Carbon-Based Nanomaterials 3.0

Carbon-Based Nanomaterials 3.0

Editor

Ana María Díez-Pascual

MDPI • Basel • Beijing • Wuhan • Barcelona • Belgrade • Manchester • Tokyo • Cluj • Tianjin

Editor
Ana María Díez-Pascual
Analytical Chemistry
Universidad de Alcalá
Madrid
Spain

Editorial Office
MDPI
St. Alban-Anlage 66
4052 Basel, Switzerland

This is a reprint of articles from the Special Issue published online in the open access journal *International Journal of Molecular Sciences* (ISSN 1422-0067) (available at: www.mdpi.com/journal/ijms/special_issues/carbon_nano_3).

For citation purposes, cite each article independently as indicated on the article page online and as indicated below:

LastName, A.A.; LastName, B.B.; LastName, C.C. Article Title. *Journal Name* **Year**, *Volume Number*, Page Range.

ISBN 978-3-0365-6551-4 (Hbk)
ISBN 978-3-0365-6550-7 (PDF)

© 2023 by the authors. Articles in this book are Open Access and distributed under the Creative Commons Attribution (CC BY) license, which allows users to download, copy and build upon published articles, as long as the author and publisher are properly credited, which ensures maximum dissemination and a wider impact of our publications.

The book as a whole is distributed by MDPI under the terms and conditions of the Creative Commons license CC BY-NC-ND.

Contents

About the Editor . . . . **vii**

Preface to "Carbon-Based Nanomaterials 3.0" . . . . **ix**

Ana M. Díez-Pascual
Carbon-Based Nanomaterials 3.0
Reprinted from: *Int. J. Mol. Sci.* **2022**, *23*, 9321, doi:10.3390/ijms23169321 . . . . **1**

Daniela Bala, Iulia Matei, Gabriela Ionita, Dragos-Viorel Cosma, Marcela-Corina Rosu and Maria Stanca et al.
Luminescence, Paramagnetic, and Electrochemical Properties of Copper Oxides-Decorated TiO_2/Graphene
Oxide Nanocomposites
Reprinted from: *Int. J. Mol. Sci.* **2022**, *23*, 14703, doi:10.3390/ijms232314703 . . . . **5**

Thi-Hoa Le, Ji-Hyeon Kim and Sang-Joon Park
"Turn on" Fluorescence Sensor of Glutathione Based on Inner Filter Effect of Co-Doped Carbon Dot/Gold Nanoparticle Composites
Reprinted from: *Int. J. Mol. Sci.* **2021**, *23*, 190, doi:10.3390/ijms23010190 . . . . **17**

Emőke Sikora, Gábor Muránszky, Ferenc Kristály, Béla Fiser, László Farkas and Béla Viskolcz et al.
Development of Palladium and Platinum Decorated Granulated Carbon Nanocomposites for Catalytic Chlorate Elimination
Reprinted from: *Int. J. Mol. Sci.* **2022**, *23*, 10514, doi:10.3390/ijms231810514 . . . . **29**

Joon-Hyuk Bang, Byeong-Hoon Lee, Young-Chul Choi, Hye-Min Lee and Byung-Joo Kim
A Study on Superior Mesoporous Activated Carbons for Ultra Power Density Supercapacitor from Biomass Precursors
Reprinted from: *Int. J. Mol. Sci.* **2022**, *23*, 8537, doi:10.3390/ijms23158537 . . . . **43**

Vito Despoja and Leonardo Marušić
Prediction of Strong Transversal s(TE) Exciton–Polaritons in C_{60} Thin Crystalline Films
Reprinted from: *Int. J. Mol. Sci.* **2022**, *23*, 6943, doi:10.3390/ijms23136943 . . . . **57**

Rene Mary Amirtha, Hao-Huan Hsu, Mohamed M. Abdelaal, Ammaiyappan Anbunathan, Saad G. Mohamed and Chun-Chen Yang et al.
Constructing a Carbon-Encapsulated Carbon Composite Material with Hierarchically Porous Architectures for Efficient Capacitive Storage in Organic Supercapacitors
Reprinted from: *Int. J. Mol. Sci.* **2022**, *23*, 6774, doi:10.3390/ijms23126774 . . . . **77**

Ting Sun, Xian Li, Xiaochuan Jin, Ziyi Wu, Xiachao Chen and Jieqiong Qiu
Function of Graphene Oxide as the "Nanoquencher" for Hg^{2+} Detection Using an Exonuclease I-Assisted Biosensor
Reprinted from: *Int. J. Mol. Sci.* **2022**, *23*, 6326, doi:10.3390/ijms23116326 . . . . **89**

Tomas Plachy, Erika Kutalkova, David Skoda and Pavlina Holcapkova
Transformation of Cellulose via Two-Step Carbonization to Conducting Carbonaceous Particles and Their Outstanding Electrorheological Performance
Reprinted from: *Int. J. Mol. Sci.* **2022**, *23*, 5477, doi:10.3390/ijms23105477 . . . . **103**

Joonwon Bae, Gyo Eun Gu, Yeon Ju Kwon, Jea Uk Lee and Jin-Yong Hong
Functionalization of Tailored Porous Carbon Monolith for Decontamination of Radioactive Substances
Reprinted from: *Int. J. Mol. Sci.* **2022**, *23*, 5116, doi:10.3390/ijms23095116 . . . . **117**

Mónica Canales, Juan Manuel Ramírez-de-Arellano, Juan Salvador Arellano and Luis Fernando Magaña
Ab Initio Study of the Interaction of a Graphene Surface Decorated with a Metal-Doped C_{30} with Carbon Monoxide, Carbon Dioxide, Methane, and Ozone
Reprinted from: *Int. J. Mol. Sci.* **2022**, *23*, 4933, doi:10.3390/ijms23094933 . . . . **131**

Leonardo Marušić, Ana Kalinić, Ivan Radović, Josip Jakovac, Zoran L. Mišković and Vito Despoja
Resolving the Mechanism of Acoustic Plasmon Instability in Graphene Doped by Alkali Metals
Reprinted from: *Int. J. Mol. Sci.* **2022**, *23*, 4770, doi:10.3390/ijms23094770 . . . . **149**

Wen-Shuo Kuo, Yen-Sung Lin, Ping-Ching Wu, Chia-Yuan Chang, Jiu-Yao Wang and Pei-Chi Chen et al.
Two-Photon–Near Infrared-II Antimicrobial Graphene-Nanoagent for Ultraviolet–Near Infrared Imaging and Photoinactivation
Reprinted from: *Int. J. Mol. Sci.* **2022**, *23*, 3230, doi:10.3390/ijms23063230 . . . . **165**

Viritpon Srimaneepong, Hans Erling Skallevold, Zohaib Khurshid, Muhammad Sohail Zafar, Dinesh Rokaya and Janak Sapkota
Graphene for Antimicrobial and Coating Application
Reprinted from: *Int. J. Mol. Sci.* **2022**, *23*, 499, doi:10.3390/ijms23010499 . . . . **177**

Manuel A. Valdés-Madrigal, Fernando Montejo-Alvaro, Amelia S. Cernas-Ruiz, Hugo Rojas-Chávez, Ramon Román-Doval and Heriberto Cruz-Martinez et al.
Role of Defect Engineering and Surface Functionalization in the Design of Carbon Nanotube-Based Nitrogen Oxide Sensors
Reprinted from: *Int. J. Mol. Sci.* **2021**, *22*, 12968, doi:10.3390/ijms222312968 . . . . **195**

About the Editor

Ana María Díez-Pascual

Ana Maria Diez-Pascual graduated in Chemistry in 2001 (awarded Extraordinary Prize) at the Complutense University (Madrid, Spain), where she completed her Ph.D. (2002–2005) on dynamic and equilibrium properties of fluid interfaces under the supervision of Prof. Rubio. In 2005, she worked at the Max Planck Institute of Colloids and Interfaces (Germany) with Prof. Miller on the rheological characterization of water-soluble polymers. During 2006–2008, she was a postdoctoral researcher at the Physical Chemistry Institute of the RWTH-Aachen University (Germany), where she worked on the layer-by-layer assembly of polyelectrolyte multilayers onto thermoresponsive microgels. Then, she moved to the Institute of Polymer Science and Technology (Madrid, Spain) and participated in a Canada–Spain joint project to develop carbon nanotube (CNT)-reinforced epoxy and polyetheretherketone composites for transport applications. Currently, she is a full professor at Alcala University (Madrid, Spain) focused on the development of polymer/nanofiller systems for biomedical applications. She has published over 200 SCI articles (97% in Q1 journals). She has an H-index of 43 and more than 4000 total citations. She has published 21 book chapters, 2 monographies, and edited 3 books, and is the first author of an international patent. She has been invited to impart seminars at prestigious international research centers (i.e., Max Planck in Germany, NRC in Canada, School of Materials in Manchester, U.K.). She was awarded the TR35 2012 prize by the Massachusetts Technological Institute (MIT) for her innovative work in the field of nanotechnology. She is member of the Young Academy of Spain.

Preface to "Carbon-Based Nanomaterials 3.0"

This Special Issue, with a collection of 12 original contributions and 2 literature reviews, offers select examples of the surface modifications of carbon nanomaterials that adapt their physicochemical properties, as well as their applications in a variety of fields, such as supercapacitors, sensors, antimicrobial coatings, bioimaging, decontamination, and so forth.

Ana María Díez-Pascual
Editor

Editorial

Carbon-Based Nanomaterials 3.0

Ana M. Díez-Pascual

Universidad de Alcalá, Facultad de Ciencias, Departamento de Química Analítica, Química Física e Ingeniería Química, Ctra. Madrid-Barcelona Km. 33.6, 28805 Alcalá de Henares, Madrid, Spain; am.diez@uah.es

Carbon-based nanomaterials are currently attracting a lot of interest in many fields, ranging from medicine and biotechnology to electronics, energy storage, and sensing applications [1,2]. They show a variety of shapes, from 0D fullerenes, nanodiamonds, and quantum dots (QDs) to 1D carbon nanotubes (CNTs), and 2D graphene (G) and its derivatives graphene oxide (GO) and reduced graphene oxide (rGO) [3]. Furthermore, new carbon-based nanomaterials are currently under investigation, such as mutated graphene-like nanomaterials, which have been found to be very effective for the removal of organic pollutants from wastewater [4], as well 3D carbon monolithics, which have great potential for the decontamination of radioactive substances [5].

One characteristic of all carbon nanomaterials is the possibility of functionalizing them through non-covalent and covalent methods [6,7], which generally modifies their hydrophilic, electronic, optical, and mechanical properties. Non-covalent approaches are attained via π–π stacking, electrostatic forces, and Van der Waals forces. On the other hand, covalent functionalization can be performed via simple oxidation, leading to oxygen-containing groups suitable for reacting with functional groups of other molecules or polymers. This Special Issue "Carbon-Based Nanomaterials 3.0", with a collection of 10 original contributions and 2 literature reviews, offers select examples of the surface modifications of carbon nanomaterials that adapt their physicochemical properties, as well as their applications in a variety of fields, such as supercapacitors, sensors, antimicrobial coatings, bioimaging, decontamination, and so forth.

With an enlarged global focus on tackling the challenges of environmental pollution, the interest in novel energy devices as an alternative to petroleum-based ones has also increased. In this regard, supercapacitors can be designed for use in environmentally friendly vehicles and new renewable energy; however, the limitation of a low energy density remains a challenge [8]. To further increase the power density of supercapacitors, mesoporous carbon nanomaterials can be used. However, activated carbons with high mesopore volumes generated via physical activation are not economically viable. Thus, to develop cost-effective high-performance supercapacitors with high energy and power densities, the preparation of novel mesoporous activated carbons should be investigated. In this regard, Bang et al. [9] have recently synthesized a kenaf-derived activated carbon (KAC) for a high-power density supercapacitor via phosphoric acid activation. Kenaf was chosen as the precursor due to its high productivity, and phosphoric acid activation was applied to create a high specific surface area and advanced mesoporous structure. The pore-growth mechanism for KAC through phosphoric acid activation was explored by analyzing the textural properties and crystal structures. The electrochemical properties of KAC were compared with commercial activated carbon, and an improvement was found both in the specific capacitance and the ion-diffusion resistance. Mesopore control of the electrode material is crucial in improving the supercapacitor resistance and output.

Hierarchical porous activated carbon (HPAC) is another interesting active material for supercapacitors due to its huge specific surface are. Preparing electrodes with high-mass loading is interesting in providing high total capacitances and gravimetric or volumetric energy densities [10]. Therefore, developing an HPAC that increases the mass loading of

Citation: Díez-Pascual, A.M. Carbon-Based Nanomaterials 3.0. *Int. J. Mol. Sci.* **2022**, *23*, 9321. https://doi.org/10.3390/ijms23169321

Received: 9 August 2022
Accepted: 17 August 2022
Published: 18 August 2022

Publisher's Note: MDPI stays neutral with regard to jurisdictional claims in published maps and institutional affiliations.

Copyright: © 2022 by the author. Licensee MDPI, Basel, Switzerland. This article is an open access article distributed under the terms and conditions of the Creative Commons Attribution (CC BY) license (https://creativecommons.org/licenses/by/4.0/).

the resulting electrode is highly desirable. In this regard, Amirtha et al. [11] prepared a novel hydrogel-derived encapsulated H-HPAC (H@H) composite material, with reduced specific surface area and pore parameters but increased proportions of nitrogen species. The free-standing and flexible H@H electrodes showed remarkable reversible capacitance, rate capability, and cycling stability and are also promising electrode materials for other energy storage fields such as metal–ion capacitors.

In another study, a novel carbonaceous material was prepared from cellulose carbonized via two-steps—hydrothermal and thermal carbonization, in order—without any chemicals [12], giving high yields after a treatment at 600 °C under an inert atmosphere. This led to nanospheres with increased specific surface areas, as confirmed by SEM, FTIR, X-ray diffraction, and Raman spectroscopy, as well as enhanced conductivity. The nanospheres were used as a dispersed phase in electrorheological fluids, displaying exceptional electrorheological effects, considerably surpassing recent state-of-the-art findings. These new carbonaceous particles prepared from renewable cellulose have further potential to be utilized in many other applications that require conducting carbonaceous structures with high specific surface area such as adsorption, catalyst, filtration, and energy storage, to mention but a few.

GO is well known for its outstanding fluorescence quenching capability. Sun et al. [13] prepared a water-soluble positively charged graphene oxide by grafting polyetherimide onto GO nanosheets via a carbodiimide reaction. Compared with conventional GO, the fluorescence quenching ability of the DNA strand of the novel positively charged one was significantly improved via an additional electrostatic interaction. The DNA probe was almost completely quenched for concentrations of the positively charged GO as low as 0.1 µg/mL. This quenching ability was used to develop a sensor for Hg^{2+} detection, leading to a linear concentration range of 0–250 nM, with a limit of detection of 3.93 nM, and it was successfully applied to real samples of pond water, leading to recoveries in the range from 99.6% to 101.1%.

Based on the fluorescence quenching ability of nitrogen- and phosphorous-doped carbon dots, a simple and selective sensor for glutathione detection was also developed [14]. The reductant potential of the doped carbon dots was used to synthesize AuNPs and to subsequently form composites, which were characterized via spectroscopic and microscopic techniques, including electrophoretic light scattering and XRD. The overlap of the fluorescence emission spectrum of the doped carbon dots and the absorption spectrum of AuNPs resulted in an effective inner filter effect in the composite material, leading to a quenching of the fluorescence intensity. In the presence of GSH, the fluorescence intensity of the composite was recovered, leading to a sensing method with a limit of detection of 0.1 µM.

Nitrogen-doped amino acid-functionalized GQDs show enhanced photoluminescence and photostability and lead to the generation of reactive oxygen species through two-photon photodynamic therapy (PDT) [15]. This amino-N-GQDs can be used as two-photon contrast probes to trail and localize analytes in in-depth two-photon imaging executed in a biological environment along with two-photon PDT to eliminate infectious or multidrug-resistant microbes.

Nitrogen oxides (NOx) are amongst the foremost atmospheric pollutants; hence, it is imperative to screen and detect their presence in the atmosphere. For such a purpose, low-dimensional carbon structures have been broadly used as NOx sensors. In particular, CNTs have been applied for sensing toxic gases due to their high specific surface area and excellent mechanical properties. Even though pristine CNTs have shown promising performance for NOx detection, several strategies have been developed such as surface functionalization and defect engineering to expand the NOx-sensing ability of pristine CNT-based sensors. In this regard, the surface modification approaches used in the recent decade to modify the sensitivity and the selectivity of CNTs to NOx have recently been reviewed [16].

Other atmospheric contaminants that threaten the environment and life include CO, CO_2, CH_4, and O_3. Canales et al. [17] have explored the use of small fullerenes such as

C$_{30}$ for the adsorption of these pollutants. They performed computational simulations to investigate their adsorption on graphene-semifullerene (C$_{30}$) surfaces, considering two C$_{30}$ geometries—hexagonal and pentagonal bases—and found that it is possible to dope all surfaces with Li, Ti, and Pt, which can be used as effective catalysts.

On the other hand, given that the control over radioactive species is currently critical, the development of functional materials for the decontamination of radioactive substances has also become imperative. In this regard, Bae and coworkers [5] have recently developed a 3D porous carbon monolith functionalized with Prussian blue particles via the removal of colloidal silica particles from exfoliated graphene/silica composite precursors. The colloidal silica acted as a template and provided enough surface area that could accommodate potentially hazardous radioactive substances by adsorption. The exceptional surface and pore structure of the novel carbon monolith was examined using SEM, XRD, FTIR, and XPS analysis. Moreover, a nitrogen adsorption/desorption study showed that surface area and pore volume increased significantly compared with the starting precursor. It was found that the novel nanomaterial had a higher adsorption capacity than that of pristine porous carbon monoliths to most radioactive ions and, hence, can be used for decontamination in many fields.

Studying acoustic plasmons (APs) in single-layer, double-layer, and multilayer graphene or in metal/dielectric/graphene superstructures is another active field of research. Although the mechanism of the formation of these plasmons in electrostatically biased graphene or at noble metal surfaces is well known, the mechanism of their formation in alkali-doped graphene is not well understood yet. In this regard, Marušic and coworkers have investigated the interplay of the p and s intraband transitions with plasmon resonance [18]. Their work illustrates the importance of understanding the nature of the chemical bonding between alkaline atoms and graphene and the perpendicular dispersivity of the dynamical response in theoretical simulations of low-energy plasmons.

On the other hand, the interaction between photons and polarization modes can result in the formation of hybrid photon polarization modes, called polaritons. The same authors [19] have shown that 2D layered nanomaterials enable the formation of well-defined exciton–polaritons even at room temperature and that the exciton–photon coupling can be manipulated simply by changing the number of single layers. These nanostructures can be applied in photonic devices, such as LED, telecommunications, or chemical and biological sensing.

As known, graphene is a versatile compound with many outstanding properties, providing a combination of a huge surface area, a high strength, and thermal and electrical properties, with a wide array of functionalization possibilities. However, the available literature on graphene-based coatings in dentistry and medical implant technology is scarce. Srimaneepong and coworkers [20] have recently provided a comprehensive and well-organized review on graphene applications in such field. Graphene displays good biocompatibility, corrosion prevention, and excellent antimicrobial properties to prevent the colonization of bacteria. Moreover, graphene coatings improve cell adhesion and osteogenic differentiation, and promote antibacterial activity to parts of titanium unaffected by the thermal treatment. Additionally, the coating can improve the surface properties of implants, which can then be used for biomedical applications. Hence, graphene and its derivatives may hold the key to the next revolution in dental and medical technology.

Funding: This research received no external funding.

Conflicts of Interest: The author declares no conflict of interest.

References

1. Diez Pascual, A.M. Carbon-Based Polymer Nanocomposites for High-Performance Applications II. *Polymers* **2022**, *14*, 870. [CrossRef] [PubMed]
2. Díez-Pascual, A.M. Carbon-Based Nanomaterials. *Int. J. Mol. Sci.* **2021**, *22*, 7726. [CrossRef] [PubMed]
3. Nanomaterials definition matters. In *Nature Nanotechnology*; Nature Publishing Group: Berlin, Germany, 2019; Volume 14, p. 193.

4. Maqbool, Q.; Barucca, G.; Sabbatini, S.; Parlapiano, M.; Ruello, M.L.; Tittarelli, F. Transformation of industrial and organic waste into titanium doped activated carbon–cellulose nanocomposite for rapid removal of organic pollutant. *J. Hazard. Mater.* **2022**, *423*, 126958. [CrossRef] [PubMed]
5. Bae, J.; Gu, G.E.; Kwon, Y.J.; Lee, J.U.; Hong, J.-Y. Functionalization of Tailored Porous Carbon Monolith for Decontamination of Radioactive Substances. *Int. J. Mol. Sci.* **2022**, *23*, 5116. [CrossRef] [PubMed]
6. Díez-Pascual, A.M. Chemical Functionalization of Carbon Nanotubes with Polymers: A Brief Overview. *Macromol* **2021**, *1*, 64–83. [CrossRef]
7. Díez-Pascual, A.M. Surface Engineering of Nanomaterials with Polymers, Biomolecules, and Small Ligands for Nanomedicine. *Materials* **2022**, *15*, 3251. [CrossRef] [PubMed]
8. Díez-Pascual, A.M.; Sánchez, J.A.L.; Capilla, R.P.; Díaz, P.G. Recent Developments in Graphene/Polymer Nanocomposites for Application in Polymer Solar Cells. *Polymers* **2018**, *10*, 217. [CrossRef] [PubMed]
9. Bang, J.-H.; Lee, B.-H.; Choi, Y.-C.; Lee, H.-M.; Kim, B.-J. A Study on Superior Mesoporous Activated Carbons for Ultra Power Density Supercapacitor from Biomass Precursors. *Int. J. Mol. Sci.* **2022**, *23*, 8537. [CrossRef] [PubMed]
10. Dong, Y.; Zhu, J.; Li, Q.; Zhang, S.; Song, H.; Jia, D. Carbon materials for high mass-loading supercapacitors: Filling the gap between new materials and practical applications. *J. Mater. Chem. A* **2020**, *8*, 21930–21946. [CrossRef]
11. Amirtha, R.M.; Hsu, H.-H.; Abdelaal, M.M.; Anbunathan, A.; Mohamed, S.G.; Yang, C.-C.; Hung, T.-F. Constructing a Carbon-Encapsulated Carbon Composite Material with Hierarchically Porous Architectures for Efficient Capacitive Storage in Organic Supercapacitors. *Int. J. Mol. Sci.* **2022**, *23*, 6774. [CrossRef] [PubMed]
12. Plachy, T.; Kutalkova, E.; Skoda, D.; Holcapkova, P. Transformation of Cellulose via Two-Step Carbonization to Conducting Carbonaceous Particles and Their Outstanding Electrorheological Performance. *Int. J. Mol. Sci.* **2022**, *23*, 5477. [CrossRef] [PubMed]
13. Sun, T.; Li, X.; Jin, X.; Wu, Z.; Chen, X.; Qiu, J. Function of Graphene Oxide as the "Nanoquencher" for Hg^{2+} Detection Using an Exonuclease I-Assisted Biosensor. *Int. J. Mol. Sci.* **2022**, *23*, 6326. [CrossRef] [PubMed]
14. Le, T.-H.; Kim, J.-H.; Park, S.-J. "Turn on" Fluorescence Sensor of Glutathione Based on Inner Filter Effect of Co-Doped Carbon Dot/Gold Nanoparticle Composites. *Int. J. Mol. Sci.* **2022**, *23*, 190. [CrossRef] [PubMed]
15. Kuo, W.-S.; Lin, Y.-S.; Wu, P.-C.; Chang, C.-Y.; Wang, J.-Y.; Chen, P.-C.; Hsieh, M.-H.; Kao, H.-F.; Lin, S.-H.; Chang, C.-C. Two-Photon–Near Infrared-II Antimicrobial Graphene-Nanoagent for Ultraviolet–Near Infrared Imaging and Photoinactivation. *Int. J. Mol. Sci.* **2022**, *23*, 3230. [CrossRef] [PubMed]
16. Valdés-Madrigal, M.A.; Montejo-Alvaro, F.; Cernas-Ruiz, A.S.; Rojas-Chávez, H.; Román-Doval, R.; Cruz-Martinez, H.; Medina, D.I. Role of Defect Engineering and Surface Functionalization in the Design of Carbon Nanotube-Based Nitrogen Oxide Sensors. *Int. J. Mol. Sci.* **2021**, *22*, 12968. [CrossRef] [PubMed]
17. Canales, M.; Ramírez-de-Arellano, J.M.; Arellano, J.S.; Magaña, L.F. Ab Initio Study of the Interaction of a Graphene Surface Decorated with a Metal-Doped C30 with Carbon Monoxide, Carbon Dioxide, Methane, and Ozone. *Int. J. Mol. Sci.* **2022**, *23*, 4933. [CrossRef] [PubMed]
18. Marušić, L.; Kalinić, A.; Radović, I.; Jakovac, J.; Mišković, Z.L.; Despoja, V. Resolving the Mechanism of Acoustic Plasmon Instability in Graphene Doped by Alkali Metals. *Int. J. Mol. Sci.* **2022**, *23*, 4770. [CrossRef] [PubMed]
19. Despoja, V.; Marušić, L. Prediction of Strong Transversal s(TE) Exciton–Polaritons in C60 Thin Crystalline Films. *Int. J. Mol. Sci.* **2022**, *23*, 6943. [CrossRef] [PubMed]
20. Srimaneepong, V.; Skallevold, H.E.; Khurshid, Z.; Zafar, M.S.; Rokaya, D.; Sapkota, J. Graphene for Antimicrobial and Coating Application. *Int. J. Mol. Sci.* **2022**, *23*, 499. [CrossRef] [PubMed]

Article

Luminescence, Paramagnetic, and Electrochemical Properties of Copper Oxides-Decorated TiO$_2$/Graphene Oxide Nanocomposites

Daniela Bala [1], Iulia Matei [2], Gabriela Ionita [2], Dragos-Viorel Cosma [3], Marcela-Corina Rosu [3], Maria Stanca [4], Carmen Gaidau [4], Maria Baleanu [5], Marian Virgolici [5] and Ioana Stanculescu [1,5,*]

[1] Physical Chemistry Department, Faculty of Chemistry, University of Bucharest, Regina Elisabeta, No. 4-12, 030018 Bucharest, Romania
[2] "Ilie Murgulescu" Institute of Physical Chemistry, 202 Splaiul Independentei, 060021 Bucharest, Romania
[3] National Institute for Research and Development of Isotopic and Molecular Technologies, 67–103 Donat Street, 400293 Cluj-Napoca, Romania
[4] Leather Research Department, National Institute for Textiles and Leather, Division Leather and Footwear Research Institute (ICPI), 93 Ion Minulescu Street, 031215 Bucharest, Romania
[5] Horia Hulubei National Institute of Research and Development for Physics and Nuclear Engineering, 30 Reactorului Str., 077125 Magurele, Romania
* Correspondence: istanculescu@nipne.ro

Abstract: The properties of newly synthesized Cu$_2$O/CuO-decorated TiO$_2$/graphene oxide (GO) nanocomposites (NC) were analyzed aiming to obtain insight into their photocatalytic behavior and their various applications, including water remediation, self-cleaning surfaces, antibacterial materials, and electrochemical sensors. The physico-chemical methods of research were photoluminescence (PL), electron paramagnetic resonance (EPR) spectroscopy, cyclic voltammetry (CV), and differential pulse voltammetry (DPV). The solid samples evidenced an EPR signal that can be attributed to the oxygen-vacancy defects and copper ions in correlation with PL results. Free radicals generated before and after UV-Vis irradiation of powders and aqueous dispersions of Cu$_2$O/CuO-decorated TiO$_2$/GO nanocomposites were studied by EPR spectroscopy using two spin traps, DMPO (5,5-dimethyl-1-pyrroline-N-oxide) and CPH (1-hydroxy-3-carboxy-2,2,5,5-tetramethylpyrrolidine), to highlight the formation of hydroxyl and superoxide reactive oxygen species, respectively. The electrochemical characterization of the NC modified carbon-paste electrodes (CPE) was carried out by CV and DPV. As such, modified carbon-paste electrodes were prepared by mixing carbon paste with copper oxides-decorated TiO$_2$/GO nanocomposites. We have shown that GO reduces the recombination process in TiO$_2$ by immediate electron transfer from excited TiO$_2$ to GO sheets. The results suggest that differences in the PL, respectively, EPR data and electrochemical behavior, are due to the different copper oxides and GO content, presenting new perspectives of materials functionalization.

Keywords: TiO$_2$ nanocomposites; EPR; photoluminescence; electrochemistry

1. Introduction

Titania (TiO$_2$)-based materials have attracted great scientific interest due to their physico-chemical properties and numerous applications. By UV photoexcitation, these materials are able to produce electron–hole pairs that can determine a series of consecutive reactions, most often involving the formation of radical reactive species [1].

TiO$_2$ nanoparticles have shown good optical, electrical, and photocatalytic properties [2]. TiO$_2$ is a substance with applications in various fields such as paints and plastics, water remediation, paper, and sensors [3–5]. The modification of TiO$_2$ using metals, non-metals, carbon-based materials may lead to an improvement of its photocatalytic as well as photoelectrochemical activity. TiO$_2$ absorbs only ultraviolet light due to its large bandgap (3.0–3.2 eV). The optical absorption performance in the visible region could be enhanced by adding copper oxides and graphene oxide (GO) to TiO_2 nanoparticles [6,7]. Various methods of obtaining TiO_2 and GO based nanocomposites and their optical and photocatalytic properties extensive characterization are reported [8–12].

The electrochemical response of TiO_2-modified electrodes can be improved by the high conductivity of TiO_2. Such electrodes were used for electrochemical measurements of guanine, adenine, and dopamine [13,14]. The TiO_2 doped in the carbon paste electrode (CPE) sensor was developed to detect methyldopa in pharmaceutical samples since it presented excellent electrochemical behavior, correlated to better electrode applicability. This electrode may promote analyte electro-oxidation, increasing method sensibility [15]. Carbon-paste electrodes modified with Cu_2O/CuO-decorated TiO_2/graphene oxide nanocomposites may be a valuable and cheap alternative to determine compounds such as neurotransmitters in drug formulae.

The EPR measurements on solid TiO_2 samples and on water suspensions of TiO_2 samples evidenced the presence of an EPR signal due to the oxygen defects and/or to the presence of cooper, as well as the formation of reactive oxygen species (ROS) in suspensions. The generation of ROS ($HO^\bullet$, $O_2^{\bullet-}$, singlet oxygen, etc.) in water titania suspensions recommends these systems as alternative oxidizing agents that can be used in the annihilation of water pollutants or can find antibacterial applications. In other fields such as cosmetics, titania-based materials should be carefully used in order to control their ROS activity. In this context, the importance of this study consists in highlighting the intimate relation between the composition, structure, and activity of Cu_2O/CuO-decorated TiO_2/graphene oxide nanocomposites by using physico-chemical methods. The significance of the work is high because the named nanocomposites may be used as advanced materials for various applications: environmental, medical textiles, self-cleaning surfaces, and electrochemical sensors.

2. Results and Discussion

2.1. Photoluminescence (PL) Data

As shown in Figure 1, the pure TiO_2 nanoparticles present clear PL emission bands: at 412 nm, corresponding to the oxygen vacancy with two trapped electrons (center F) [16]; at 426 nm, attributed to the recombination of self-trapped excitons (STE) or free excitons [17,18]; at 451 nm, 468 nm, and 484 nm, assigned to electrons' trapping in shallow traps resulting from oxygen vacancies of TiO_2 [18]; and at 493 nm, corresponding to emissions from the TiO_2 surface states [18].

As a result of TiO_2 decoration with Cu_2O and CuO species (identified by XPS in the previous study [9]), a lower PL intensity of TC1, TC2, and TC3 was observed, indicating an efficient charge–carrier separation. This finding is in good agreement with the data reported by M. Janczarek and E. Kowalska in their comprehensive review that presents the Cu_2O and CuO as active species in TiO_2 photocatalytic system being efficient electron trappers to prevent the recombination of the photogenerated electron–hole pairs [19]. This trend is more pronounced after graphene oxide addition, confirming that GO reduces the recombination process in TiO_2 by immediate electron transfer from excited TiO_2 to GO sheets [20,21].

2.2. EPR Spectroscopy Data

2.2.1. EPR Spectra of Solid Samples

The EPR spectra of the solid TiO_2 samples are presented in Figure 2. The g factors, calculated from the values of the microwave frequency (ν) and magnetic field (B) as shown in ref. [22], are given in Table 1. As can be observed from Figure 2, the EPR spectra of the copper oxides-decorated TiO_2/graphene oxide samples present a broad line corresponding to copper (II) centers, with g factors in the range 2.1464–2.1516.

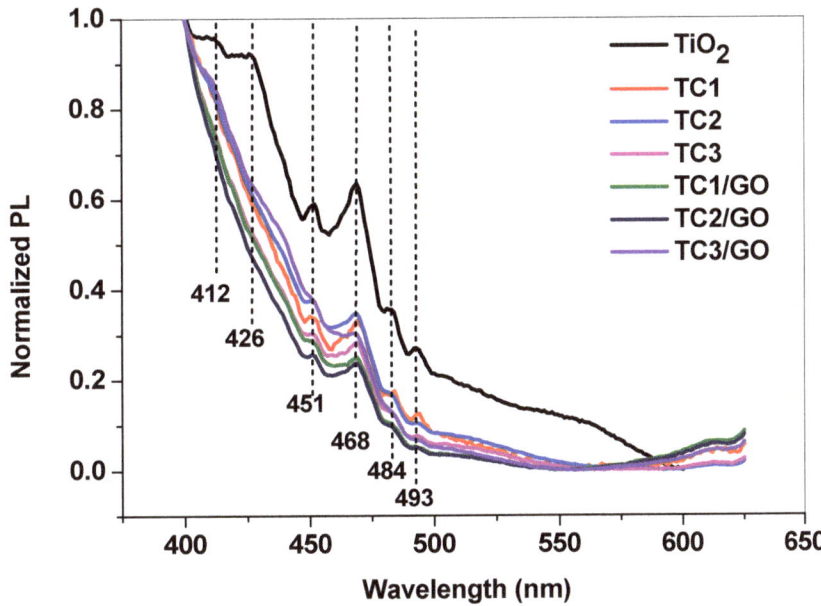

Figure 1. PL spectra of prepared nanocomposites compared to those of TiO$_2$ nanoparticles.

Table 1. The g factors of the solid samples investigated.

Sample	ν (GHz)	B (mT)	g
TC1	9.046638	300.419	2.1516
TC1-GO	9.047555	299.914 310.639	2.1554 2.0810
TC2	9.046588	301.141	2.1464
TC2-GO	9.047630	298.793 310.079	2.1635 2.0848
TC3	9.047059	300.865	2.1485
TC3-GO	9.046954	298.233 309.038	2.1674 2.0916

In the case of Cu$_2$O/CuO-decorated TiO$_2$/graphene oxide, it can be noticed that the line attributed to copper (II) becomes asymmetric, and this is due in fact to the contribution of the EPR line of the free electron due to defects present in the carbon nanomaterial and to copper (II) centers. The g values attributed to these signals are also included in Table 1 and range from 2.0810 to 2.0916.

2.2.2. Spin-Trapping Measurements

The spin-trapping method was employed in order to investigate whether the TC3 and TC3-GO samples generate ROS. Since HO$^\bullet$ and O$_2^{\bullet-}$ are the radical species most commonly reported in TiO$_2$ systems [11,12,23,24], two spin traps were used: 5,5-dimethyl-1-pyrroline N-oxide (DMPO), sensitive to the HO$^\bullet$ radical, and 1-hydroxy-3-carboxy-2,2,5,5-tetramethylpyrrolidine (CPH), having high affinity for the O$_2^{\bullet-}$ radical.

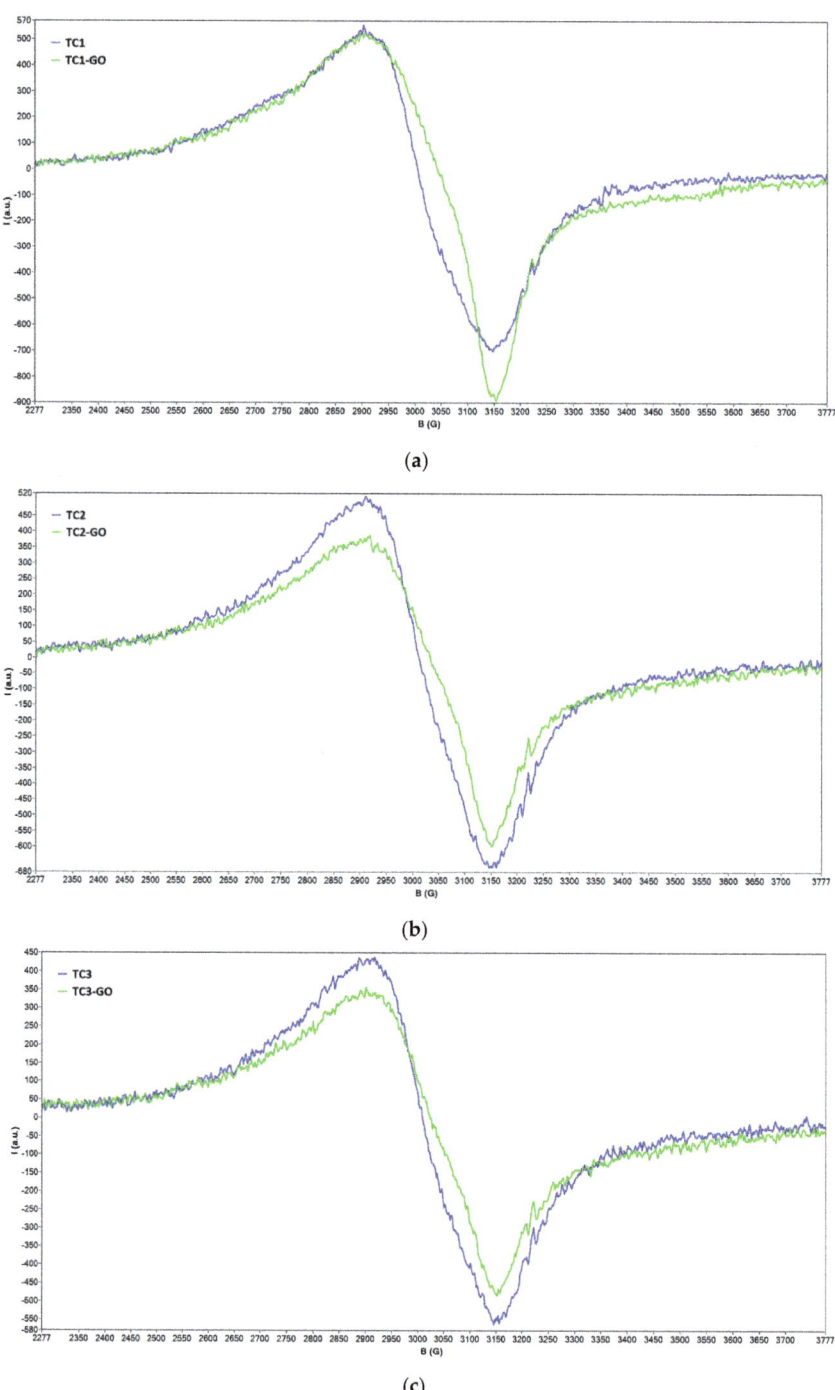

Figure 2. Solid state EPR spectra of the TC1/TC1-GO (**a**), TC2/TC2-GO (**b**), and TC3/TC3-GO pairs (**c**).

The EPR spectra of the TiO$_2$ aqueous samples in the presence of spin traps are shown in Figure 3. One may observe the characteristic 1:1:1 triplet signal of the stable 3-carboxy-carproxyl nitroxide formed by oxidation of CPH by ROS [25] (Figure 3a) and the 1:2:2:1 quartet signal characteristic to the •DMPO-OH spin adduct [26] (Figure 3b). The hyperfine coupling constant of the stable nitroxide was determined from the experimental spectrum as a_N = 16.18 G, typical for a nitroxide. The hyperfine coupling constants of the •DMPO-OH spin adduct, obtained by spectral simulation, are a_N = 15.00 G and a_H^β = 14.56 G, in accordance to the data reported in ref. [27]. Signal intensities are slightly lower for the sample containing graphene oxide. The weak signal recorded for the •DMPO-OH spin adduct may indicate a fast consumption of the HO• radical in these systems.

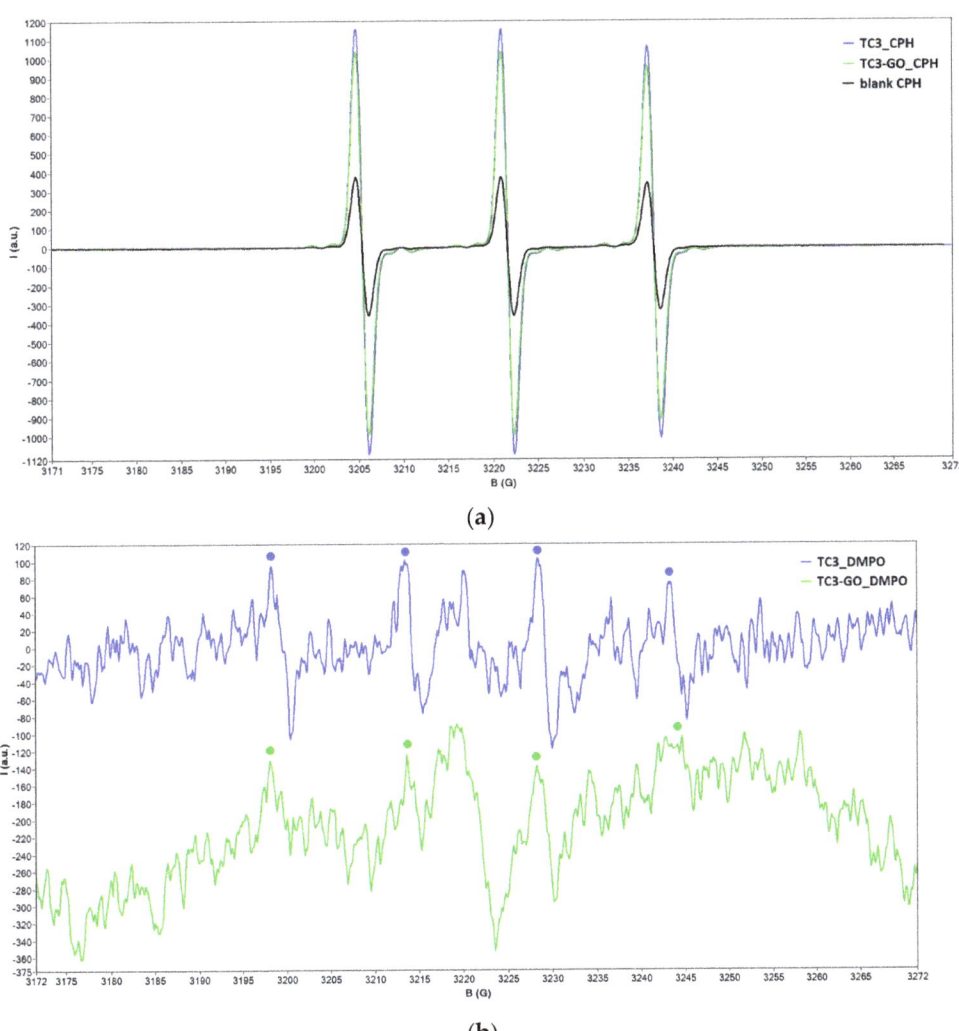

Figure 3. The EPR spectra of the TC3 and TC3-GO samples in the presence of (**a**) CPH and (**b**) DMPO.

2.3. Electrochemical Characterization

2.3.1. Cyclic Voltammetry Study

The results show that the anodic peaks increase when the sweep rate is increased and there is a move to positive potentials. Additionally, by increasing the sweep rate, the peak shape does not change, which leads to the conclusion that the modified electrode is sensitive regarding the electrochemical investigation of the ferri/ferro process. In Figure 4a, cyclic voltammograms for TC2 are presented.

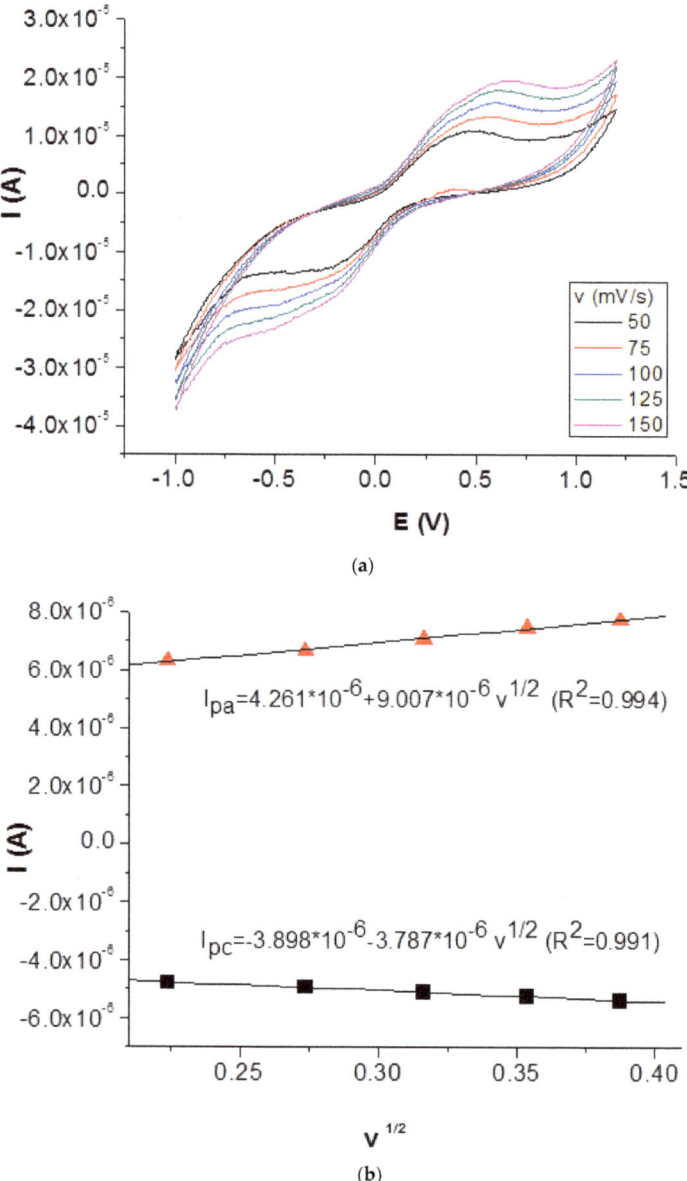

Figure 4. (**a**) Cyclic voltammograms for 1.0 mM $K_3[Fe(CN)_6]$ in 0.1 M KCl solution on TC2 modified carbon paste electrode, v = 50–150 mV s^{-1}) and (**b**) I vs. $v^{\frac{1}{2}}$ plot.

An analysis of the voltammetric peak height as a function of the square-root of the scan rate reveals a highly linear response, with good correlation factors, as observed in Figure 4b. This response indicates a diffusion-controlled electrochemical process.

The same experiments were performed for all modified electrodes. A comparison of CV measurements at 100 mV/s for all electrodes is presented in Figure 5.

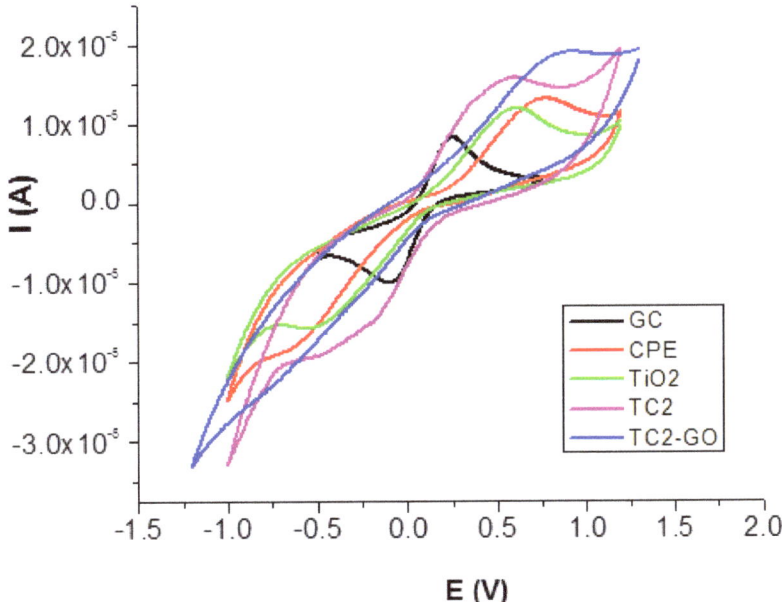

Figure 5. Overlay of the cyclic voltammograms for selected TC2 modified electrodes for the redox process of 1.0 mM $K_3[Fe(CN)_6]$ in 0.1 M KCl solution (scan rate was 100 mV s^{-1}).

Electrochemical CV data for bare and modified carbon-paste electrodes are presented in Table 2.

Table 2. Electrochemical data from CV measurements at 100 mV/s; I_a and I_c represent the anodic and cathodic peak currents, and E_a and E_c represent the anodic and cathodic peak potentials.

	E_c (V)	I_c (A)	E_a (V)	I_a (A)	ΔE (V)
CPE	−0.568	-2.69×10^{-6}	0.706	6.22×10^{-6}	1.274
TiO$_2$	−0.424	-4.69×10^{-6}	0.567	7.28×10^{-6}	0.991
TC2	−0.203	-5.00×10^{-6}	0.466	7.08×10^{-6}	0.669
TC2-GO	−0.557	-1.07×10^{-6}	0.768	6.04×10^{-6}	1.325

Both anodic and cathodic peak potentials are shifted for TiO$_2$ and TC2 when compared with the potential of bare carbon paste electrode. An increase in the peak currents and a decrease in the separation between the peak potentials (ΔE_p) at 100 mV·s^{-1} were observed for these two modified electrodes in comparison to the bare CPE, indicating that the electron transfer reaction was kinetically and thermodynamically favored at the copper oxides-decorated TiO$_2$-modified electrode surface. Enhanced electron transfer capacity was also found by CV by Mirza-Aghayan et al. for the CuO/rGO/TiO2 system [12]. In the case of the electrode modified with copper oxide-decorated TiO$_2$/graphene oxide, peak currents and potentials decreased. By increasing the scan rate, the intensity of the peak increases not only in the anodic direction but also in the cathodic side. The parameter of most significant importance is represented by the position of the voltammetric peak rather than the magnitude of the wave. In the case of the metal-doped graphene modified electrodes, the larger peak current is likely due to a slightly larger surface area at the electrode.

Large band gap narrowing of Cu_2O/CuO-decorated TiO_2/graphene oxide nanocomposites: TC1, 2.90 eV, TC2, 2.94 eV, TC3, 2.86 eV, TC1-GO, 2.75 eV, TC2-GO, 2.56 eV, TC3-GO, and 2.76 eV as compared to pure TiO2, 3.2 eV reported previously [9] may explain their enhanced electron transfer capacity.

2.3.2. Differential Pulse Voltammetry Results

To get a better understanding of the redox behavior at the modified electrodes, DPV measurements were performed. Differential pulse voltammetry is a more sensitive technique than cyclic voltammetry. The DPV traces are presented in Figure 6.

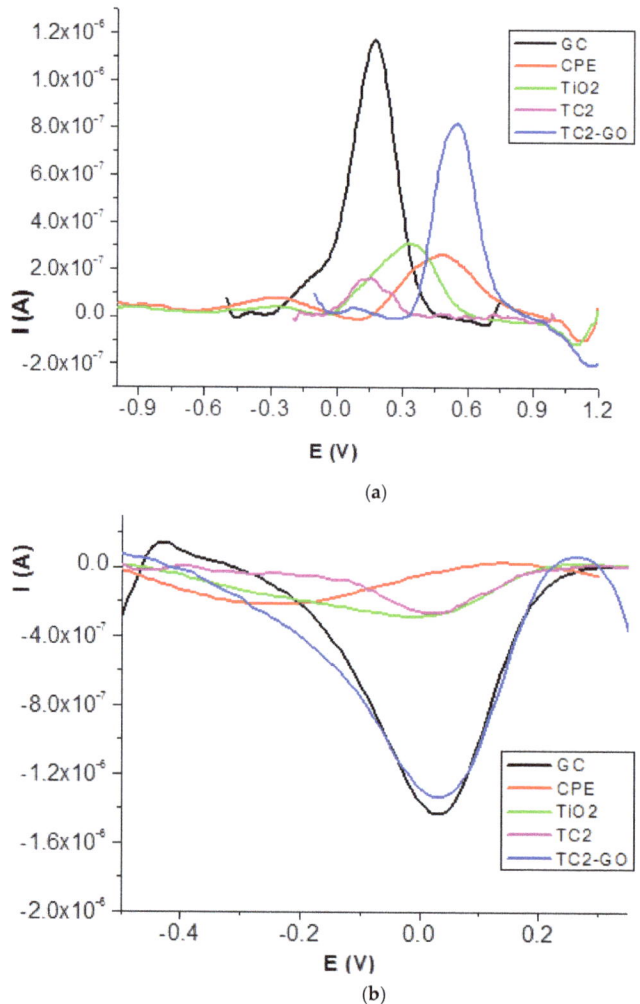

Figure 6. Differential pulse voltammograms for TC2 modified electrodes for the redox process of 1.0 mM $K_3[Fe(CN)_6]$ in 0.1 M KCl solution (with step potential 10 mV and modulation amplitude 25 mV): (**a**) anodic and (**b**) cathodic waves.

Anodic peak potentials shifted positively for all carbon paste bare and modified electrodes when compared with glassy carbon (0.175 V). The best response was obtained for the TC2-GO electrode, as stated by the highest peak current values (increased by 2.5 to 5.5). For the reduction process, the peak potential and peak current at TC2-GO (0.031 V and -1.41×10^{-6} A) were almost the same as those at glassy carbon (0.035 V and -1.46×10^{-6} A). For all other electrodes, the peak potentials shifted cathodically, and the peak currents were 4–5 times smaller. The electrodes investigated may be a useful and cheap alternative for the determination of redox active compounds contained in drug formulae.

Reported scientific data showed that the electrochemical response of graphene modified electrodes can be improved by increasing the amount of graphene in the electrode [28].

3. Materials and Methods

3.1. Nanomaterials Preparation

Cu_2O/CuO-decorated TiO_2/graphene oxide nanocomposites were prepared by the liquid impregnation method as previously described [29]. Briefly, TiO_2 powder (P25 Evonik) was dispersed in the appropriate solutions of copper (II) nitrate trihydrate under magnetic stirring. Subsequently, the dispersing medium was evaporated, and the resulting powders were calcinated in argon atmosphere at 450 °C, then in argon/hydrogen (10% H_2) at 280 °C. For high homogeneity, all three powders were dispersed in double-distilled water, frozen, and freeze dried. The resulting powders were denoted TC1, TC2, and TC3 according to the copper content 1%, 2%, and 3%, respectively. The TC(1,2,3)-graphene oxide nanocomposites were also prepared using the freeze-drying procedure by mixing TC1, TC2, and TC3 powders with graphene oxide (GO) in a weight ratio of 10:1. The final powders were denoted TC1-GO, TC2-GO, and TC3-GO. Graphene oxide was synthesized according to an improved version of Hummer's method that was reported elsewhere [30].

3.2. Photoluminescence and EPR Spectroscopy Characterization

Photoluminescence (PL) spectra of the nanocomposites were recorded using a Jasco FP-6500 spectrofluorimeter equipped with a 150 W Xenon lamp. The excitation wavelength used was 320 nm. The experiments were performed in triplicate.

The EPR spectra of the solid probes were recorded on a JEOL FA100 spectrometer equipped with a cylindrical-type resonator TE011 using the following parameters: frequency modulation 100 kHz, microwave power 0.998 mW, sweep time 1800 s, modulation amplitude 1 G, time constant 1 s, and magnetic field scan range 1500 G. Each solid sample was placed in a glass capillary and introduced in the spectrometer's cavity.

For the spin-trapping measurements, two spin traps, 5,5-dimethyl-1-pyrroline N-oxide (DMPO) and 1-hydroxy-3-carboxy-2,2,5,5-tetramethylpyrrolidine (CPH), purchased from Sigma-Aldrich (St. Louis, MO, USA) and ENZO Life Sciences, Inc. (Lausen, Switzerland), respectively, were used. The parameter settings of the EPR spectrometer for spin-trapping experiments were frequency modulation 100 kHz, microwave power 0.998 mW, sweep time 60 s, modulation amplitude 1 G, time constant 0.1 s, and magnetic field scan range 100 G.

Aqueous dispersions containing TiO_2-based samples (1 mg/mL), DMPO (25 mM), and hydrogen peroxide (125 mM) were prepared and incubated for 10 min at 37 °C in the dark. The samples were vortexed during the last 5 min of incubation and then centrifuged at 15.000 rpm for 1 min. The supernatant was immediately transferred into a capillary tube and exposed for 10 min to UVA radiation (370 nm, mercury arc lamp, 500 W, LOT-Quantum Design, Darmstadt, Germany) directly in the spectrometer's cavity; then, the EPR spectrum was recorded.

A stock solution of CPH (10 mM) was prepared in phosphate buffer of pH 7.4. To this solution, the chelating agent deferoxamine mesylate (100 µM) was added in order to prevent the oxidation of CPH that is catalyzed by traces of transition metal ions [31]. Aqueous dispersions containing TiO_2 (1 mg/mL) and CPH (0.5 mM) were prepared and treated similarly to the case of the DMPO-containing samples. The experiments were performed in duplicate.

The simulation of the EPR spectra of spin adducts formed by the DMPO and CPH spin traps with the ROS generated by TiO_2 samples was performed using the WinSim program [32,33].

3.3. Preparation of the Carbon Paste Electrodes

Graphite powder (GP) (<20 μm, synthetic, Sigma-Aldrich), mineral oil (MO) (Sigma-Aldrich), and copper oxides-TiO_2 graphene oxide powders were used for the preparation of the electrodes. A certain amount of graphite powder was placed in a mortar and pestle and was thoroughly hand mixed for 40 min with paraffin oil, until a consistent uniformly wetted paste was obtained. The GP:MO ratio was approximately 3:1 (w/w). The obtained paste was placed into a plastic syringe of 1.0 mL. The electrical contact was made by forcing a copper wire down into the syringe and into the back of the graphite paste. The surface of the electrodes was obtained by polishing it on a weighing paper and, when it was necessary, a renewed surface was made by pushing a small excess of the paste out of the tube and polishing it again. The bare carbon paste electrode will be denoted by CPE. The modified electrodes were prepared by mixing certain amounts of carbon paste with Cu_2O/CuO-decorated TiO_2 graphene oxide nanocomposites (97:3 w/w) and were denoted by TC1, TC2, TC3, TC1-GO, TC2-GO, and TC3-GO. The obtained materials were pressed at the end of carbon paste from syringes. For comparison, a few electrodes were also prepared as follows: one was left unmodified (CPE), one modified with graphene oxide (denoted by GO), and one modified with TiO_2 powder (denoted TiO_2). The surface of all electrodes was smoothed by polishing on a piece of weighing paper. All electrodes were kept in distilled water before and after measurements.

Electrochemical measurements were carried out in duplicate on a potentiostat galvanostat system AutoLabPGStat 12, controlled by a general purpose electrochemical system (GPES) with interface for Windows (version 4.9.007). Three electrodes in a one compartment cell (10 mL) were used in all experiments. A glassy carbon electrode (Metrohm, 3 mm in diameter) and carbon-paste electrodes (unmodified and modified) served as working electrodes. The counter electrode was a Pt wire of large area. All experimental potentials were referred to Ag/AgCl/KClsat as reference electrodes.

3.4. Testing the Modified Carbon Paste Electrodes

The electrochemical characterization of the modified CPE electrodes was carried out by cyclic voltammetry (CV) and differential pulse voltammetry (DPV). The CV experiments were recorded in 0.1 mol L^{-1} KCl solution containing 1.0 mmol L^{-1} K_3[Fe(CN)$_6$] in the potential range of (-1) to ($+1.2$) V at scan rates of 50 to 150 mV s^{-1}. DPV curves were recorded on the same potential domains at step potential (SP) 10 mV and modulation amplitude (MA) 25 mV.

All modified electrodes were tested for the redox process of 1 mM potassium ferrocyanide(III) using 0.1 M KCl as electrolyte, being an one-electron reversible redox system.

$$[Fe(CN)_6]^{3-} + e^- \leftrightarrow [Fe(CN)_6]^{4-}$$

A control experiment was first performed utilizing a bare carbon paste electrode. The voltammetric profile of bare CPE and copper oxides-TiO_2 ± graphene oxide-modified electrodes was explored by sweep rate variation from 50 to 150 mV/s.

4. Conclusions

The enhanced free radical generation and electrochemical response of Cu_2O/CuO-decorated TiO_2/graphene oxide nanocomposites are associated with modifications of transition metal oxides. Their electron-accepting properties may enhance the oxidation of analyte when anodic scans are performed. The potential application of the demonstrated electrochemical properties of modified electrodes with nanostructured oxides may increase the efficiency of drug detection in electroanalysis through electro-catalytic effects. Additionally, it was shown for the first time that the obtained modified TiO_2 nanocomposites transfer electrons under UV irradiation and generate hydroxyl and superoxide radicals reactive oxygen species (ROS), as emphasized by EPR spectroscopy. Further applications of these new nanomaterials could be bacterial inactivation; obtaining self-cleaning surfaces; sensors; and various uses in environmental remediation.

Author Contributions: Conceptualization I.S. and G.I.; methodology, M.S. and C.G.; validation, D.B. and M.V.; formal analysis, M.B.; investigation, D.-V.C. and M.-C.R.; writing—original draft preparation, M.-C.R., I.M., G.I. and D.B.; writing—review and editing, I.S.; supervision, C.G.; and funding acquisition, M.S. All authors have read and agreed to the published version of the manuscript.

Funding: This work was supported by grants of the Romanian Ministry of Research and Innovation, UEFISCDI, contract no. 44PCCDI/2018 (**PhysForTeL**), project **no.** PN-III-P1-1.1-PD-2021-0189 (**GAMMA-COLL**), contract **no.** 71/2022; project no. 4N/2019–PN 19 17 01 02/2022 (**CREATIV_PIEL**) and project no. 10N/2019 (Interdisciplinary applications of gamma irradiation) National Program Nucleus. The APC was funded by all the authors.

Institutional Review Board Statement: Not applicable.

Informed Consent Statement: Not applicable.

Data Availability Statement: Data are available on request from the authors.

Conflicts of Interest: The authors declare no conflict of interest. The funders had no role in the design of the study; in the collection, analyses, or interpretation of data; in the writing of the manuscript, or in the decision to publish the results.

References

1. Brezová, V.; Billik, P.; Vrecková, Z.; Plesch, G. Photoinduced formation of reactive oxygen species in suspensions of titania mechanochemically synthesized from $TiCl_4$. *J. Mol. Catal. A Chem.* **2010**, *327*, 101–109. [CrossRef]
2. Do Kim, K.; Lee, T.J.; Kim, H.T. Optimal conditions for synthesis of TiO_2 nanoparticles in semi-batch reactor. *Colloids Surf. A* **2003**, *224*, 1–9. [CrossRef]
3. Kim, I.; Kim, G.; Choi, Y.; Lee, W.; Smith, C., Jr.; Kim, Y. Method for Synthesizing Nano-Sized Titanium Dioxide Particles. US Patent WO 2006/044495 A1, 27 April 2006.
4. Yang, J.; Mei, S.; Ferreira, J. Hydrothermal synthesis of TiO_2 nanopowders from tetraalkylammonium hydroxide peptized sols. *Mater. Sci. Eng. C* **2001**, *15*, 183–185. [CrossRef]
5. Zhang, C.J.; Anasori, B.; Seral-Ascaso, A.; Park, S.H.; McEvoy, N.; Shmeliov, A.; Dues-berg, G.S.; Coleman, J.N.; Gogotsi, Y.; Nicolosi, V. Transparent, flexible, and conductive 2D titanium carbide (MXene) films with high volumetric capacitance. *Adv. Mater.* **2017**, *29*, 1702678. [CrossRef] [PubMed]
6. Ajmala, A.; Majeed, I.; Malik, R.N.; Iqbal, M.; Arif Nadeem, M.; Hussain, I.; Yousaf, S.; Zeshan, M.G.; Zafar, M.I.; Amtiaz Nadeem, M. Photocatalytic degradation of textile dyes on Cu_2O-CuO/TiO_2 anatase powders. *J. Environ. Chem. Eng.* **2016**, *4*, 2138–2146. [CrossRef]
7. Padmanabhan, N.T.; Thomas, N.; Louis, J.; Mathew, D.T.; Ganguly, P.; John, H.; Pillai, S.C. Graphene coupled TiO_2 photocatalysts for environmental applications: A review. *Chemosphere* **2021**, *271*, 129506. [CrossRef]
8. Darabi, H.; Adelifard, M.; Rajabi, Y. Characterization of nonlinear optical refractive index for graphene oxide–silicon oxide nanohybrid composite. *J. Nonlinear Opt. Phys. Mater.* **2019**, *28*, 1950005. [CrossRef]
9. Safa, M.; Rajabi, Y.; Ardyanian, M. Influence of preparation method on the structural, linear, and nonlinear optical properties of TiN nanoparticles. *J. Mate-Rials Sci. Mater. Electron.* **2021**, *32*, 19455–19477. [CrossRef]
10. Dadkhah, S.; Rajabi, Y.; Zare, E.N. Thermal Lensing Effect in Laser Nanofluids Based on Poly (aniline-co-ortho phenylenediamine) @TiO_2 Interaction. *J. Electron. Mater.* **2021**, *50*, 4896–4907. [CrossRef]
11. Altin, I. CuO-TiO_2/graphene ternary nanocomposite for highly efficient visible-light-driven photocatalytic degradation of bisphenol A. *J. Mol. Struct.* **2022**, *1252*, 132199. [CrossRef]
12. Mirza-Aghayan, M.; Saeedi, M.; Boukherroub, R. An efficient $CuO/rGO/TiO_2$ photocatalyst for the synthesis of benzopyranopyrimidine compounds under visible light irradiation. *New J. Chem.* **2022**, *46*, 3817–3830. [CrossRef]
13. Fan, Y.; Huang, K.-J.; Niu, D.-J.; Yang, C.-P.; Jing, Q.-S. TiO_2-graphene nanocomposite for electrochemical sensing of adenine and guanine. *Electrochim. Acta* **2011**, *56*, 4685–4690. [CrossRef]
14. Liu, A.; Wei, M.D.; Honma, I.; Zhou, H. Biosensing properties of titanate nanotube films: Selective detection of dopamine in the presence of ascorbate and uric acid. *Adv. Funct. Mater.* **2006**, *16*, 371–376. [CrossRef]
15. Tavakkoli, N.; Soltani, N.; Salavati, H.; Talakoub, M. New carbon paste electrode modified with graphene/TiO_2/V_2O_5 for electrochemical measurement of chlorpromazine hydrochloride. *J. Taiwan Inst. Chem. Eng.* **2018**, *83*, 50–58. [CrossRef]

16. Choudhury, B.; Dey, M.; Choudhury, A. Shallow and deep trap emission and luminescence quenching of TiO_2 nanoparticles on Cu doping. *Appl. Nanosci.* **2014**, *4*, 499–506. [CrossRef]
17. Kernazhitsky, L.; Shymanovska, V.; Gavrilko, T.; Naumov, V.; Fedorenko, L.; Kshnyakin, V.; Baran, J. Room temperature photoluminescence of anatase and rutile TiO_2 powders. *J. Lumin.* **2014**, *146*, 199–204. [CrossRef]
18. Nair, R.V.; Gayathri, P.K.; Gummaluri, V.S.; Nambissan, P.M.G.; Vijayan, C. Large bandgap narrowing in rutile TiO_2 aimed towards visible light applications and its correlation with vacancy-type defects history and transformation. *J. Phys. D Appl. Phys.* **2018**, *51*, 045107. [CrossRef]
19. Janczarek, M.; Kowalska, E. On the origin of enhanced photocatalytic activity of copper-modified titania in the oxidative reaction systems. *Catalysts* **2017**, *7*, 317. [CrossRef]
20. Yadav, H.M.; Kim, J.S. Solvothermal synthesis of anatase TiO_2-graphene oxide nanocomposites and their photocatalytic performance. *J. Alloys Compd.* **2016**, *688*, 123–129. [CrossRef]
21. Liu, Y.; Wang, L.; Xue, N.; Wang, P.; Pei, M.; Guo, W. Ultra-highly efficient removal of methylene blue based on graphene oxide/TiO_2/bentonite sponge. *Materials* **2020**, *13*, 824. [CrossRef] [PubMed]
22. Hagen, W.R. EPR spectroscopy as a probe of metal centres in biological systems. *Dalton Trans.* **2006**, *37*, 4415–4434. [CrossRef] [PubMed]
23. Dvoranová, D.; Barbieriková, Z.; Brezová, V. Radical intermediates in photoinduced reactions on TiO_2 (An EPR spin trapping study). *Molecules* **2014**, *19*, 17279–17304. [CrossRef]
24. Carter, E.; Carley, A.F.; Murphy, D.M. Evidence for O_2-radical stabilization at surface oxygen vacancies on polycrystalline TiO_2. *J. Phys. Chem. C* **2007**, *111*, 10630–10638. [CrossRef]
25. Hellack, B.; Nickel, C.; Schins, R.P.F. Oxidative potential of silver nanoparticles measured by electron paramagnetic resonance spectroscopy. *J. Nanopart. Res.* **2017**, *19*, 404. [CrossRef]
26. Pou, S.; Ramos, C.L.; Gladwell, T.; Renks, E.; Centra, M.; Young, D.; Cohen, M.S.; Rosen, G.M. A Kinetic Approach to the Selection of a Sensitive Spin Trapping System for the Detection of Hydroxyl Radical. *Anal. Biochem.* **1994**, *217*, 76–83. [CrossRef] [PubMed]
27. Buettner, G.R. Spin trapping: ESR parameters of spin adducts. *Free Radic. Biol. Med.* **1987**, *3*, 259–303. [CrossRef]
28. Lin, W.J.; Liao, C.S.; Jhang, J.H.; Tsai, Y.C. Graphene modified basal and edge plane pyrolytic graphite electrodes for electrocatalytic oxidation of hydrogen peroxide and β-nicotinamide adenine dinucleotide. *Electrochem. Commun.* **2009**, *11*, 2153. [CrossRef]
29. Cosma, D.; Urda, A.; Radu, T.; Rosu, M.C.; Mihet, M.; Socaci, C. Evaluation of the Photocatalytic Properties of Copper Oxides/Graphene/TiO_2 Nanoparticles Composites. *Molecules* **2022**, *27*, 5803. [CrossRef]
30. Pogacean, F.; Socaci, C.; Pruneanu, S.; Biris, A.R.; Coros, M.; Magerusan, L.; Katona, G.; Turcu, R.; Borodi, G. Graphene based nanomaterials as chemical sensors for hydrogen peroxide–A comparison study of their intrinsic peroxidase catalytic behavior. *Sens. Actuators B Chem.* **2015**, *213*, 474–483. [CrossRef]
31. Dikalov, S.S.; Grigor'ev, I.A.; Voinov, M.; Bassenge, E. Detection of superoxide radicals and peroxynitrite by 1-hydroxy-4-phosphonooxy-2,2,6,6-tetramethylpiperidine: Quantification of extracellular superoxide radicals formation. *Biochem. Biophys. Res. Commun.* **1998**, *248*, 211–215. [CrossRef]
32. Duling, D.R. *PEST Winsim, version 0.96*; National Institute of Environmental Health Sciences: Triangle Park, NC, USA, 1996.
33. Duling, D.R. Simulation of multiple isotropic spin-trap EPR spectra. *J. Magn. Reson. B* **1994**, *104*, 105–110. [CrossRef] [PubMed]

"Turn on" Fluorescence Sensor of Glutathione Based on Inner Filter Effect of Co-Doped Carbon Dot/Gold Nanoparticle Composites

Thi-Hoa Le, Ji-Hyeon Kim and Sang-Joon Park *

Department of Chemical and Biological Engineering, Gachon University, Seongnam 13120, Korea; hoale2907@gmail.com (T.-H.L.); jihyeon@gachon.ac.kr (J.-H.K.)
* Correspondence: psj@gachon.ac.kr

Abstract: Glutathione (GSH) is a thiol that plays a significant role in nutrient metabolism, antioxidant defense and the regulation of cellular events. GSH deficiency is related to variety of diseases, so it is useful to develop novel approaches for GSH evaluation and detection. In this study we used nitrogen and phosphorus co-doped carbon dot-gold nanoparticle (NPCD–AuNP) composites to fabricate a simple and selective fluorescence sensor for GSH detection. We employed the reductant potential of the nitrogen and phosphorus co-doped carbon dots (NPCDs) themselves to form AuNPs, and subsequently NPCD–AuNP composites from Au^{3+}. The composites were characterized by using a range of spectroscopic and electron microscopic techniques, including electrophoretic light scattering and X-ray diffraction. The overlap of the fluorescence emission spectrum of NPCDs and the absorption spectrum of AuNPs resulted in an effective inner filter effect (IFE) in the composite material, leading to a quenching of the fluorescence intensity. In the presence of GSH, the fluorescence intensity of the composite was recovered, which increased proportionally to increasing the GSH concentration. In addition, our GSH sensing method showed good selectivity and sensing potential in human serum with a limit of detection of 0.1 μM and acceptable results.

Keywords: glutathione; NPCD–AuNP composites; inner filter effect (IFE); "turn on" fluorescence

1. Introduction

Glutathione (GSH) is a tripeptide compound produced by the liver that is composed of three different amino acids: cysteine, glutamic acid and glycine (ɣ-Glu-Cys-Gly) [1]. GSH acts as an antioxidant and is involved in many processes in the body, including tissue building and repair, cysteine transport and storage, signal transduction modulation, cell propagation control and immune system response regulation [2,3]. Abnormal levels of GSH have been reported to contribute to oxidative stress, which plays a key role in aging, and in a variety of diseases, including cancer, neurotoxicity, psoriasis, liver damage, Parkinson's disease, and Alzheimer's disease. Therefore, GSH is considered a universal biomarker in the diagnosis and therapeutic monitoring of cancers [4,5].

Many techniques have been used for the identification and quantification of GSH, such as HPLC [6], electrochemistry [7], fluorescence [8], electrochemiluminescence [9] and surface enhanced Raman scattering (SERS) [10].

Fluorescence-based methods are widely used for sensing different analytes based on photoinduced electron transfer (PET), photoinduced charge transfer (PCT), fluorescence resonance energy transfer (FRET) and the inner filter effect (IFE). PET and PCT are intramolecular processes, whereas IFE and FRET are intermolecular processes, all of which involve energy transfer between at least two independent molecules [11–14]. We developed FRET-based sensors in previous studies and designed detection methods that reply via IFE in this study. The IFE is a decrease in the fluorescence intensity of a fluorophore (donor) due to the absorption of the excitation/emission light by an absorber (acceptor). This occurs when the absorption spectrum of the absorber overlaps with the fluorescence excitation or emission spectrum of the fluorophore [14]. The IFE is classified as primary when the excitation beam is attenuated by the volume of the sample, and secondary when the emission of the fluorophore is absorbed by a molecule present in the solution [15]. Unlike FRET-based sensors, IFE-based sensors do not provide site-specific information and cannot be used to determine donor–acceptor interactions. However, the IFE has been shown to be useful for converting analytical absorption signals into fluorescence signals, which has been proven to enhance sensitivity and selectivity [16,17]. Moreover, IFE-based sensors do not require any surface modification, or covalent linking between a donor and acceptor, which makes probe fabrication simpler and flexible [18].

Recently, the combination of carbon dots (CDs) and gold nanoparticles (AuNPs) has been shown to be an effective IFE pair because of the outstanding properties of both materials. CDs are a promising class of fluorescent carbon materials that have attracted considerable interest because of their distinctive properties, such as excellent solubility, low toxicity, good biocompatibility, high photostability, tunable surface functionalities and cell membrane permeability [19,20]. AuNPs have been widely used in fundamental studies for a variety of applications. It is worth noting that the AuNP solutions have different colors depending on their size distributions and morphologies, which is advantageous for the recognition of different target molecules [21–23]. In addition, AuNPs are ideal fluorescent absorbers because of their high extinction coefficients, and extensive absorption spectrum that can easily overlap with the excitation/emission spectra of CDs to produce an IFE process [24,25].

Many previous studies have used CD/AuNP pair-based IFE mechanisms for sensing analytes, such as hyaluronidase [26], aldicarb [27], metformin hydrochloride [28], cyanide ion [29] and biothiols. For the analytes, AuNPs were obtained via the reduction of chloroauric acid using tri-sodium citrate as a reductant and surfactant. This results in several negatively charged citrate ions on the surfaces of AuNPs, which may affect the IFE process. Most fluorescence-based sensors have focused on the fluorescence properties of CDs. However, it is beneficial to take full advantage of CDs, including their reducing and stabilizing potential. CDs synthesized using the hydrothermal method contain a large number of carboxyl and hydroxyl groups [30]. Negatively charged carboxyl groups on the CDs can stabilize metal particles in solution [31]. In addition, the hydroxyl groups allow CDs to act as green reducing agents in the synthesis of metal nanoparticles [32]. Hence, the CDs have two important roles: (1) a fluorescence source, and (2) a reducing agent and stabilizer for AuNPs in the fabrication of CD/AuNP pair-based IFE sensors.

Another novelty of this research is the use of nitrogen and phosphorus co-doped carbon dots (NPCDs) in an IFE-based sensor for GSH detection. NPCDs possess enhanced fluorescence properties and more negatively charged functional groups and amine groups on the surface compared to pristine CDs [19]. The amine and thiol groups show strong binding affinity to the surfaces of AuNPs [33,34], which can increase the effectiveness of the IFE process between NPCDs and AuNPs. NPCDs can successfully reduce $HAuCl_4$ salt to AuNPs without adding any other reductant and surfactant. Once AuNPs are formed, NPCDs act as fluorescence donors that trigger the IFE process with AuNP acceptors, which leads to the quenching of the fluorescence of NPCDs. The fluorescence of NPCDs is "turned on" after GSH is added. The introduction of GSH to the AuNP system can induce aggregation of AuNPs [24]. In addition, because of its specific multidentate and steric structure, GSH molecules are able to preferentially enclose AuNPs and isolate them from NPCDs. Hence, the interaction between NPCDs and AuNPs can be disrupted by GSH, leading to a decrease in the IFE and a restoration in the fluorescence of NPCDs. The mechanism of "turn on" fluorescence for detecting GSH is illustrated in Scheme 1. Compared to "turn off" fluorescence sensors, the "turn on" fluorescence sensors are preferable because the likelihood of false positives is reduced.

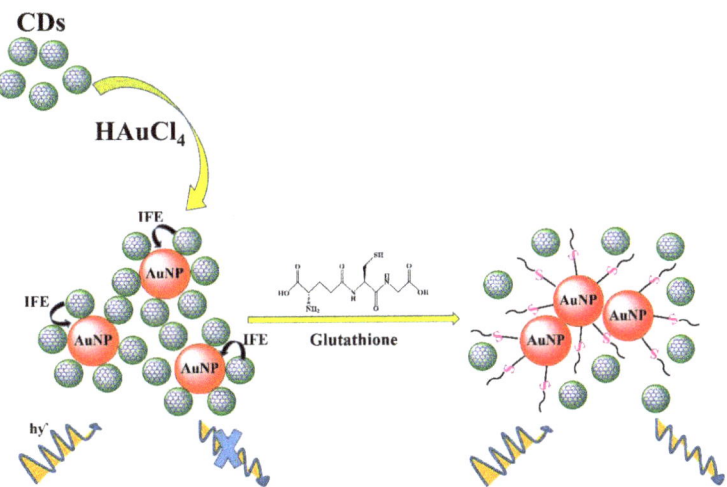

Scheme 1. Demonstration of the glutathione detection based on inner filter effect happening in NPCD–AuNP composite material.

2. Chemicals and Experiments

2.1. Chemicals

All chemicals including $(NH_4)_2HPO_4$, citric acid monohydrate, $HAuCl_4 \cdot 3H_2O$, glutathione, cysteine, homocysteine, alanine, glycine, lysine, tryptophan, methionine, glutamic acid, human serum and deionized (DI) water were obtained from Sigma-Aldrich.

2.2. Instruments

Photoluminescence (PL) and UV–Vis spectra were obtained using a QuantaMaster TM 50 PTI spectrofluorometer (Photon Technology International, San Diego, CA, USA) and a G1103A UV–Vis spectrophotometer (Agilent, Santa Clara, CA, USA). Size distribution measurements were performed using electrophoretic light scattering (ELS; Photal Otsuka Electronics, ELS 8000, Osaka, Japan). The structural and morphological characterization of the samples was carried out by scanning electron microscopy (SEM; S-4700, Hitachi Ltd., Tokyo, Japan) and transmission electron microscopy (TEM; Tecnai, F30S-Twin, Hillsboro, OR, USA). X-ray photoelectron spectroscopy (XPS) and powder X-ray diffraction (XRD) were performed using an X-ray photoelectron spectrometer (PHI 5000, Chigasaki, Kanagawa Prefecture, Japan) and X-ray source (Rigaku/Smartlab, Tokyo, Japan), respectively.

2.3. Preparation of NPCDs

NPCDs were fabricated according to the method presented in our previous work [19]. In particular, $(NH_4)_2HPO_4$ (5 g) and citric acid (2 g) monohydrate (molar ratio of 4:1) were mixed with DI water (30 mL). Subsequently, the mixture was transferred into a 50 mL Teflon-lined stainless-steel autoclave and heated at 180 °C for 4 h. After cooling, the solution was purified by filtering through a 0.22-μm polyethersulfone membrane into a dialysis bag (MWCO: 1000 Da) to dialyze for 48 h. The obtained light-violet colored solution was lyophilized to collect a powdery substance.

2.4. Synthesis of NPCD–AuNP Composites

The NPCD solution (1.2 mL, 0.075 g/mL) was added to DI water (10 mL). The solution was heated to boiling before 1% $HAuCl_4$ (100 μL) was dropped into the solution. The color of the solution quickly changed to purple and then burgundy after 10 min, indicating the successful formation of AuNPs.

2.5. Fluorescence Sensing of GSH

GSH sensing was carried out at room temperature. First, NPCD–AuNP solution (300 µL) was added to DI water (2700 µL) to obtain a solution concentration of 9.05×10^{-4} g/mL. Subsequently, different volumes of GSH solution were added via dropping into the NPCD–AuNP solution to obtain various concentrations from 0.1 to 300 µM. The solutions were further diluted to the mark using DI water and kept in equilibrium for 60 min before fluorescence measurements were made with an excitation wavelength of 340 nm. Each concentration was measured three times to determine the standard deviation.

2.6. Detection of GSH in Human Serum

Human serum (30 µL) was added to DI water (2670 µL) followed by NPCD–AuNP solution (300 µL). A quantity of GSH was dropped into the solution, followed by dilution to the mark with DI water before fluorescence analysis without pretreatment. The sensing of GSH in human serum samples was performed using the method described above.

2.7. Selectivity

To study the GSH specificity of the NPCD–AuNP-based fluorescence sensing approach, the fluorescence responses of other biothiols, including HCys, Cys and some amino acids, were investigated.

3. Results and Discussion

3.1. Characterization of NPCDs-AuNPs Composites

UV–Vis absorption and fluorescence spectroscopy were used to characterize the optical properties of the NPCDs and NPCD–AuNP composites. The UV–Vis spectrum of the aqueous NPCD solution (Figure 1) shows a peak appears at 234 nm and another at 334 nm, which originate from the surface states [20,35,36]. There is a peak located at 521 nm in the spectrum of AuNP solution synthesized using sodium citrate as a reductant. The maximum absorption peak of AuNPs can redshift with an increase in size of AuNPs [37]. The spectrum of an aqueous solution of NPCD–AuNP composites shows the same peak of NPCDs at 334 nm and a specific peak at 548 nm, which confirms the successful reduction of NPCDs to form AuNPs.

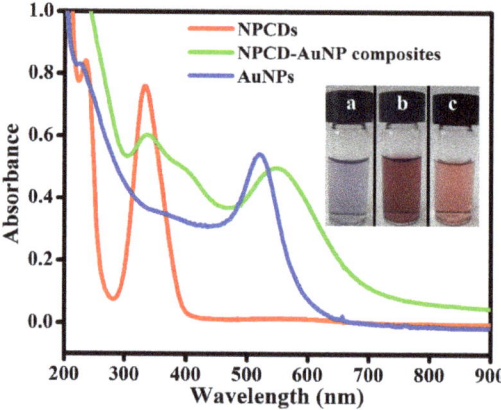

Figure 1. UV–Vis spectra of NPCDs, NPCD–AuNP composites and AuNPs (inset: aqueous solution of NPCDs (**a**), NPCD–AuNP composites (**b**) and AuNPs (**c**)).

The NPCD and NPCD–AuNP composites' fluorescence emission peaks appeared at 447 and 442 nm under excitation wavelengths of 355 and 340 nm, respectively (Figure 2). Therefore, there was a blue shift in the emission and excitation peaks for the NPCD–AuNP composites. In comparison to the fluorescence intensity of the NPCDs, that of the NPCD–AuNP composites was considerably quenched. As shown in Figure 1, the absorption spectrum of the NPCD–AuNP composites presents a broad peak between 440 and 650 nm with a peak maximum at 548 nm because of the presence of AuNP components, and the emission peak is located at 442 nm due to the NPCD components. Hence, an overlap between the absorption and emission peaks in the composite material easily produced an IFE process, leading to the quenching of fluorescence intensity.

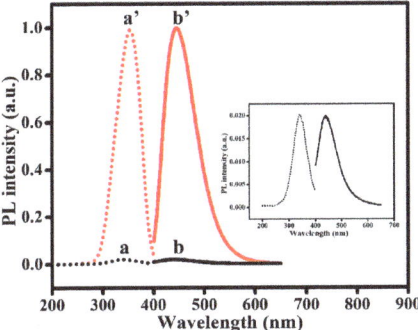

Figure 2. Excitation (a, a') and emission (b, b') fluorescence spectra of the NPCDs (a', b') and NPCD–AuNP composites (a, b) (inset: enlarged fluorescence spectrum of NPCD–AuNP composites).

The morphology and size of the NPCD–AuNP composites were analyzed by SEM, TEM and DLS. Under the influence of NPCDs, the synthesized AuNPs were relatively monodispersed and spherical. They had a narrow size distribution range of 19.2 to 51.5 nm and an average size of 32.5 nm (Figure 3). A high-resolution (HR) TEM image (inset of Figure 3D) shows a crystal of the AuNPs, which was further characterized by powder XRD.

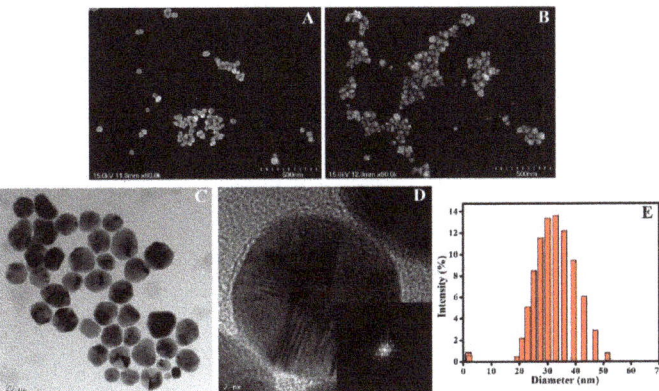

Figure 3. SEM images of NPCD–AuNP composites in the (**A**) absence and (**B**) presence of GSH. (**C**) TEM image. (**D**) High-resolution TEM image (inset: fast Fourier transform (FFT) pattern). (**E**) DLS analysis of NPCD–AuNP composites.

The powder XRD pattern of the NPCD–AuNP composites is shown in Figure 4. The composites exhibited six planes of (111), (200), (220), (311), (222) and (400), which were assigned to sharp peaks at 38.2°, 44.4°, 64.5°, 77.5°, 81.7° and 98.1° and lattice fringe distances of 2.36, 2.04, 1.44, 1.23, 1.18 and 1.02 Å, respectively. Hence, the pattern closely matches with the XRD pattern of pure crystalline Au° with face-centered cubic crystal structures (PDF 04-0784).

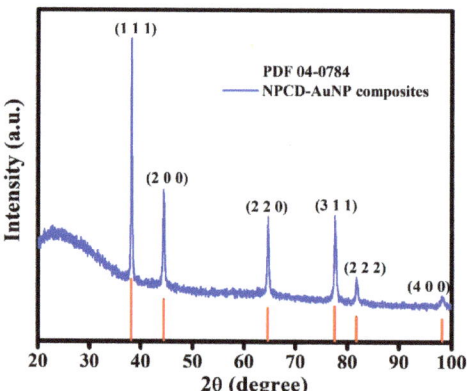

Figure 4. Powder XRD pattern of NPCD–AuNP composites.

The XPS profile (Figure 5A) presents typical peaks at 83.97, 133.71, 284.92, 336.14, 354.3, 400.5 and 532.33 eV, corresponding to Au4f, P2p, C1s, Au4d (Au4d$_{5/2}$ and Au4d$_{3/2}$), N1s and O1s, respectively. The atomic percentages of Au, C, O, N and P were found to be 5.35%, 39.18%, 45.61%, 6.98% and 2.87%, respectively. The high-resolution XPS profile of the Au4f peak is shown in Figure 5B. Two peaks of Au4f$_{5/2}$ and Au4f$_{7/2}$ located at 84.4 and 87.7 eV, respectively, correspond to the metallic state of Au^0. Figure 5C shows that the C1s peak can be parsed to four peaks at 284.2, 285, 286 and 288.4 eV, which are assigned as C-C/C=C, C-N/C-P, C-O and C=O bonds, respectively. These peaks originate from the sp^2 graphitic structure [38] and several carboxyl, hydroxyl, amine and phosphate groups on the surfaces of the NPCDs. These assignments can be further confirmed from the P2p and N1s spectra, which are similar to those measured in the our previous study [19].

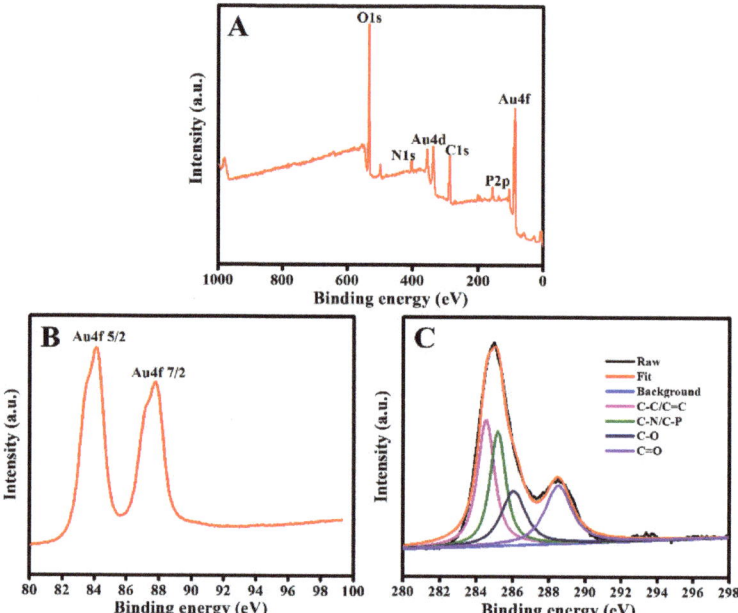

Figure 5. (**A**) XPS survey spectrum of NPCD–AuNP composites. High resolution XPS spectra of Au4f (**B**) and C1s (**C**).

3.2. Detection of GSH

3.2.1. Mechanism of Sensing

As previously discussed, the overlap of the absorption spectrum of the AuNP components and the emission spectrum of the NPCD components led to a significant quenching in the fluorescence intensity of the NPCD–AuNP composites compared to that of pure NPCDs. This overlap is illustrated in Figure 6A. However, the intensity of the NPCD–AuNP composites recovered in the presence of GSH. This phenomenon did not occur with NPCDs. In other words, GSH cannot affect the fluorescence of NPCDs but can turn on that of the NPCD–AuNP composites (Figure 7). This is explained by the fact that GSH can bind to AuNP surfaces via Au–S bonding and subsequently induce the aggregation of AuNPs [34,39]. This can be clearly seen in the SEM images of the NPCD–AuNP composites in the presence and absence of GSH (Figure 3A,B). Moreover, the color of the composite solution changed from red to purple blue, and the absorbance peak of the composites redshifted (Figure 6B), further indicating the aggregation of AuNPs in the composites after the addition of GSH. Aggregation may limit the ability of AuNPs to act as acceptors in the IFE process. Furthermore, GSH shows a strong affinity to bind AuNPs through its multidentate ligands, and has a steric structure that can keep the bound AuNPs separate from the NPCDs, which results in fluorescence recovery of NPCDs.

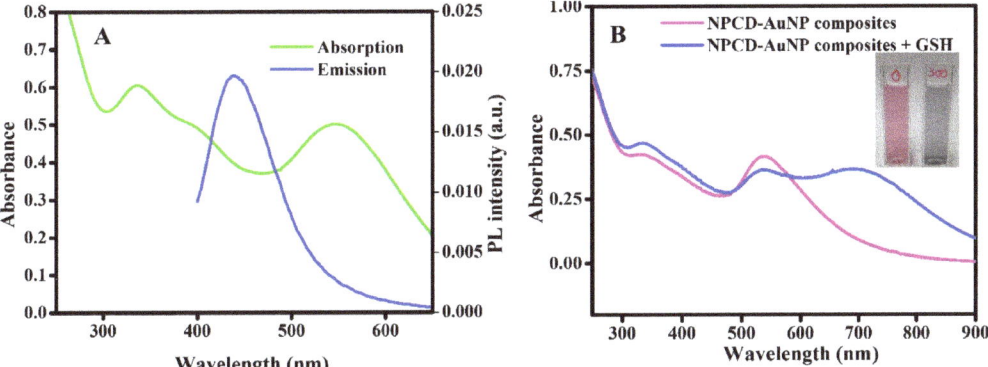

Figure 6. (**A**) Overlap of absorption peaks and emission peaks of NPCD–AuNP composites. (**B**) Absorption spectra (inset: image of colored solution) of NPCD–AuNP composites in the presence and absence of GSH (300 µM).

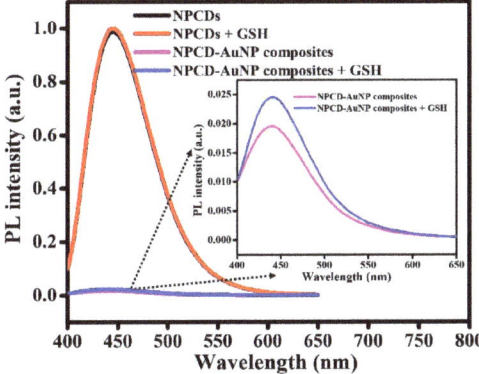

Figure 7. Variation in fluorescence intensity of NPCDs and NPCD–AuNP composites (inset: enlarged fluorescence spectra of NPCD–AuNP composites) in the presence and absence of GSH (40 µM).

After adding GSH to a fixed concentration of NPCD–AuNP composites, the fluorescence intensity was recorded every 10 min to investigate the kinetics of the IFE process. As shown in Figure 8, the intensity of the composites increased steadily in the presence of GSH from 0 to 60 min. Thereafter, the intensity remained relatively stable. Therefore, 60 min was chosen as the optimal time for the next step.

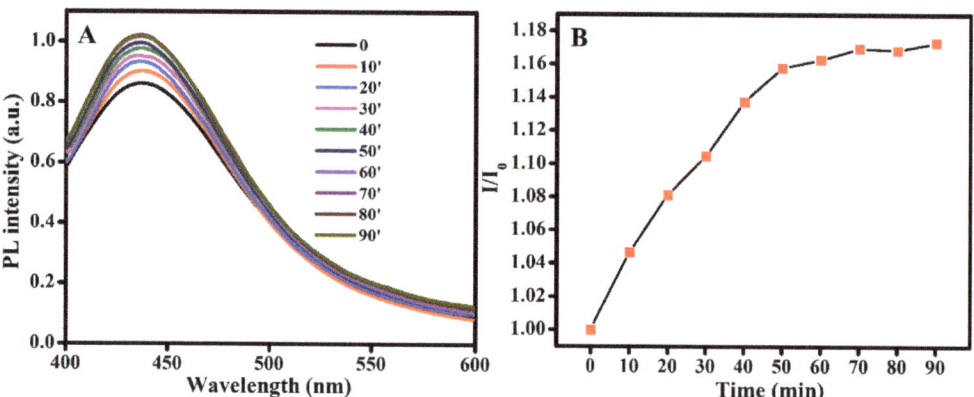

Figure 8. (**A**) Fluorescence intensity of NPCD–AuNP composites with GSH over time. (**B**) Relative fluorescence (I/I_0) of NPCD–AuNP composites over time with 20 μM GSH.

3.2.2. GSH Sensing

Figure 9 clearly shows the fluorescence recovery of the NPCD–AuNP composites in the presence of GSH. The fluorescence intensity of the composites increased steadily when the GSH concentration was increased from 0 to 300 μM. When the GSH level was within the range of 10–120 μM, there was a linear relationship between the NPCD–AuNP composite intensity and the GSH concentration, $I/I_0 = 1.22872 + 0.00303 C_M$, with a correlation coefficient (R^2) value of 0.99716. The measurement was performed three times, and the data were plotted with standard deviations. In addition, the limit of detection (LOD) value was determined to be approximately 0.1 μM. A comparison of the detection results using our approach with those of other approaches is presented in Table 1.

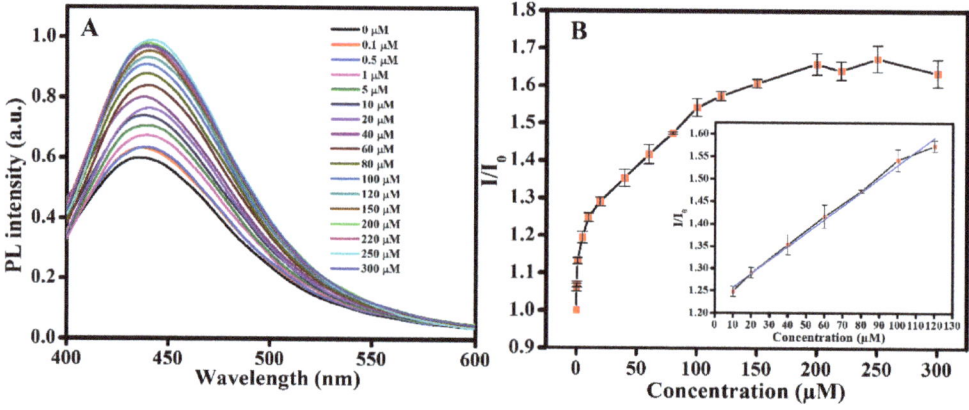

Figure 9. (**A**) Fluorescence spectra. (**B**) Relative fluorescence of NPCD–AuNP composites in the presence of various concentrations of GSH (inset: linear relationship between I/I_0 and GSH concentration in the range of 10 to 120 μM).

Table 1. Comparison of the GSH detection results of our method to those of other methods.

Method	Linear Range	LOD	Reference
Colorimetric		0.03 mM	[40]
Fluorescence	0.07–70 µM	48 nM	[41]
Fluorescence	0.5–300 µM	0.26 µM	[42]
Fluorescence	0–50 µM	0.19 µM	[43]
Electrochemical	10–250 µM	25 µM	[44]
Electrochemical	10 µM–500 mM	703 nM	[45]
Electrochemical	100 nM–10 mM	41.9 nM	[46]
Photoelectrochemical	1–10 µM	0.8 µM	[47]
Fluorescence	10–20 µM	0.1 µM	Our method

3.2.3. Sensing in Serum

To further investigate the practicality and reliability of the NPCD–AuNP composite-based sensor, we tested this material for GSH detection in human serum. The results are displayed in Table 2. The recovery range of GSH was 90.74–107.04%. The experiment was repeated three times, and all relative standard deviations (RSDs) were less than 5%. These results indicate that the method has potential applications for practical the detection of GSH.

Table 2. Determination of GSH in human serum.

Sample	Added (µM)	Founded (µM)	Recovery (%)	RSD (n = 3)
1	10	9.07	90.74	2.27
2	30	31.49	104.96	4.54
3	80	78.19	97.73	2.75
4	100	95.43	95.43	4.35
5	120	128.45	107.04	1.55

3.2.4. Selectivity

High specificity is a necessary condition for a good sensor. To confirm that NPCD–AuNP composites are specific to sensing GSH, the "turn on" fluorescence responses of other biothiols—including Cys; HCys; and amino acids, such as glycine (Gly), alanine (Ala), lycine (Lys), trytophan (Trp), glutamic acid (Glu) and methionine (Met)—were studied in the absence and presence of GSH. As shown in Figure 10, all biothiols, such as GSH, Cys and Hcys, enhanced the fluorescence intensity of the composites. On the other hand, the addition of an amino acid did not increase the fluorescence intensity of the composites or affect the role of the fluorescence "turn on" of GSH. The introduction of biothiols to the fluorescent NPCD–AuNP composites can induce the aggregation of AuNPs in the order of HCys >> Cys > GSH [48]. The intensity was the strongest in the presence of GSH, followed by HCys and Cys. This is explained by the high denticity of GSH, which is typically two or more parts of the molecule (–SH and –COO$^-$). This chelation makes the GSH–metal atom interactions stronger than interactions of metal atoms with the other two biothiols or amino acids that possess either a single –SH, weak amine or –COO$^-$ binding group [49]. In addition, the large steric hindrance effect created by GSH is greater than that of Cys or HCys, so it is much more able to coordinate stably to the metal atoms or ions [50]. In addition, many studies indicate that GSH is superior to Cys and HCys in the role of a capping ligand for semiconductor nanocrystals [51]. For these reasons, the interaction between GSH and AuNPs is much stronger than the interactions of the other molecules studied. This results in significant enhancement in the fluorescence intensity of the composites only when GSH is added, which ensures that the sensor based on NPCD–AuNP composite material is highly selective for GSH sensing.

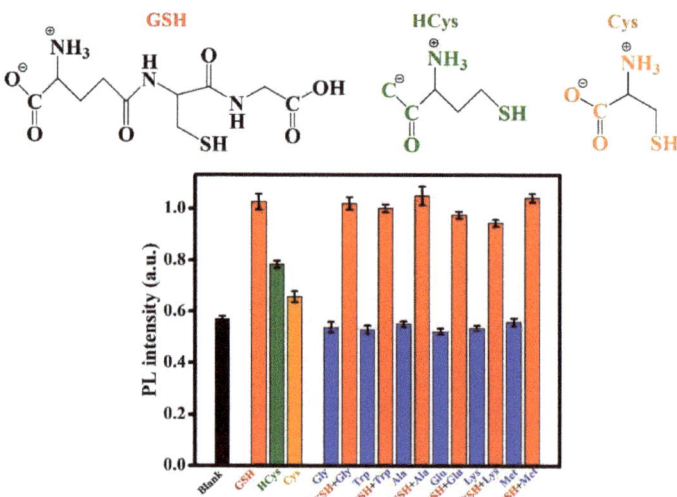

Figure 10. Fluorescence response of NPCD–AuNP composites on adding 200 µM of different biothiols and amino acids.

4. Conclusions

In summary, we successfully synthesized NPCD–AuNP composites, which are ideal IFE pairs for GSH sensing applications. We effectively utilized the enhanced fluorescence properties of NPCDs and their reducing nature to synthesize AuNPs and the subsequent composites from an Au^{3+} salt. The fluorescence of the NPCD–AuNP composites recovered steadily with increasing concentrations of GSH. A linear relationship of fluorescence against GSH concentration from 10 to 120 µM was determined, with an R^2 value of 0.99716 and LOD value of 0.1 µM. In comparison to the results of some previous studies, our sensing results were not as sensitive. However, our NPCD–AuNP composite-based GSH sensor does not require a complicated design or expensive instruments. Its simplicity, sensitivity, selectivity and effectivity make it a potentially viable sensor for GSH detection. Thus, we have introduced a new strategy for development of simple sensors for biomedical applications.

Author Contributions: Conceptualization, S.-J.P.; Data curation, J.-H.K.; Investigation, T.-H.L.; Methodology, J.-H.K.; Writing—original draft, T.-H.L.; Writing—review & editing, S.-J.P. All authors have read and agreed to the published version of the manuscript.

Funding: This research was supported by the Basic Science Capacity Enhancement Project through Korea Basic Science Institute (National Research Facilities and Equipment Center) grant funded by the Ministry of Education (grant number 2019R1A6C1010016).

Conflicts of Interest: The authors declare no conflict of interest.

References

1. Saydam, N.; Kirb, A.; Demir, O.; Hazan, E.; Oto, O.; Saydam, O.; Guner, G. Determination of glutathione, glutathione reductase, glutathione peroxidase and glutathione S-transferase levels in human lung cancer tissues. *Cancer Lett.* **1997**, *119*, 13–19. [CrossRef]
2. Tsiasioti, A.; Tzanavaras, P.D. Determination of glutathione and glutathione disulfide using zone fluidics and fluorimetric detection. *Talanta* **2021**, *222*, 121559. [CrossRef]
3. Sen, C.K. Nutritional biochemistry of cellular glutathione. *J. Nutr. Biochem.* **1997**, *8*, 660–672. [CrossRef]
4. Torres, S.; Matías, N.; Baulies, A.; Nuñez, S.; Alarcon-Vila, C.; Martinez, L.; Nuño, N.; Fernandez, A.; Caballeria, J.; Levade, T.; et al. Mitochondrial GSH replenishment as a potential therapeutic approach for niemann pick type C disease. *Redox Biol.* **2017**, *11*, 60–72. [CrossRef]

5. Kong, F.; Liang, Z.; Luan, D.; Liu, X.; Xu, K.; Tang, B. A glutathione (GSH)-responsive near-infrared (NIR) theranostic prodrug for cancer therapy and imaging. *Anal. Chem.* **2016**, *88*, 6450–6456. [CrossRef]
6. Giustarini, D.; Dalle-Donne, I.; Colombo, R.; Milzani, A.; Rossi, R. An improved HPLC measurement for GSH and GSSG in human blood. *Free Radic. Biol. Med.* **2003**, *35*, 1365–1372. [CrossRef]
7. Noh, H.-B.; Chandra, P.; Moon, J.O.; Shim, Y.-B. In vivo detection of glutathione disulfide and oxidative stress monitoring using a biosensor. *Biomaterials* **2012**, *33*, 2600–2607. [CrossRef] [PubMed]
8. Yang, C.; Deng, W.; Liu, H.; Ge, S.; Yan, M. Turn-on fluorescence sensor for glutathione in aqueous solutions using carbon dots–MnO_2 nanocomposites. *Sens. Actuators B* **2015**, *216*, 286–292. [CrossRef]
9. Niu, W.-J.; Zhu, R.-H.; Cosnier, S.; Zhang, X.-J.; Shan, D. Ferrocyanide-ferricyanide redox couple induced electrochemiluminescence amplification of carbon dots for ultrasensitive sensing of glutathione. *Anal. Chem.* **2015**, *87*, 11150–11156. [CrossRef]
10. Saha, A.; Jana, N.R. Detection of cellular glutathione and oxidized glutathione using magnetic—Plasmonic nanocomposite-based "turn-off" surface enhanced raman scattering. *Anal. Chem.* **2013**, *85*, 9221–9228. [CrossRef]
11. Suzuki, Y.; Yokoyama, K. Development of functional fluorescent molecular probes for the detection of biological substances. *Biosensors* **2015**, *5*, 337–363. [CrossRef]
12. Sauer, M. Single-molecule-sensitive fluorescent sensors based on photoinduced intramolecular charge transfer. *Angew. Chem. Int. Ed.* **2003**, *42*, 1790–1793. [CrossRef]
13. Ma, Y.Q.; Pandzic, E.; Nicovich, P.R.; Yamamoto, Y.; Kwiatek, J.; Pageon, S.V.; Benda, A.; Rossy, J.; Gaus, K. An intermolecular FRET sensor detects the dynamics of T cell receptor clustering. *Nat. Commun.* **2017**, *8*, 15100. [CrossRef]
14. Al-Hashimi, B.; Omer, K.M.; Rahman, H.S. Inner filter effect (IFE) as a simple and selective sensing platform for detection of tetracycline using milk-based nitrogen-doped carbon nanodots as fluorescence probe. *Arab. J. Chem.* **2020**, *13*, 5151–5159. [CrossRef]
15. Panigrahi, S.K.; Mishra, A.K. Inner filter effect in fluorescence spectroscopy: As a problem and as a solution. *J. Photochem. Photobiol. C Photochem. Rev.* **2019**, *41*, 100318. [CrossRef]
16. Chang, H.C.; Ho, J.A.A. Gold nanocluster-assisted fluorescent detection for hydrogen peroxide and cholesterol based on the inner filter effect of gold nanoparticles. *Anal. Chem.* **2015**, *87*, 10362–10367. [CrossRef]
17. Yan, X.; Li, H.X.; Han, X.S.; Su, X.G. A ratiometric fluorescent quantum dots based biosensor for organophosphorus pesticides detection by inner-filter effect. *Biosens. Bioelectron.* **2015**, *74*, 277–283. [CrossRef]
18. He, H.R.; Li, H.; Mohr, G.; Kovacs, B.; Werner, T.; Wolfbeis, O.S. Novel type of ion-selective fluorosensor based on the inner filter effect—An optrode for potassium. *Anal. Chem.* **1993**, *65*, 123–127. [CrossRef]
19. Le, T.H.; Lee, H.J.; Kim, J.H.; Park, S.J. Detection of ferric ions and catecholamine neurotransmitters via highly fluorescent heteroatom co-doped carbon dots. *Sensors* **2020**, *20*, 3470. [CrossRef]
20. Ming, F.L.; Hou, J.Z.; Hou, C.J.; Yang, M.; Wang, X.F.; Li, J.W.; Huo, D.Q.; He, Q. One-step synthesized fluorescent nitrogen doped carbon dots from thymidine for Cr (VI) detection in water. *Spectroc. Acta Part A-Molec. Biomolec. Spectr.* **2019**, *222*, 117165. [CrossRef]
21. Vilela, D.; González, M.C.; Escarpa, A. Sensing colorimetric approaches based on gold and silver nanoparticles aggregation: Chemical creativity behind the assay. A review. *Anal. Chim. Acta* **2012**, *751*, 24–43. [CrossRef]
22. Kailasa, S.K.; Koduru, J.R.; Desai, M.L.; Park, T.J.; Singhal, R.K.; Basu, H. Recent progress on surface chemistry of plasmonic metal nanoparticles for colorimetric assay of drugs in pharmaceutical and biological samples. *TrAC Trends Anal. Chem.* **2018**, *105*, 106–120. [CrossRef]
23. Kateshiya, M.R.; George, G.; Rohit, J.V.; Malek, N.I.; Kumar Kailasa, S. Ractopamine as a novel reagent for the fabrication of gold nanoparticles: Colorimetric sensing of cysteine and Hg^{2+} ion with different spectral characteristics. *Microchem. J.* **2020**, *158*, 105212. [CrossRef]
24. Qin, X.; Yuan, C.; Chen, Y.; Wang, Y. A fluorescein—Gold nanoparticles probe based on inner filter effect and aggregation for sensing of biothiols. *J. Photochem. Photobiol. B* **2020**, *210*, 111986. [CrossRef]
25. Chen, S.; Yu, Y.-L.; Wang, J.-H. Inner filter effect-based fluorescent sensing systems: A review. *Anal. Chim. Acta* **2018**, *999*, 13–26. [CrossRef]
26. Lu, H.Z.; Quan, S.; Xu, S.F. Highly sensitive ratiometric fluorescent sensor for trinitrotoluene based on the inner filter effect between gold nanoparticles and fluorescent nanoparticles. *J. Agric. Food. Chem.* **2017**, *65*, 9807–9814. [CrossRef] [PubMed]
27. Sajwan, R.K.; Lakshmi, G.; Solanki, P.R. Fluorescence tuning behavior of carbon quantum dots with gold nanoparticles via novel intercalation effect of aldicarb. *Food Chem.* **2021**, *340*, 127835. [CrossRef] [PubMed]
28. Zhang, G.Q.; Zhang, X.Y.; Luo, Y.X.; Li, Y.S.; Zhao, Y.; Gao, X.F. A flow injection fluorescence "turn-on" sensor for the determination of metformin hydrochloride based on the inner filter effect of nitrogen-doped carbon dots/gold nanoparticles double-probe. *Spectrochim. Acta A Mol. Biomol. Spectrosc.* **2021**, *250*, 119384. [CrossRef]
29. Zhang, J.; Dong, L.; Yu, S.H. A selective sensor for cyanide ion (CN-) based on the inner filter effect of metal nanoparticles with photoluminescent carbon dots as the fluorophore. *Sci. Bull.* **2015**, *60*, 785–791. [CrossRef]
30. Singh, I.; Arora, R.; Dhiman, H.; Pahwa, R. Carbon quantum dots: Synthesis, characterization and biomedical applications. *Turkish J. Pharm. Sci.* **2018**, *15*, 219–230. [CrossRef] [PubMed]
31. Li, M.X.; Chen, T.; Gooding, J.J.; Liu, J.Q. Review of carbon and graphene quantum dots for sensing. *ACS Sens.* **2019**, *4*, 1732–1748. [CrossRef]

32. Ma, J.L.; Yin, B.C.; Wu, X.; Ye, B.C. Simple and cost-effective glucose detection based on carbon nanodots supported on silver nanoparticles. *Anal. Chem.* **2017**, *89*, 1323–1328. [CrossRef] [PubMed]
33. Kumar, N.; Seth, R.; Kumar, H. Colorimetric detection of melamine in milk by citrate-stabilized gold nanoparticles. *Anal. Biochem.* **2014**, *456*, 43–49. [CrossRef] [PubMed]
34. Gao, Q.; Zheng, Y.; Song, C.; Lu, L.Q.; Tian, X.K.; Xu, A.W. Selective and sensitive colorimetric detection of copper ions based on anti-aggregation of the glutathione-induced aggregated gold nanoparticles and its application for determining sulfide anions. *RSC Adv.* **2013**, *3*, 21424–21430. [CrossRef]
35. Amjadi, M.; Hallaj, T.; Manzoori, J.L.; Shahbazsaghir, T. An amplified chemiluminescence system based on Si-doped carbon dots for detection of catecholamines. *Spectrochim. Acta A Mol. Biomol. Spectrosc.* **2018**, *201*, 223–228. [CrossRef]
36. Wang, W.J.; Peng, J.W.; Li, F.M.; Su, B.Y.; Chen, X.; Chen, X.M. Phosphorus and chlorine co-doped carbon dots with strong photoluminescence as a fluorescent probe for ferric ions. *Microchim. Acta* **2019**, *186*, 32. [CrossRef] [PubMed]
37. Borse, V.; Konwar, A.N. Synthesis and characterization of gold nanoparticles as a sensing tool for the lateral flow immunoassay development. *Sens. Int.* **2020**, *1*, 100051. [CrossRef]
38. Xu, Q.; Li, B.F.; Ye, Y.C.; Cai, W.; Li, W.J.; Yang, C.Y.; Chen, Y.S.; Xu, M.; Li, N.; Zheng, X.S.; et al. Synthesis, mechanical investigation, and application of nitrogen and phosphorus co-doped carbon dots with a high photoluminescent quantum yield. *Nano Res.* **2018**, *11*, 3691–3701. [CrossRef]
39. Basu, S.; Pal, T. Glutathione-induced aggregation of gold nanoparticles: Electromagnetic interactions in a closely packed assembly. *J. Nanosci. Nanotechnol.* **2007**, *7*, 1904–1910. [CrossRef] [PubMed]
40. Vobornikova, I.; Pohanka, M. Smartphone-based colorimetric detection of glutathione. *Neuro Endocrinol. Lett.* **2016**, *37*, 139–143.
41. Wang, Q.; Li, L.F.; Wang, X.D.; Dong, C.; Shuang, S.M. Graphene quantum dots wrapped square-plate-like MnO_2 nanocomposite as a fluorescent turn-on sensor for glutathione. *Talanta* **2020**, *219*, 121180. [CrossRef]
42. Zheng, C.; Ding, L.; Wu, Y.N.; Tan, X.H.; Zeng, Y.Y.; Zhang, X.L.; Liu, X.L.; Liu, J.F. A near-infrared turn-on fluorescence probe for glutathione detection based on nanocomposites of semiconducting polymer dots and MnO_2 nanosheets. *Anal. Bioanal. Chem.* **2020**, *412*, 8167–8176. [CrossRef] [PubMed]
43. Chu, S.Y.; Wang, H.Q.; Du, Y.X.; Yang, F.; Yang, L.; Jiang, C.L. Portable smartphone platform integrated with a nanoprobe-based fluorescent paper strip: Visual monitoring of glutathione in human serum for health prognosis. *ACS Sustain. Chem. Eng.* **2020**, *8*, 8175–8183. [CrossRef]
44. Anik, U.; Cubukcu, M.; Ertas, F.N. An effective electrochemical biosensing platform for the detection of reduced glutathione. *Artif. Cells Nanomed. Biotechnol.* **2016**, *44*, 971–977. [CrossRef]
45. Rawat, B.; Mishra, K.K.; Barman, U.; Arora, L.; Pal, D.; Paily, R.P. Two-dimensional MoS_2-based electrochemical biosensor for highly selective detection of glutathione. *IEEE Sens. J.* **2020**, *20*, 6937–6944. [CrossRef]
46. Barman, U.; Mukhopadhyay, G.; Goswami, N.; Ghosh, S.S.; Paily, R.P. Detection of glutathione by glutathione-S-transferase-nanoconjugate ensemble electrochemical device. *IEEE Trans. Nanobiosci.* **2017**, *16*, 271–279. [CrossRef] [PubMed]
47. Tian, J.; Zhao, P.; Zhang, S.S.; Huo, G.A.; Suo, Z.C.; Yue, Z.; Zhang, S.M.; Huang, W.P.; Zhu, B.L. Platinum and iridium oxide Co-modified TiO_2 nanotubes array based photoelectrochemical sensors for glutathione. *Nanomaterials* **2020**, *10*, 522. [CrossRef]
48. Li, Y.; Wu, P.; Xu, H.; Zhang, H.; Zhong, X.H. Anti-aggregation of gold nanoparticle-based colorimetric sensor for glutathione with excellent selectivity and sensitivity. *Analyst* **2011**, *136*, 196–200. [CrossRef]
49. Díaz-Cruz, M.S.; Mendieta, J.; Tauler, R.; Esteban, M. Cadmium-binding properties of glutathione: A chemometrical analysis of voltammetric data. *J. Inorg. Biochem.* **1997**, *66*, 29–36. [CrossRef]
50. Han, B.; Yuan, J.; Wang, E. Sensitive and selective sensor for biothiols in the cell based on the recovered fluorescence of the cdte quantum dots−Hg(II) system. *Anal. Chem.* **2009**, *81*, 5569–5573. [CrossRef] [PubMed]
51. Qian, H.F.; Dong, C.Q.; Weng, J.F.; Ren, J.C. Facile one-pot synthesis of luminescent, water-soluble, and biocompatible glutathione-coated CdTe nanocrystals. *Small* **2006**, *2*, 747–751. [CrossRef] [PubMed]

Article

Development of Palladium and Platinum Decorated Granulated Carbon Nanocomposites for Catalytic Chlorate Elimination

Emőke Sikora [1,*], Gábor Muránszky [1], Ferenc Kristály [2], Béla Fiser [1,3,4], László Farkas [5], Béla Viskolcz [1,3] and László Vanyorek [1,*]

1. Institute of Chemistry, University of Miskolc, H-3515 Miskolc, Hungary
2. Institute of Mineralogy and Geology, University of Miskolc, H-3515 Miskolc, Hungary
3. Higher Education and Industrial Cooperation Centre, University of Miskolc, H-3515 Miskolc, Hungary
4. Ferenc Rakoczi II Transcarpathian Hungarian College of Higher Education, 90200 Beregszász, Transcarpathia, Ukraine
5. BorsodChem Ltd., Bolyai tér 1., H-3700 Kazincbarcika, Hungary
* Correspondence: kemsik@uni-miskolc.hu (E.S.); kemvanyi@uni-miskolc.hu (L.V.)

Abstract: Granulated carbon nanotube-supported palladium and platinum-containing catalysts were developed. By using these, remarkable catalytic activity was achieved in chlorate ion hydrogenation. Nitrogen-doped bamboo-like carbon nanotubes (N-BCNTs) loaded gel beads were prepared by using Ca^{2+}, Ni^{2+} or Fe^{3+} ions as precursors for cross-linking of sodium alginate. The gel beads were carbonized at 800 °C, and these granulated carbon nanocomposites (GCNC) were used as supports to prepare palladium and platinum-containing catalysts. All in all, three catalysts were developed and, in each case, >99 n/n% chlorate conversion was reached in the aqueous phase by using the Pd-Pt containing GCNCs, moreover, these systems retained their catalytic activity even after repeated use.

Keywords: GCNC; N-BCNT; hydrogenation; gel beads; sodium alginate

1. Introduction

Carbon has one of the most versatile allotropic modifications on Earth. From powder to fibers, there is a wide range of varieties, including activated carbon, fullerenes, carbon nanotubes, carbon fibers, graphene and of course diamond. These carbon variants provide special electrical, mechanical, and optical properties that led to previously unimaginable applications including catalysis [1–6]. Carbon-based materials play an exceptional role in the development of catalytic systems.

To achieve an environmentally friendly chemical industry, catalytic processes are essential and need to be applied as widely as possible. Therefore, to increase the catalytic activity and selectivity of the catalysts and to reduce the cost of catalyst production, the necessary heat and amount of chemicals are inevitable [7]. In parallel with the development of new active phases, intensive research is being carried out by both academic and industrial groups to develop new catalyst supports that can modify the catalytic activity and selectivity of existing active phases [8–13]. Besides cost-effectiveness, carbonaceous solid catalysts have the unique ability of keeping their activity not only in organic solutions but also in aqueous solutions [14]. It was revealed that heat treatment can improve the oxidation resistance of porous graphene more efficiently than conventional porous carbon [6]. These unique properties make this material promising catalyst support for catalytic reactions conducted under oxidative conditions [6].

There are several examples in the literature where carbon-based catalysts have been used for halogenate reduction by catalytic hydrogenation [15–18]. Nitrogen-doped bamboo-like carbon nanotubes (N-BCNTs) and non-doped multi-walled carbon nanotubes were compared in terms of their activity and the former was better because electron-rich N atoms promote electron transfer processes [19,20]. N-BCNTs are special types of carbon nanotubes, made from nitrogen-containing carbon compounds. Due to the nitrogen-doping, defect sites are also appearing in the structure of the nanotubes, which are potential binding sites for the catalytically active metal particles [21].

The drawback of the most promising carbon forms is that they often fall into the nano range in more dimensions, which makes their use difficult on an industrial scale (especially in fixed-bed catalytic reactions) and, last but not least, makes them dangerous to handle (e.g., inhalation). Consequently, it is important to search for and develop methods that allow the large-scale synthesis of carbon nanostructures and their direct macroscopic sizing for immobilization [6,22–25]. Structured catalyst support with oxygenated functional groups from commercial graphite felt (OFG) raw material was prepared and applied in Pd/OFG catalyst synthesis, which was tested through the liquid-phase hydrogenation of cinnamaldehyde and it was found that the system expressed excellent catalytic activity, and stability as well as recyclability [26]. Furthermore, monolithic biochar-based catalysts decorated with graphitic carbon-covered metal nanoparticles were synthesized [27]. This catalyst with the incorporation of Ni-Co alloy nanoparticles achieved a synergistic effect of Co and Ni nanoparticles in tar decomposition, exhibiting higher activity and better stability than the catalyst incorporating pure Co and Ni nanoparticles for both toluene cracking and steam reforming processes [27]. Nanodiamond-based monolith within an N-doped mesoporous carbon matrix was prepared [28]. Cheap food-grade components were used as glue for the dispersion of nanodiamonds and this metal-free composite system served as a highly stable and well-performing catalyst for the conversion of ethylbenzene to styrene [28]. It has also been found that the incorporated nitrogen in this monolith contributes to increased styrene selectivity [28]. The use of activated carbon cloth as platinum catalyst support has also been investigated and used in catalytic hydrogenation [29]. In addition to their activity and the particular shape of the cloth support, the possibility of being used in a wide range of temperatures is very advantageous [29]. Granulated nitrogen-doped graphene oxide aerogels (N-dGOA) were synthesized and showed encouraging electro-catalytic activity in oxygen reduction, which makes them applicable in areas such as hydrogen and thermal energy storage [30]. In addition, granulated platinum-decorated carbon nanotubes were found to have much better catalytic activity in the liquid phase hydrogenation of nitrobenzene than their platinum-decorated activated carbon (AC) counterparts [31]. The granulated CNTs had larger pores than the AC particles, which gave a faster mass transfer rate of H_2 and helped to produce aniline with high selectivity [31]. Nitrogen-doped carbon nanotubes (N-CNT) supported on a macroscopic structure of SiC have shown to be an active and selective metal-free catalyst for the low-temperature oxidation of H_2S into elemental sulfur [22]. The macroscopic shaping allows us to avoid the problems with the handling and transport of the nanoscopic system, and this hybrid metal-free catalyst with controlled macroscopic shape can be efficiently employed in a fixed-bed configuration without facing the problem with the pressure drop across the catalytic bed [22].

To combine the above-mentioned positive effects of carbon-based materials, N-BCNT was used to prepare a granulated catalyst support and by using it sodium alginate nanotube-loaded gel beads were created. These beads were carbonized and used as supports to develop palladium and platinum catalysts and tested in catalytic chlorate hydrogenation.

2. Materials and Methods

2.1. Materials

For the CCVD synthesis of N-BCNTs, *n*-butylamine (Sigma Aldrich, St. Louis, MO, USA) was used as a carbon source and nickel(II) nitrate hexahydrate ($Ni(NO_3)_2 \cdot 6H_2O$, ThermoFisher GmbH, 76870 Kandel, Germany) was applied as a catalyst on magnesium oxide (Sigma Aldrich). Nitrogen (99.995%) was used as carrier gas (Messer Hungary Ltd, Budapest, Hungary). Sodium alginate (Sigma-Aldrich, 3050 Saint Louis, MO 63103, USA) was applied to prepare the N-BCNT gelatine beads along with calcium-chloride ($CaCl_2 \cdot 2H_2O$, Merck Ltd, Darmstadt, Germany), nickel(II) nitrate hexahydrate ($Ni(NO_3)_2 \cdot 6H_2O$, ThermoFisher GmbH, 76870 Kandel, Germany), and iron(III) nitrate nonahydrate

(Fe(NO$_3$)$_3$ · 9H$_2$O, VWR Int. Ltd., B-3001 Leuven, Belgium). To prepare the final catalyst, palladium(II) nitrate dihydrate (Pd(NO$_3$)$_2$ · 2H$_2$O, Merck Ltd.) and hexachloroplatinic acid (H$_2$PtCl$_6$, Merck Ltd.) were applied as Pd and Pt precursors, respectively. During the chlorate hydrogenation tests, potassium iodide (KI, Merck), 35 wt% hydrochloric acid (HCl, VWR), and potassium chlorate (KClO$_3$) were applied.

2.2. Characterization Techniques

High-resolution transmission electron microscopy (HRTEM, FEI Technai G2 electron microscope, 200 kV) was used to characterize the nanoparticles. The sample preparation was carried out by using the aqueous suspension of the nanoparticles, which was dropped on 300 mesh copper grids (Ted Pella Inc, Redding, CA, USA). The surface of the samples was further examined by a Helios G4 PFIB CXe Plasma Focused Ion Beam Scanning Electron Microscope (PFIB-SEM) equipped with an EDAX Octane Elect EDS System and APEX Analysis Software. Carbon tape was used for sample preparation. EDS maps were created with a 1024 × 800 resolution, and 1 frame was recorded with a 1000 µs collecting time. The particle diameters of the nanoparticles were manually scaled using the ImageJ program, based on the scale bar of SEM images. The qualitative and quantitative analysis of the different metals and metal-oxide forms was carried out by using X-ray diffraction (XRD) measurements with Rietveld analysis. Bruker D8 diffractometer (Cu-Kα source) in parallel beam geometry (Göbel mirror) with Vantec detector was applied. The metal (palladium and platinum) content of the catalyst samples was determined by using a Varian 720 ES inductively coupled optical emission spectrometer (ICP-OES). For the ICP-OES measurements, the samples were burned in air, after the remaining ash was solved in aqua regia. The specific surface area of the catalysts was also measured by CO$_2$ adsorption experiments using Micromeritics ASAP 2020 sorptometer and the calculations were carried out based on the Dubinin-Ashtakov isotherm. The types of the incorporated nitrogen atoms were determined by X-ray photoelectron spectroscopy (XPS), using a Kratos XSAM-800 XPS instrument. The MgKα X-ray source was operated with 120 W (12 kV, 10 mA). Samples were examined on double-sided carbon tape. Survey spectra were collected with a pass energy of 80 eV and 1 eV step size. The N-BCNTs were examined by using Raman microscopy (WITECH 3112973 instrument with HeNe laser, λ = 632.92 nm). The structural defects were quantified by calculating the ratio of the intensities of the defect peak (D-peak, ~1340 cm^{-1}) and the graphite peak (G-peak, ~1580 cm^{-1}) (ID/IG).

2.3. Synthesis of the Nitrogen-Doped Bamboo-Like Carbon Nanotubes (N-BCNTs)

N-BCNTs were synthesized from *n*-butylamine by using the catalytic chemical vapor deposition (CCVD) method in a quartz reactor, which was placed in a tube furnace. Butylamine was dosed (16.2 mL h^{-1}) into the quartz reactor with a syringe pump. The carrier gas was nitrogen (100 scm), and the carbon nanotubes were synthesized at 750 °C for 20 min by using of 5 wt% nickel-containing MgO catalyst (2.5 g). After the N-BCNT synthesis, the catalyst was removed by cc. hydrochloric acid.

2.4. Preparation of Catalyst Supports and Pd-Pt Containing Catalysts

The preparation of the catalyst supports is similar to a method applied in our previous work, with some modifications [32]. A mixture of 100 mL distilled water, 0.75 g sodium alginate, and 1 g N-BCNT were prepared by using a Hielscher homogenizer. This mixture was added dropwise to 300 mL 5.5 g CaCl$_2$ or 7.5 g Ni(NO$_3$)$_2$ or 10.86 g Fe(NO$_3$)$_3$ solution by using a syringe pump. After the preparation, the remaining solution was removed and the spheres were washed with distilled water and then, dried at 370 K for 24 h. The prepared beads were calcinated at 800 °C under a nitrogen flow for 60 min. For the calcium-containing samples, a concentrated hydrochloric acid wash was also applied prior to heat treatment. All in all, three different GCNC (granulated carbon nanocomposite) supports were prepared.

Palladium and platinum were added to the supports by using an impregnation method. A total of 1.15 g catalyst support was added to a 50 mL metal solution containing 0.1 g of $Pd(NO_3)_2 \cdot 2 H_2O$ and 0.01 g of H_2PtCl_6. This was followed by vacuum evaporation and reduction at 673 K in the H_2 stream for 30 min.

2.5. Catalytic Hydrogenation of Chlorate Ions

The prepared catalysts were tested in catalytic chlorate hydrogenation. To test the catalytic activity of each system, potassium chlorate (200 mg/dm^3) was hydrogenated in the presence of a 1 g catalyst in aqueous media. Gas supply was provided (40 sccm nitrogen and 100 sccm hydrogen) during the experiments and the temperature was set to 80 °C by using a Julabo circulator. The solution was placed in a side-inlet gas washing bottle with a fritted disc. The hydrogenation was carried out for 3 h in each case, and sampling took place at 0, 5, 15, 30, 45, 60, 90, 120, 150, and 180 min. Thereafter, distilled water was used to wash the catalysts which then, were dried at 105 °C overnight. UV-6300PC spectrophotometer was applied at 351 nm to determine the chlorate concentration in the collected samples. During the measurements, the following redox reaction was considered between iodide and chlorate ions:

$$KClO_3 + 6 KI + 6 HCl \rightarrow 3 H_2O + 3 I_2 + 7 KCl \qquad (1)$$

The intensity of the color, and thus, the absorbance of the samples changed due to iodine formation, from which the chlorate concentration can be determined by appropriate calibration. Potassium chlorate solutions with different concentrations (0, 50, 100, 150, and 200 mg/dm^3) were prepared for calibration. A total of 100 mg potassium iodide and 1 mL HCl were added to 1 mL sample, and then, it was diluted by using 50 mL distilled water, and this was measured by the spectrophotometer.

3. Results and Discussion

3.1. Characterization of the Catalyst Supports

By using transmission electron microscopy, the fibrous structure of the prepared N-BCNTs was verified (Figure 1A). The graphene layers dividing the nanotubes into segments are also visible at higher resolution (Figure 1B).

At the edge of the before-mentioned graphene layers, the carbon atoms are easily oxidized, and thus, several hydroxyl and carboxyl functional groups can be formed on the surface of the N-BCNTs. Due to the deprotonation of these functional groups in aqueous dispersions, negative charges appear, which lead to the decreasing electrokinetic potential of the N-BCNTs. This can be used effectively to anchor the catalytically metal ions (i.e., palladium- or nickel-ions) on the surface of the nanotubes by ion exchange adsorption and electrostatic interactions. On the other hand, this can also promote the continuous, and homogenous coverage of the surface of the nanotubes by metal nanoparticles. Moreover, the interaction between the nanoparticles and the N-BCNTs will be favored. Thus, the extraordinary bamboo-like structure formed due to the incorporation of nitrogen atoms in the graphitic structure of the nanotubes is beneficial.

On the deconvoluted N1s band of the N-BCNTs' XPS spectrum, three peaks were identified at 398.4 eV, 401.2 eV, and 404.9 eV binding energy, which are associated with the pyridinic, and graphitic nitrogen atom types, and nitrogen oxides, respectively (Figure 1C). The C1s band of the N-BCNTs' XPS spectrum was also deconvoluted and three peaks were identified at 284.5 eV, 287.5 eV, and 291.2 eV binding energy which can be associated with carbon atoms of the graphitic character (C=C- and -C-C- bonds), C=O, and C atoms in carboxyl functional groups.

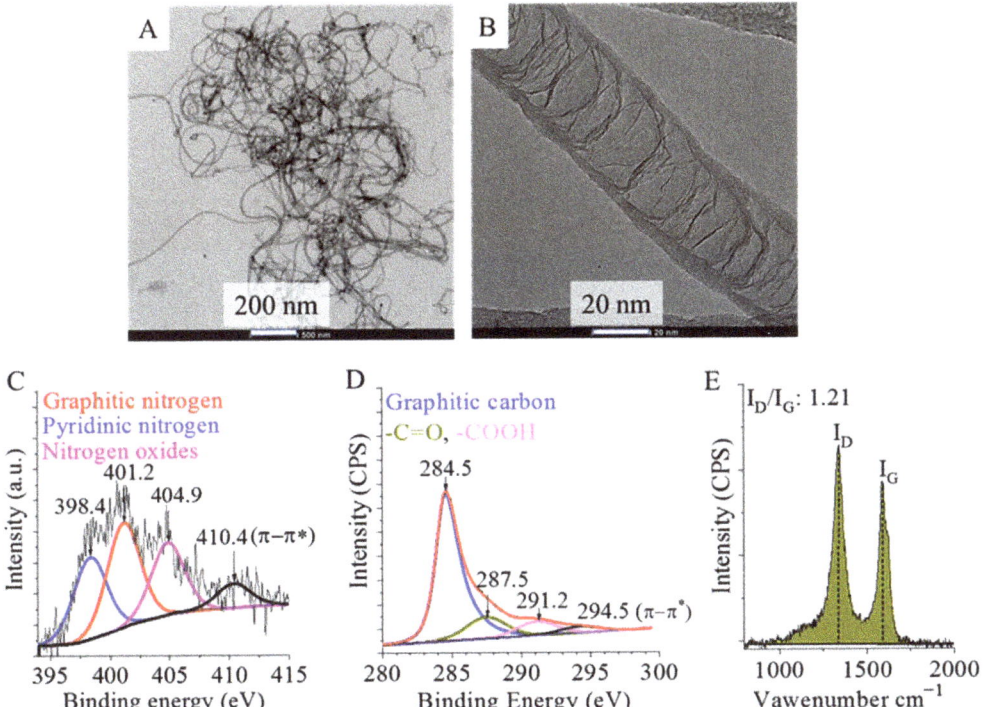

Figure 1. HRTEM images (**A**,**B**), deconvoluted N1s (**C**) and C1s band (**D**) on the XPS spectrum and Raman spectrum (**E**) of the synthesized nitrogen-doped bamboo-like carbon nanotubes (NBCNTs).

The characteristic bamboo-like structure of the nanotubes is achieved by the nitrogen incorporation into the system. This also increases crystal lattice defects and disorder in the graphitic structure, which makes the N-BCNTs less graphitic compared to conventional non-doped multi-walled carbon nanotubes (MWCNTs). Raman spectroscopy was applied to quantify the disordered character of the structure of the nanotubes (Figure 1E). The ratio of the intensities of the defect peak (D-peak, ~1340 cm^{-1}) and the graphite peak (G-peak, ~1580 cm^{-1}) (ID/IG) was calculated [33]. The impurities or disorder in the nanotubes are defined by the D-peak, while the G-peak is associated with carbon-carbon bond stretching. The ID/IG ratio is relatively high, 1.21 in the case of the studied N-BCNTs. (For reference, Kim et al used heat treatment to improve the degree of CNT crystallinity, resulting a ratio of 0.4 [34]). Thus, due to the presence of lattice defects, the nitrogen-doped nanotubes are prone to modification, and various functional groups (-COOH, -OH) could occur on their surfaces which led to high dispersibility in polar solvents.

The N-BCNTs were embedded into alginate gel beads. Ca^{2+}, Ni^{2+}, or Fe^{3+} ions were applied as precursors for cross-linking sodium alginate, and thus, three different N-BCNT loaded gel beads (or granulated carbon nanocomposites, GCNC) were created, GCNC, Ni-GCNC, and Fe-GCNC, respectively. The advantage of transition metals compared to calcium is that they can show catalytic activity or influence the catalytic behavior of the noble metal-containing hydrogenation catalysts. The nickel and iron-containing, N-BCNT loaded alginate gel beads were carbonized at 800 °C in nitrogen flow.

On the SEM images of the carbonized supports (Supplementary Materials Figure S1A,B), the fibrous structure of the N-BCNTs is visible and decorated with spherical objects, which are nickel nanoparticles that remained from the CCVD synthesis. More nanoparticles are visible on the SEM images of the Ni- and Fe-GCNC samples (Figure 2A,B) which contain not only the residual Ni from the CCVD synthesis, but also Ni and Fe particles that have replaced Na from Na-alginate.

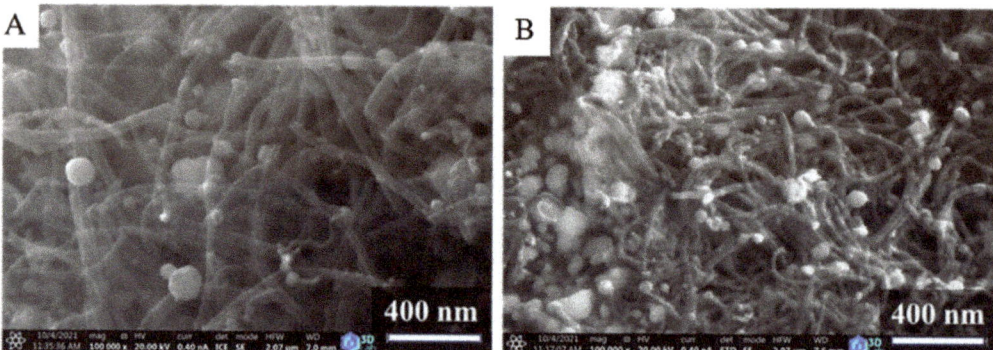

Figure 2. SEM images of the nickel (**A**) and magnetite (**B**) containing N-BCNT loaded gel beads.

To determine the composition of the nickel and iron-containing beads, XRD measurements were carried. On the diffractogram of the Ni-GCNC, peaks were identified at 44.4° and 51.8° two theta degrees, which can be associated with (111) and (200) reflexions (PDF 04-0850), and characteristic to the metallic nickel phase (Figure 3A). The reflexions at 25.8° (002) and 43.1° (100) indicate the presence of a carbon phase as well (PDF 75-1621). In the case of the Fe-GCNC sample, the (111) and (200) reflexions at 43.7° and 50.9° 2 θ degrees (PDF 06-0696) belong to the α-Fe phase (Figure 3B). Furthermore, the reflexion at 44.6° (110) confirmed that the carbonization of the gel beads led to the formation of a γ-Fe phase. The presence of elemental metallic particles can be explained by the reducing effect of carbon at high temperatures (800 °C).

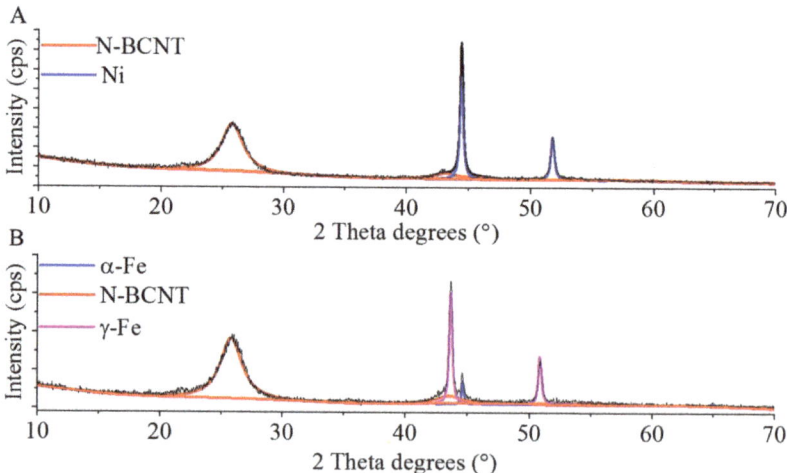

Figure 3. XRD pattern of the nickel (**A**) and magnetite (**B**) containing granulated carbon nanocomposites (GCNC).

3.2. Characterization of the Noble Metal Containing Catalysts

The GCNC, Ni-GCNC and Fe-GCNC supports were used for catalyst preparation, which was impregnated with the solutions of palladium and platinum precursors. The dried impregnated supports were activated in a hydrogen atmosphere at 400 °C. The surface of the grains is richly covered by metal nanoparticles in the case of the Pd-Pt/Ni-GCNC and Pd-Pt/Fe-GCNC catalysts (Figure 4A,C). A significant amount of these nanoparticles is aggregated, but the fibrous structure of the N-BCNTs is still visible at higher resolution (Figure 4B,D). The Pd-Pt/GCNC sample is similar in terms of surface coverage and aggregation (Supplementary Materials Figure S1C,D). In the case of the Pd-Pt/Fe-GCNC catalyst, the particle size of the anchored Pd and Pt crystallites on the N-BCNT surface is smaller compared to those in the case of the Pd-Pt/Fe-GCNC (Figure 4). To confirm this, the SEM images were used to prepare size distribution diagrams (Supplementary Materials Figure S2). The smallest average particle size was measured in the case of Pd-Pt/Ni-GCNC (14.4 ± 7.0 nm), while the largest value corresponds to Pd-Pt/Fe-GCNC (105.2 ± 58.1 nm) and Pd-Pt/Ca-GCNC is in the middle in terms of this property (46.5 ± 14.3 nm). The median values were 12.5 nm, 89.1 nm, and 46.5 nm for Pd-Pt/Ni-GCNC, Pd-Pt/Fe-GCNC, and Pd-Pt/Ca-GCNC, respectively. Based on the mean and standard deviation values, the nickel-containing sample was the most homogeneous with the smallest particles and the most uniform in terms of particle size.

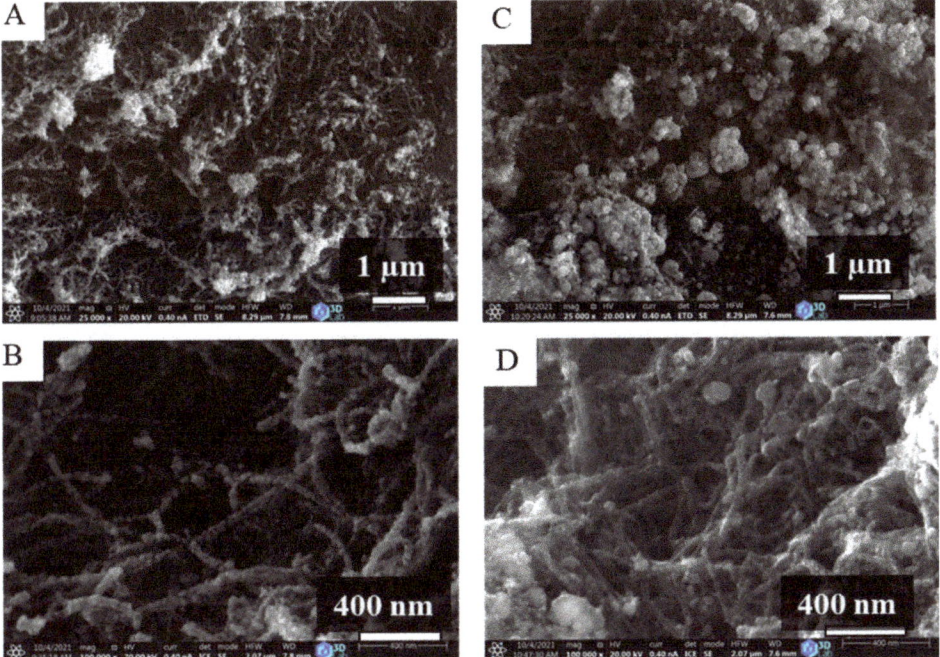

Figure 4. SEM images of the nickel (**A**,**B**) and magnetite (**C**,**D**) containing, palladium and platinum decorated granulated carbon nanocomposites (GCNC). Pd-Pt/Ni-GCNC—(**A**,**B**) and Pd-Pt/Fe-GCNC (**C**,**D**).

Elemental mapping was carried out on the Pd-Pt containing catalysts, to examine the metal distribution and coverage of the catalyst support. Elemental maps of the Pd-Pt/Ni-GCNC system indicates that the distribution of the nickel particles is homogenous and the bigger particle aggregates are mainly based on palladium, but where the Pd is enriched, Pt is also found in greater amount (Figure 5A). In the case of the Pd-Pt/Fe-GCNC catalyst, the large aggregates contain mostly iron, while the palladium and platinum nanoparticles are located in homogenous distribution on the surface (Figure 5B).

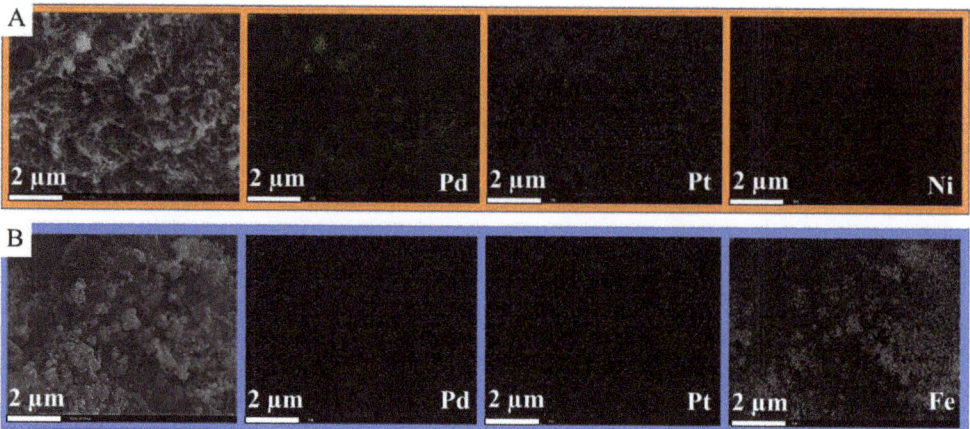

Figure 5. Elemental maps of the nickel (**A**) and magnetite (**B**) containing, palladium and platinum decorated granulated carbon nanocomposites (GCNC).

On the diffractogram of the iron and nickel-free Pd-Pt/GCNC catalyst, the reflexions of the palladium were identified at 40.1° (111) and 46.8° (200) two theta degrees (Figure 6A) (PDF 046–1043). The presence of elemental platinum was also verified by peaks located at 40.3° (111) and 46.0° (200) 2 θ degrees (PDF 04-0802).

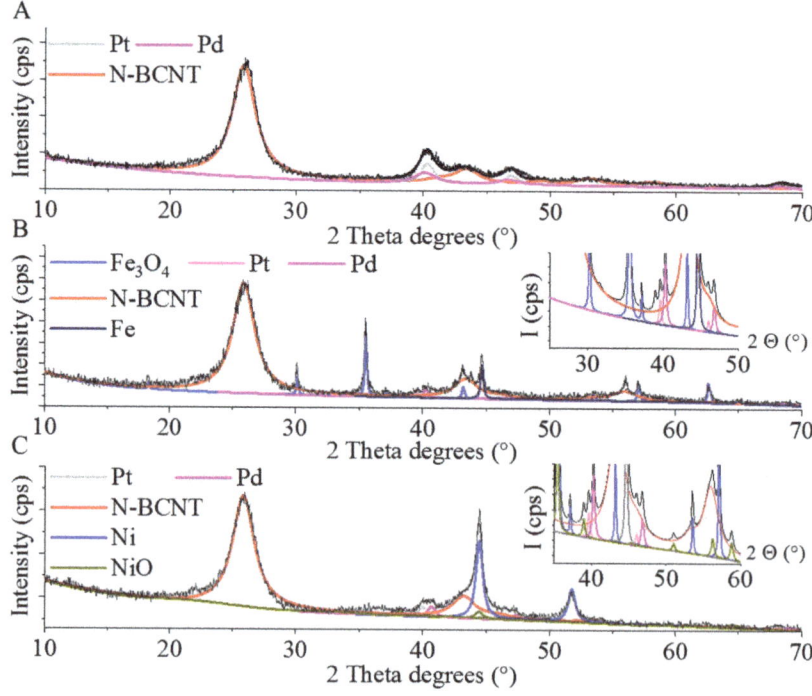

Figure 6. XRD patterns of the palladium and platinum decorated calcium (Pd-Pt/GCNC) (**A**), magnetite (Pd-Pt/Fe-GCNC) (**B**), and nickel (Pd-Pt/Ni-GCNC) (**C**) containing granulated carbon nanocomposites.

In the case of the Pd-Pt/Fe-GCNC catalyst, reflexions at 18.4° (111), 30.1° (220), 35.5° (311), 43.1° (400), 53.4° (422), 57.1° (511), and 62.6° (440) two Theta degrees were located, and these can be associated with Fe_3O_4 (Figure 6B) (PDF 89-0691). The presence of elemental iron was also verified by the peak at 44.6° (110) two theta degrees (PDF 06-0696). The (111) and (200) reflexions of palladium and platinum are found at 40.3° and 46.7° as well as 40.1° and 46.0° two theta degrees, respectively (PDF 46-1043 and PDF 04-0802). Reflexions indicating the presence of carbon are visible at 25.8° (002) and 43.1° (100) 2 θ degrees (PDF 75-1621).

The presence of elemental Ni and Ni oxide in the nickel-containing system, Pd-Pt/Ni-GCNC, was verified (Figure 6C). The characteristic reflexions of Ni are found at 44.4° (111) and 51.7° (200) two theta degrees (PDF 04-0850). Peaks of the NiO phase are found at 37.2° (111) and 43.3° (200) two theta degrees (PDF 47-1049). Elemental palladium and platinum are also found in the sample.

The reductive treatment of the samples in a hydrogen atmosphere at 400 °C was enough to form palladium and platinum nanoparticles. However, despite the reductive atmosphere, due to the decomposition of the nitrate salt of palladium, nitrogen oxides were developed, which led to the formation of NiO and Fe_3O_4 nanoparticles. The reduction time at 400 °C was not enough to produce only the metallic form Ni and Fe, because of the difficulty to access NiO and Fe_3O_4 particles. Moreover, due to the palladium and platinum precursors being in hydrated form, water was present which further retards the reduction of oxides [35].

The specific surface area of the catalysts was determined before and after applying them in catalytic chlorate hydrogenation (Table 1). All in all, the surface area of the catalytic systems only slightly changed after using them five times. The largest deviation occurred in the case of the nickel-containing sample, while the smallest in the case of the Pd-Pt/Fe-GCNC.

Table 1. Specific surface area of the developed granulated carbon nanocomposite supported Pd-Pt decorated catalysts before and after using them 5 times.

Sample	SSA (m^2/g)
Pd-Pt/GCNC	75
Pd-Pt/GCNC 5× used	81
Pd-Pt/Ni-GCNC	214
Pd-Pt/Ni-GCNC 5× used	192
Pd-Pt/Fe-GCNC	177
Pd-Pt/Fe-GCNC 5× used	179

Based on the ICP results, the GCNC sample contained 2.66 wt % Ca despite acid washing (Table 2). In the case of the Ni-GCNC and Fe-GCNC samples, ~22 wt % nickel and ~18 wt % Fe were found, respectively. All samples contain nickel, even the GCNC and Fe-GCNC supports include ~3.5 wt %, which is due to the fact, that the carbon nanotubes contain a small amount of nickel since a nickel-containing catalyst was used to synthesize the N-BCNTs. The Pd content of the noble metal-containing catalysts varied between 2.38 and 3.03 wt %, while the Pt content was between 0.28 and 0.45 wt %.

Table 2. Quantitative analysis of the granulated carbon nanocomposite supported Pd-Pt decorated catalysts based on ICP-OES measurements.

	Weight %				
	Ca	Fe	Ni	Pd	Pt
GCNC	2.66 ± 0.22	-	3.87 ± 0.21	-	-
Ni-GCNC	-	-	21.66 ± 0.77	-	-
Fe-GCNC	-	17.76 ± 0.36	3.23 ± 0.11	-	-
Pd-Pt/GCNC	2.11 ± 0.13	-	3.56 ± 0.10	3.03 ± 0.18	0.38 ± 0.03
Pd-Pt/GCNC5× used	0.24 ± 0.01	-	3.16 ± 0.16	3.54 ± 0.22	0.42 ± 0.02
Pd-Pt/Ni-GCNC	-	-	21.07 ± 0.55	2.68 ± 0.42	0.45 ± 0.01
Pd-Pt/Ni-GCNC5× used	-	-	12.72 ± 0.86	1.41 ± 0.11	0.15 ± 0.01
Pd-Pt/Fe-GCNC	-	13.69 ± 4.27	3.19 ± 0.25	2.38 ± 0.28	0.28 ± 0.03
Pd-Pt/Fe-GCNC5× used	-	9.20 ± 0.89	3.22 ± 0.12	1.64 ± 0.16	0.20 ± 0.02

3.3. Comparison of the Catalytic Activity of the Developed Catalysts in Chlorate Hydrogenation

In the first experiments, the Pd-Pt/Ni-GCNC (Figure 7A) and the Pd-Pt/Fe-GCNC (Figure 7B) catalysts showed better performance, converting almost all of the chlorate to chloride in 60 min. The Pd-Pt/GCNC sample (Figure 7C) took about twice as long to achieve this level of conversion. Although some decline after the first use is experienced, significant change in the catalytic behavior of the Pd-Pt/Ni-GCNC and Pd-Pt/GCNC catalysts was not detected (Figure 7A,C). In the case of the Pd-Pt/Fe-GCNC catalyst, after the first cycle, the catalytic activity decreased to a greater extent (Figure 7B). Nonetheless, the Pd-Pt/Fe-GCNC retains this decreased catalytic activity until the end of the fifth cycle.

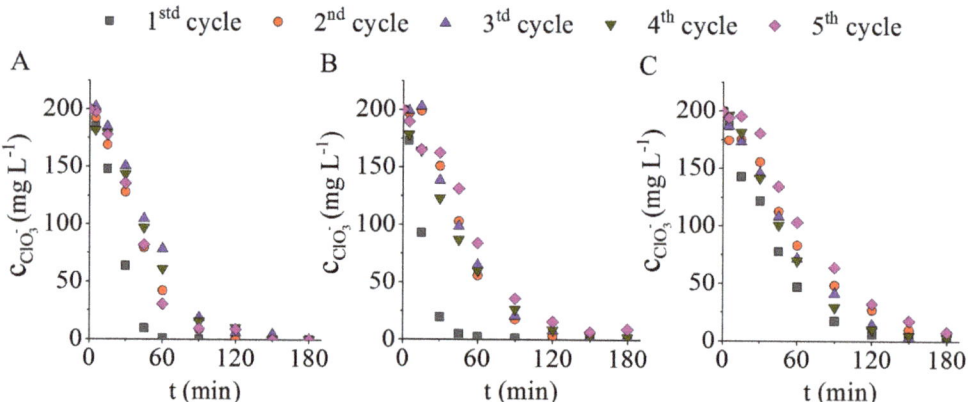

Figure 7. Decreasing chlorate concentration depending on the hydrogenation time by applying the developed catalysts over five cycles: Pd-Pt/Ni-GCNC (**A**), Pd-Pt/Fe-GCNC (**B**) and Pd-Pt/GCNC (**C**).

ICP measurements were also performed after the reuse tests (5×) of the catalysts (Table 2). The results show a significant reduction in the precious metal content after the fifth cycle in the case of the Pd-Pt/Ni-GCNC and Pd-Pt/Fe-GCNC samples. As the iron and nickel contents also significantly decreased, it is likely that the palladium and platinum crystallized on these metals have also leached from the samples' surface during the catalytic process. This metal leaching may have caused the decrease in activity experienced during the catalytic tests (Figure 7). The Pd-Pt/GCNC sample only shows a significant decrease in its Ca content. There is no decrease in its precious metal content, but even a small increase can be seen due to sample inhomogeneity.

4. Conclusions

Three different catalyst supports were developed by embedding nitrogen-doped bamboo-like carbon nanotubes into alginate gel beads. Ca^{2+}, Ni^{2+} or Fe^{3+} ions were applied as precursors for cross-linking of alginate. The prepared granulated carbon nanocomposites were carbonized and used as supports to create Pd and Pt-containing catalysts. All in all, three catalysts (Pd-Pt/GCNC, Pd-Pt/Ni-GCNC, and Pd-Pt/Fe-GCNC) were prepared by the impregnation method and successfully tested in chlorate hydrogenation. >99 n/n% chlorate conversion was achieved in each case. Reuse tests were also carried out, and it was found that the systems retained their activity after repeated use. Although, in the case of the nickel and iron-containing samples, a not too significant decrease in their activity was observed compared to the first cycle, which does not affect substantially their applicability. Therefore, the developed noble metal-containing catalysts are highly efficient and well applicable in chlorate hydrogenation.

Supplementary Materials: The following supporting information can be downloaded at: https://www.mdpi.com/article/10.3390/ijms231810514/s1.

Author Contributions: Conceptualization, L.F. and B.V.; Investigation, E.S. and F.K.; Methodology, G.M.; Supervision, L.F. and L.V.; Writing—original draft, E.S. and L.V.; Writing—review and editing, B.F. All authors have read and agreed to the published version of the manuscript.

Funding: This research was funded by Doctoral Student Scholarship Program of the Co-Operative Doctoral Program of the Ministry of Innovation and Technology financed by the National Research, Development, and Innovation Fund.

Institutional Review Board Statement: Not applicable.

Informed Consent Statement: Not applicable.

Acknowledgments: This study was prepared with the professional support of the Doctoral Student Scholarship Program of the Co-Operative Doctoral Program of the Ministry of Innovation and Technology financed by the National Research, Development, and Innovation Fund. We would like to thank János Lakatos for the SSA measurements. Furthermore, special thanks to Dániel Koncz-Horváth and to the 3D Lab at the University of Miskolc for the electron microscopic images. Special thanks to Wanhua-BorsodChem Ltd. for their cooperation and support of our research.

Conflicts of Interest: The authors declare no conflict of interest.

References

1. Sharma, P.; Krishnapriya, R.; Sharma, P.R.; Sharma, R.K. Recent Advances in Synthesis of Metal-Carbon Nanocomposites and Their Application in Catalytic Hydrogenation Reactions. *ACS Symp. Ser.* **2020**, *1359*, 403–458. [CrossRef]
2. Li, H.; He, X.; Wu, T.; Jin, B.; Yang, L.; Qiu, J. Synthesis, modification strategies and applications of coal-based carbon materials. *Fuel Process. Technol.* **2022**, *230*, 107203. [CrossRef]
3. Chen, Z.; Zhao, J.; Cao, J.; Zhao, Y.; Huang, J.; Zheng, Z.; Li, W.; Jiang, S.; Qiao, J.; Xing, B.; et al. Opportunities for graphene, single-walled and multi-walled carbon nanotube applications in agriculture: A review. *Crop Des.* **2022**, *1*, 100006. [CrossRef]
4. Sridharan, R.; Monisha, B.; Kumar, P.S.; Gayathri, K.V. Carbon nanomaterials and its applications in pharmaceuticals: A brief review. *Chemosphere* **2022**, *294*, 133731. [CrossRef] [PubMed]
5. Esteves, L.M.; Oliveira, H.A.; Passos, F.B. Carbon nanotubes as catalyst support in chemical vapor deposition reaction: A review. *J. Ind. Eng. Chem.* **2018**, *65*, 1–12. [CrossRef]
6. He, L.; Fan, Y.; Bellettre, J.; Yue, J.; Luo, L. A review on catalytic methane combustion at low temperatures: Catalysts, mechanisms, reaction conditions and reactor designs. *Renew. Sustain. Energy Rev.* **2020**, *119*, 109589. [CrossRef]
7. Sikora, E.; Koncz-Horváth, D.; Muránszky, G.; Kristály, F.; Fiser, B.; Viskolcz, B.; Vanyorek, L. Development of nickel-and magnetite-promoted carbonized cellulose bead-supported bimetallic Pd–Pt catalysts for hydrogenation of chlorate ions in aqueous solution. *Int. J. Mol. Sci.* **2021**, *22*, 11846. [CrossRef]
8. Ledoux, M.J.; Pham-Huu, C. Carbon nanostructures with macroscopic shaping for catalytic applications. *Catal. Today* **2005**, *102–103*, 2–14. [CrossRef]

9. Pathak, S.; Upadhyayula, S. A review on the development of supported non-noble metal catalysts for the endothermic high temperature sulfuric acid decomposition step in the Iodine–Sulfur cycle for hydrogen production. *Int. J. Hydrog. Energy* **2022**, *47*, 14186–14210. [CrossRef]
10. Patel, K.D.; Subedar, D.; Patel, F. Design and development of automotive catalytic converter using non-nobel catalyst for the reduction of exhaust emission: A review. *Mater. Today Proc.* **2022**, *57*, 2465–2472. [CrossRef]
11. Zhang, Z.; Tian, J.; Li, J.; Cao, C.; Wang, S.; Lv, J.; Zheng, W.; Tan, D. The development of diesel oxidation catalysts and the effect of sulfur dioxide on catalysts of metal-based diesel oxidation catalysts: A review. *Fuel Process. Technol.* **2022**, *233*, 107317. [CrossRef]
12. Qin, G.X.; Hao, Y.; Wang, S.; Dong, Y. Bin Rapid formation of nitrogen-doped carbon foams by self-foaming as metal-free catalysts for selective oxidation of aromatic alkanes. *Appl. Catal. A Gen.* **2020**, *591*, 117400. [CrossRef]
13. Li, D.D.; Zhang, J.W.; Cai, C. Chemoselective hydrogenation of nitroarenes catalyzed by cellulose-supported Pd NPs. *Catal. Commun.* **2018**, *103*, 47–50. [CrossRef]
14. Strelko, V.V.; Stavitskaya, S.S.; Gorlov, Y.I. Proton catalysis with active carbons and partially pyrolyzed carbonaceous materials. *Chin. J. Catal.* **2014**, *35*, 815–823. [CrossRef]
15. Van Santen, R.; Klesing, A.; Neuenfeldt, G.; Ottmann, A. Method for removing chlorate ions from solutions. U.S. Patent No. 6,270,682, 7 August 2001.
16. Liu, J.; Chen, X.; Wang, Y.; Strathmann, T.J.; Werth, C.J. Mechanism and Mitigation of the Decomposition of an Oxorhenium Complex-Based Heterogeneous Catalyst for Perchlorate Reduction in Water. *Environ. Sci. Technol.* **2015**, *49*, 12932–12940. [CrossRef] [PubMed]
17. Chen, X.; Huo, X.; Liu, J.; Wang, Y.; Werth, C.J.; Strathmann, T.J. Exploring beyond palladium: Catalytic reduction of aqueous oxyanion pollutants with alternative platinum group metals and new mechanistic implications. *Chem. Eng. J.* **2017**, *313*, 745–752. [CrossRef]
18. Palomares, A.E.; Franch, C.; Yuranova, T.; Kiwi-Minsker, L.; García-Bordeje, E.; Derrouiche, S. The use of Pd catalysts on carbon-based structured materials for the catalytic hydrogenation of bromates in different types of water. *Appl. Catal. B Environ.* **2014**, *146*, 186–191. [CrossRef]
19. Ayala, P.; Arenal, R.; Rümmeli, M.; Rubio, A.; Pichler, T. The doping of carbon nanotubes with nitrogen and their potential applications. *Carbon N. Y.* **2010**, *48*, 575–586. [CrossRef]
20. Chizari, K.; Janowska, I.; Houllé, M.; Florea, I.; Ersen, O.; Romero, T.; Bernhardt, P.; Ledoux, M.J.; Pham-Huu, C. Tuning of nitrogen-doped carbon nanotubes as catalyst support for liquid-phase reaction. *Appl. Catal. A Gen.* **2010**, *380*, 72–80. [CrossRef]
21. Vanyorek, L.; Bánhidi, O.; Muránszky, G.; Sikora, E.; Prekob, Á.; Boros, Z.; Farkas, L.; Viskolcz, B. Chlorate Elimination by Catalytically Hydrogenation, Catalyst Development and Characterization. *Catal. Lett.* **2019**, *149*, 196–202. [CrossRef]
22. Cuong, D.V.; Truong-Phuoc, L.; Tran-Thanh, T.; Nhut, J.M.; Nguyen-Dinh, L.; Janowska, I.; Begin, D.; Pham-Huu, C. Nitrogen-doped carbon nanotubes decorated silicon carbide as a metal-free catalyst for partial oxidation of H2S. *Appl. Catal. A Gen.* **2014**, *482*, 397–406. [CrossRef]
23. Rogozhnikov, V.N.; Salanov, A.N.; Potemkin, D.I.; Pakharukova, V.P.; Stonkus, O.A.; Glotov, A.P.; Boev, S.S.; Zasypalov, G.O.; Melnikov, D.P.; Snytnikov, P.V. Structured composite catalyst Pd/Ce$_{0.75}$Zr$_{0.25}$O$_{2-x}$/θ-Al$_2$O$_3$/FeCrAlloy for complete oxidation of methane. *Mater. Lett.* **2022**, *310*, 131481. [CrossRef]
24. Pauletto, G.; Vaccari, A.; Groppi, G.; Bricaud, L.; Benito, P.; Boffito, D.C.; Lercher, J.A.; Patience, G.S. FeCrAl as a Catalyst Support. *Chem. Rev.* **2020**, *120*, 7516–7550. [CrossRef] [PubMed]
25. Ho, P.H.; Ambrosetti, M.; Groppi, G.; Tronconi, E.; Palkovits, R.; Fornasari, G.; Vaccari, A.; Benito, P. Structured Catalysts-Based on Open-Cell Metallic Foams for Energy and Environmental Applications. *Stud. Surf. Sci. Catal.* **2019**, *178*, 303–327. [CrossRef]
26. Xu, Z.; Duong-Viet, C.; Liu, Y.; Baaziz, W.; Li, B.; Nguyen-Dinh, L.; Ersen, O.; Pham-Huu, C. Macroscopic graphite felt containing palladium catalyst for liquid-phase hydrogenation of cinnamaldehyde. *Appl. Catal. B Environ.* **2019**, *244*, 128–139. [CrossRef]
27. Tian, B.; Dong, K.; Guo, F.; Mao, S.; Bai, J.; Shu, R.; Qian, L.; Liu, Q. Catalytic conversion of toluene as a biomass tar model compound using monolithic biochar-based catalysts decorated with carbon nanotubes and graphic carbon covered Co-Ni alloy nanoparticles. *Fuel* **2022**, *324*, 124585. [CrossRef]
28. Ba, H.; Luo, J.; Liu, Y.; Duong-Viet, C.; Tuci, G.; Giambastiani, G.; Nhut, J.M.; Nguyen-Dinh, L.; Ersen, O.; Su, D.S.; et al. Macroscopically shaped monolith of nanodiamonds @ nitrogen-enriched mesoporous carbon decorated SiC as a superior metal-free catalyst for the styrene production. *Appl. Catal. B Environ.* **2017**, *200*, 343–350. [CrossRef]
29. Macías Pérez, M.C.; Salinas Martínez De Lecea, C.; Linares Solano, A. Platinum supported on activated carbon cloths as catalyst for nitrobenzene hydrogenation. *Appl. Catal. A Gen.* **1997**, *151*, 461–475. [CrossRef]
30. Baskakov, S.A.; Manzhos, R.A.; Lobach, A.S.; Baskakova, Y.V.; Kulikov, A.V.; Martynenko, V.M.; Milovich, F.O.; Kumar, Y.; Michtchenko, A.; Kabachkov, E.N.; et al. Properties of a granulated nitrogen-doped graphene oxide aerogel. *J. Non-Cryst. Solids* **2018**, *498*, 236–243. [CrossRef]
31. Jin, S.; Qian, W.; Liu, Y.; Wei, F.; Wang, D.; Zhang, J. Granulated carbon nanotubes as the catalyst support for Pt for the hydrogenation of nitrobenzene. *Aust. J. Chem.* **2010**, *63*, 131–134. [CrossRef]
32. Sikora, E.; Prekob, Á.; Halasi, G.; Vanyorek, L.; Pekker, P.; Kristály, F.; Varga, T.; Kiss, J.; Kónya, Z.; Viskolcz, B. Development and Application of Carbon-Layer-Stabilized, Nitrogen-Doped, Bamboo-Like Carbon Nanotube Catalysts in CO$_2$ Hydrogenation. *ChemistryOpen* **2018**, *7*, 789–796. [CrossRef] [PubMed]

33. Berciaud, S.; Ryu, S.; Brus, L.E.; Heinz, T.F. Probing the Intrinsic Properties of Exfoliated Graphene: Raman Spectroscopy of Free-Standing Monolayers. *Nano Lett.* **2009**, *9*, 346–352. [CrossRef] [PubMed]
34. Kim, B.J.; Kim, J.P.; Park, J.S. Effects of Al interlayer coating and thermal treatment on electron emission characteristics of carbon nanotubes deposited by electrophoretic method. *Nanoscale Res. Lett.* **2014**, *9*, 1–6. [CrossRef] [PubMed]
35. Lee, D.S.; Min, D.J. A Kinetics of Hydrogen Reduction of Nickel Oxide at Moderate Temperature. *Met. Mater. Int.* **2019**, *25*, 982–990. [CrossRef]

Article

A Study on Superior Mesoporous Activated Carbons for Ultra Power Density Supercapacitor from Biomass Precursors

Joon-Hyuk Bang [1,2,†], Byeong-Hoon Lee [1,3,†], Young-Chul Choi [1], Hye-Min Lee [1,*] and Byung-Joo Kim [4,*]

1. Convergence Research Division, Korea Carbon Industry Promotion Agency (KCARBON), Jeonju 54853, Korea; lusidsoul@naver.com (J.-H.B.); bhlee@kcarbon.or.kr (B.-H.L.); youngchoi@kcarbon.or.kr (Y.-C.C.)
2. Department of Chemistry, Kunsan National University, Kunsan 54150, Korea
3. Department of Organic Materials & Fiber Engineering, Chonbuk Natial University, Jeonju 54896, Korea
4. Department of Carbon-Nanomaterials Engineering, Jeonju University, Jeonju 55069, Korea
* Correspondence: leehm@kcarbon.or.kr (H.-M.L.); kimbyungjoo@jj.ac.kr (B.-J.K.); Tel.: +82-63-220-3293 (B.-J.K.)
† These authors contributed equally to this work.

Abstract: A kenaf-derived activated carbon (KAC) for a high-power density supercapacitor was developed in this study through phosphoric acid activation. The N_2/77K isothermal adsorption–desorption curve was used to estimate the textural properties of KAC based on BET and BJH and the pore size distribution based on NLDFT. The electrochemical properties of KAC were analyzed by using the coin-type cell applying 1 M $SPBBF_4$/PC electrolyte, and the specific surface area and total pore volume were 1490–1942 m^2/g and 1.18–3.18 cm^3/g, respectively. The pore characteristics of KAC varied according to the activation temperature, and most KAC showed a mesoporous structure. As the activation temperature increased, the mesopore volume increased up to 700 °C, then decreased. The mesoporous structure of KAC resulted in a substantial decrease in the Warburg impedance as the ion diffusion resistance decreased. Hence, the specific capacitance of KAC decreased from 82.9 F/g to 59.48 F/g as the charge–discharge rate increased from 1 mA/g to 10 mA/g, with the rate of reduction at approximately 30%. The rate of reduction of KAC's specific capacitance was 50% lower compared with commercial activated carbon; hence, KAC is a more suitable electrode-active material for high power density supercapacitors.

Keywords: supercapacitor; electric double-layer capacitor (EDLC); activated carbon (AC); power density

1. Introduction

With an increased global focus on addressing the challenge of environmental pollution, the interest in new energy devices that can replace petroleum has also increased [1]. Lithium ion batteries (LiB) and fuel cells serve as the main energy-storage devices based on high energy density; however, the limitation of low energy density remains a challenge [2]. Supercapacitors are characterized by semi-permanent life and high output characteristics, which makes them suitable for use as auxiliary power with LiB or fuel cells [3,4]. Supercapacitors contain a hybrid system with LiB or fuel cells to support regenerative breaking in environmentally friendly vehicles or a pitch peak and voltage dip in new renewable energy [5,6]. Numerous studies have shown that supercapacitors are beneficial as the auxiliary power in the hybrid system for generating the auxiliary output as well as securing the system's stability and enhancing battery life [7–9]. In the past, supercapacitors were mainly applied in electronic devices, and the focus was on improving the energy density; however, with the recent increase in their utilization in environmentally friendly vehicles and new renewable energy, there is an urgent need to improve the power density.

In supercapacitors, the energy-storage mechanism involves the adsorption of ions to the pores of activated carbon such that the electrochemical performance relies on the characteristics of the pores of activated carbon [10]. Previous studies on supercapacitors mostly focused on improving the energy density, and as a result, the specific capacitance of supercapacitors was shown to have a strong correlation with the volume of the pore that can store the ions [11–14]. The market demand for supercapacitors with high power density has recently increased, which has prompted research that is focused on decreasing the impedance of activated carbon to increase both energy and power densities [15,16]. The electrode materials, such as the 3D building blocks, 3D activated graphene, and 3D carbon frameworks, all contain a high specific surface area that ensures outstanding energy density, whereas the high conductivity (production with nanocarbon) and low-ion diffusion resistance (high mesopore volume) were found to result in outstanding electrochemical performance (high energy density & power density) [17–19]. The drawbacks regarding these electrode materials, however, are the requirements for a complex synthesis, production process, and the consequential high process cost. In our previous study, a bamboo-based activated carbon for a high power density supercapacitor was produced, which had a mesoporous structure [20]. The ion diffusion resistance was found to decrease with an increase in mesopore volume. The results showed that the mesopore volume of bamboo-based activated carbon was approximately 26% higher with an approximately 11% increase in specific capacitance, compared with commercial activated carbons. To further increase the power density of supercapacitors, the mesopore volume of activated carbon should be increased further. However, if the activated carbon with a higher mesopore volume is produced by using the physical activation based on a mesopore growth mechanism via the oxidation of crystal grain boundaries, the activation yield falls below 5%, causing low economic feasibility. Thus, to develop a high-performance supercapacitor with high energy and power densities, the production of novel mesoporous activated carbons should be investigated.

Phosphoric acid activation is a manufacturing technology for producing activated carbon having a higher mesopore ratio than the physical activation method. In particular, phosphoric acid activation of biomass precursors has a pore-development mechanism in which phosphate linkages formed by crosslinking with phosphoric acid are decomposed at 723 K to form mesopores [21]. Therefore, in order to increase the mesopore ratio of the biomass-based activated carbon prepared by phosphoric acid activation, a high cellulose content required for the crosslinking reaction is required. Kenaf is a very useful biomass precursor of activated carbon with a high yield per unit area, fast growth rate, and high fixed carbon. In particular, kenaf has a high cellulose content, so it has a high utilization value as a precursor of mesoporous activated carbon by phosphoric acid activation [22].

In this study, a mesoporous kenaf-derived activated carbon (KAC) with a high specific surface area was prepared to enhance the diffusion resistance of supercapacitors. Kenaf was chosen in place of activated carbon as the precursor based on its high productivity, and phosphoric acid activation was applied to create a high specific surface area and advanced mesoporous structure. The pore-growth mechanism for KAC through phosphoric acid activation was examined by analyzing the textural properties and crystal structures. The electrochemical properties of KAC were comparatively analyzed with reference to commercial activated carbons.

2. Results and Discussion

The isothermal adsorption–desorption curve is the most effective method for analyzing the characteristics of porous carbon materials. The isothermal adsorption–desorption curves and textural properties for activated carbons (KAC and commercial activated carbons) are given in Figure 1 and Table 1, respectively.

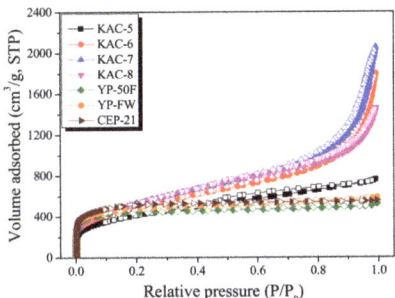

Figure 1. N$_2$/77K isothermal adsorption-desorption curves of kenaf-derived activated carbons as a function of activation temperature.

Table 1. Textural properties of kenaf-derived activated carbons as a function of activation temperature.

Samples	S$_{BET}$ [a] (m^2/g)	V$_{Total}$ [b] (cm^3/g)	V$_{Micro}$ [c] (cm^3/g)		V$_{Meso}$ [d] (cm^3/g)	Activation Yield [e] (%)
			DR Method	αs Method		
KAC-5	1490	1.18	0.54	1.06	0.82	31
KAC-6	1800	2.78	0.62	1.17	2.42	30
KAC-7	1940	3.18	0.65	1.37	2.80	27
KAC-8	1940	2.25	0.64	1.36	1.86	25
YP-50F	1780	0.83	0.69	0.76	0.17	-
YP-FW	1820	0.91	0.70	0.86	0.26	-
CEP-21	2230	0.86	0.83	0.96	0.14	-

[a] S$_{BET}$: Specific surface area; BET method $\frac{P}{v(P_0-P)} = \frac{1}{v_m c} + \frac{c-1}{v_m c}\frac{P}{P_0}$. [b] V$_{Total}$: Total pore volume; BET method. [c] V$_{Micro}$: Micropore volume. [d] V$_{Meso}$: Mesopore volume; BJH method $r_p = r_k + t$, (r_p = actual radius of the pore, t = thickness of the adsorbed film). [e] Activation yield: $\frac{\text{Weight of activated sample}}{\text{Weight of carbonized or stabilized sample input}} \times 100$.

In Figure 1, KAC and commercial activated carbons display entirely different types of isothermal adsorption–desorption curves based on the IUPAC criteria [23]. The isothermal adsorption–desorption curve for KAC showed that only KAC-5 had the type I curve, whereas all others had the type II curve. That is, the KAC activated at 500 °C mainly experiences micropore growth and the pore structure mostly comprises micropores, whereas the KAC activated at a higher temperature experiences mesopore growth and the respective pore structure comprises both micropore and mesopore. However, for commercial activated carbons (YP-50F, YP-FW, and CEP-21), the isothermal adsorption–desorption curves were all type I based on the IUPAC criteria. That is, the pore structure of commercial activated carbons mostly comprises micropores, which is presumed to be for storing a greater amount of ions to ensure a high energy density [23]. YP-FW among commercial activated carbons had the volume adsorbed fall in the range between YP-50F and CEP-21 at a relative pressure of 0.05. As the relative pressure increased, the continuous increase in the volume adsorbed led to the highest level among commercial activated carbons at relative pressure 0.99. YP-FW, therefore, showed the largest total pore volume due to high mesopore volume despite the micropore volume falling between YP-50F and CEP-21.

Table 1 shows the textural properties of activated carbon. Textural properties such as specific surface area, micropore volumes, mesopore volume, and pore size distribution of porous carbon materials are important factors to be considered when designing new materials as well as in their applications. Nitrogen adsorption has shown to be adequate to determine the textural properties of microporous and mesoporous carbon materials. However, assessing microporosity is not an easy task. As a result, several methods and models such as the Brunauer–Emmett–Teller (BET) method, the Dubinin–Radushkevich

(DR) method, alpha-s plot (αs-plot), the Barrett–Joyner–Halenda (BJH) method, and the non-linear density functional theory (NLDFT) have been proposed, with each method presenting different assumptions, physical criteria, or application ranges, resulting in underestimation or overestimation of the final results in some cases. In conclusion, because KACs have a mesoporous pore structure, the two methods (DR plot and as-plot) determine the micropore volume differently depending on different assumptions, physical criteria, or application ranges in Table 1.

In Table 1, the activation yield of KAC showed a continuous fall as the activation temperature increased, which is presumed to be due to the increase in crystal oxidation with an increase in activation temperature. For KAC, the micropore volume increased until 700 °C and stabilized, but the mesopore volume increased until 700 °C and then decreased. The specific surface area and total pore volume of KAC were 1490–1940 m^2/g and 1.18–3.18 cm^3/g, respectively. Nonetheless, the specific surface area and total pore volume of commercial activated carbons were 1780–1950 m^2/g and 0.83–0.91 cm^3/g, respectively. Thus, for all except KAC-5, the specific surface area was similar to commercial activated carbons, whereas the total pore volume was two to three times higher. The results indicated that, with the exception of KAC-5, a micropore volume of KAC was similar to commercial activated carbons, whereas the mesopore volume was 5–10 times greater, so that the total pore volume became 2–3 times higher than commercial activated carbons.

The completely different pore structures between the KAC and commercial activated carbons may be attributed to the different pore growth mechanism through the activation process. The pore characteristics of activated carbons are known to be influenced by the precursor, stabilization and carbonization processes, and the activation process, the last factor having the most significant effect [24,25]. The coconut-based activated carbon (YP-50F and YP-FW) and cokes-based activated carbon (CEP-21) were each produced through the steam activation and KOH activation processes, which share the pore-growth mechanism through the crystal oxidation mediated by CO_2 or H_2O despite differences in the chemical activation due to intercalated metallic K. Thus, the activated carbons produced through the steam and KOH activation processes mostly display a microporous structure, whereas mesopore growth is preceded by the fall in the micropore volume, which is caused by low activation yield or pore destruction [26,27]. Therefore, it is generally very difficult to produce activated carbons with a high level of mesopores through the steam activation and KOH activation processes.

Kenaf-derived activated carbons exhibit a different mechanism of pore growth via phosphoric acid activation. The dehydration, degradation, and condensation (crosslinking) of kenaf occur due to phosphoric acid. The phosphorous compound forms an ester bond with –OH groups at 200 °C for the crosslinking in polymer chains, and thermal degradation of phosphate linkages at 450 °C or above causes the expansion of the crystal structure to increase the mesopore volume [28]. Hence, the unique pore-growth mechanism of KAC compared to commercial activated carbons is thought to underlie the high proportion of mesopores at approximately 40–60% in spite of the high activation yield.

The Non-Localized Density Functional Theory (NLDFT) is a method for monitoring the pore-size distribution in activated carbons based on the thermodynamic data. The pore-distribution curves for KAC and commercial activated carbons of EDLC are shown in Figure 2. First, compared to KAC, commercial activated carbons present narrower and smaller pore diameters on the pore-distribution curves. CEP-21, YP-50F, and YP-FW each show a curve with main pore diameters of 1.14 nm, 1.2 nm, and 1.66 nm, respectively. CEP-21, in particular, as it is formed through KOH activation, displays the narrowest pore distribution curve with the smallest pore diameter. Although YP-50F and CEP-21 have a microporous structure with pore distribution curve of <3 nm pore diameter, the mesoporous structure was most distinct for YP-FW among the commercial activated carbons with the pore growth of up to a 100-nm pore diameter.

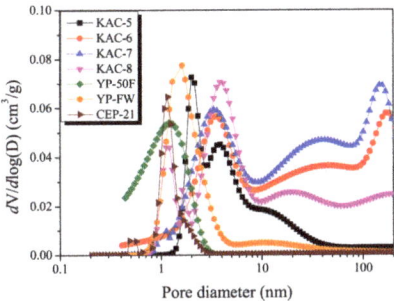

Figure 2. Pore size distribution curves of kenaf-derived activated carbons as a function of activation temperature.

However, a highly broad pore-distribution curve with more than two peaks was observed for KAC. The center of the peak was also observed at 3.0–4.0 nm. In line with the previously described isothermal adsorption–desorption curve of KAC, the mesopore volume with ≥10 nm pore diameter increased as the activation temperature increased up to 700 °C, then decreased. Thus, despite the similar specific surface area, KAC and commercial activated carbons showed completely different pore ranges on the micropore distribution curve, whereas KACs were found to have the pore growth with relatively larger pore diameters.

XRD is the most useful analytic method for the monitoring of the crystal structure of porous carbons. The XRD curve for KAC is presented in Figure 3. The XRD spectra agree well with the standard International Centre of Diffraction Data (ICDD) file (ICDD-PDF #411487), which confirmed the formation of crystalline structures of the graphitic carbon. The obtained peaks are indexed to the corresponding peaks for C(002), C(100), and C(101) crystal planes of graphitic carbon. However, the C(100) and C(101) crystal planes of porous carbon are marked as C(10*l*) because they are difficult to distinguish due to their low crystallinity and isotropic crystal structure. In the diffraction pattern of KAC, an Al(PO$_3$)$_3$ peak was observed at approximately 20° and 25°, which is considered to be formed by the reaction between phosphoric acid and an alumina tray.

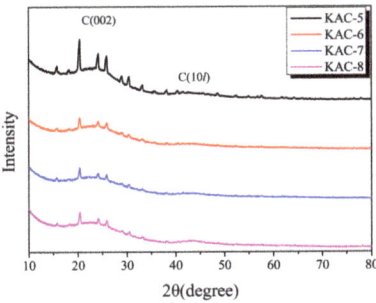

Figure 3. X-ray diffraction patterns of the kenaf-derived activated carbons as a function of activation conditions.

The peak fitting of the XRD curve in Figure 3 is used to show the changes in the L_a (width) and L_c (height) is shown in Figure 4. The L_a and L_c increased as the activation temperature increased as shown in Figure 5. In the XRD analysis, an increase in the L_a of a carbon material is observed due to two reasons: (1) the crystal growth, and (2) the oxidation of a non-crystalline substance. It is a well-established fact that the crystal of carbon material grows at a high temperature [29]. The activation of KAC was mediated without carbonization after stabilization at 200 °C. Thus, the crystal growth is presumed to have occurred as the temperature increased in the activation process. As the XRD shows the mean values after statistical analyses, the oxidation of non-crystalline parts is observed with a relative increase in crystal growth [29]. As shown in Table 1, an increase in the activation temperature caused an increase in the specific surface area; this means that with the increase in activation temperature, the oxidation of non-crystalline parts increased to form micropores. As a result, the L_a of KAC is determined to be increased due to crystal growth, and oxidation of amorphous material at a high activation temperature.

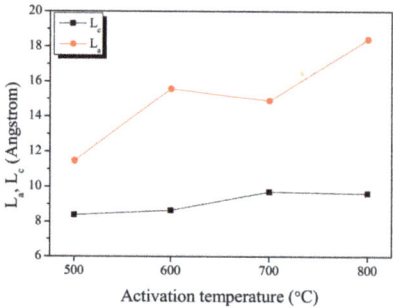

Figure 4. Structural characteristics of kenaf-derived activated carbons as a function of activation temperature.

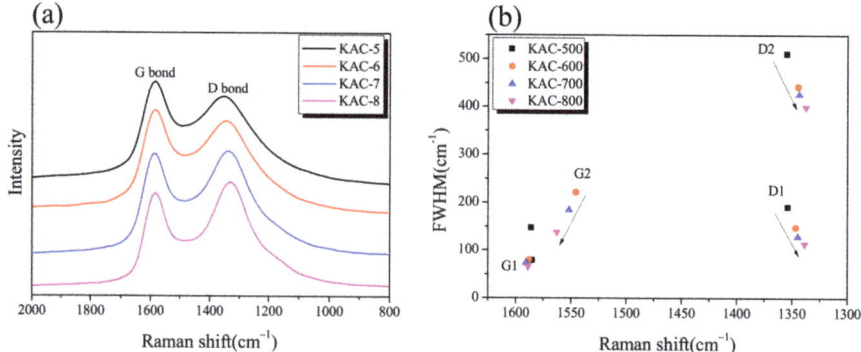

Figure 5. (a) Raman spectrum of kenaf-derived activated carbons as a function of activation temperature. (b) Band parameters derived from raw spectra decompositions.

The Raman spectrum and band parameters for KACs are shown in Figure 5. As can be seen in Figure 5a, a clear separation of G and D bands was observed with a gradual increase in ID/IG in direct proportion to the activation temperature. In our previous study, an increase in ID/IG was observed upon the oxidation of non-crystalline substances during the physical activation process [30]. Hence, the oxidation is thought to prioritize non-crystalline over crystalline substances during the chemical activation by phosphoric acid, in conformance with the previously discussed XRD result. The G1, G2, D1, and D2 calculated through peak fitting in the Raman spectrum are shown in Figure 5b. G1 is the in-plane bond stretching motion of sp2 hybridized carbon atoms pairs, D1 indicates defects/disorder, G2 indicates amorphous carbon, and D2 indicates a disordered graphitic lattice [31]. With an increase in activation temperature, a negligible change was observed in G1 across all KAC samples, whereas the FWHM decreased in G2, D2, and D1. Phosphate activation has a pore-development mechanism similar to that of physical activation in which crystal edges are oxidized by oxygen generated by the decomposition of phosphate linkages [21]. Therefore, negligible change was observed in G1, unlike G2, D1, and D2, because the crystal structure change due to the oxidation reaction of phosphoric acid activation is mainly made at the edge of the crystal. In addition, as the activation temperature increases, the oxidation due to phosphoric acid activation is likely to be more active in amorphous carbons or at the crystal grain boundaries.

CV curves give a useful method for monitoring the charge–discharge behavior of EDLC. The CV curve for KAC at a 5–400 mV/s scan rate is shown in Figure 6. It is known that the CV curve for EDLC takes a rectangular shape as the most ideal form, with constant current in accordance with changes in voltage. The CV curves for all KAC samples, as shown in Figure 6, display the change to a leaf form from a rectangular form as ion-diffusion resistance increased with an increase in the charge–discharge rate. In Figure 6a, the CV curves of all KAC was observed in a rectangular form due to low ion-diffusion resistance at the slowest charge–discharge rate of 5 mV/s. The CV curve area was also highly similar across all KAC types. Thus, at slow charge–discharge rates, all KAC are presumed to have a highly similar specific capacitance.

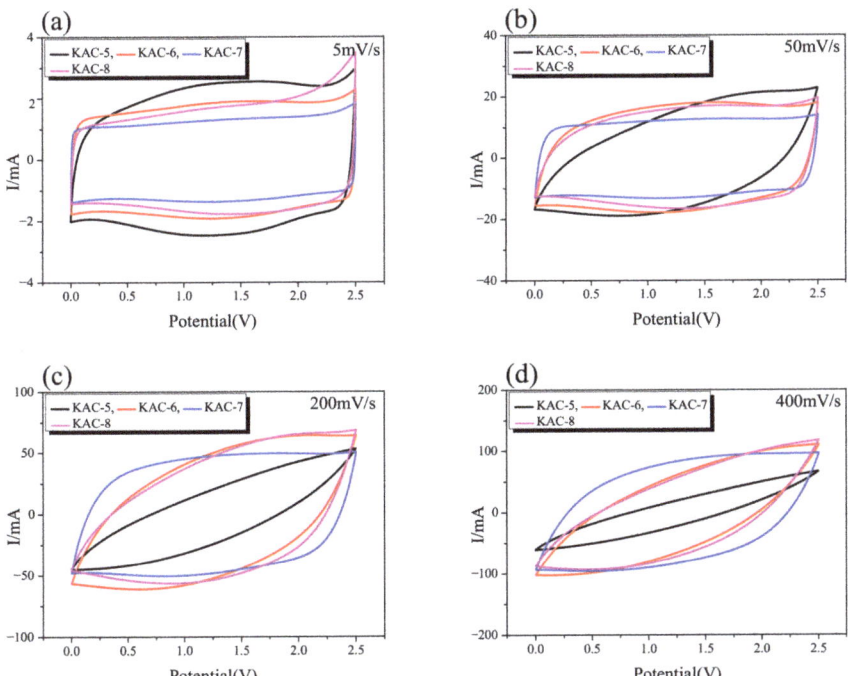

Figure 6. Cyclic voltammetry curves up to (**a**) scan rate 5 mV/s, (**b**) 50 mV/s, (**c**) 100 mV/s, and (**d**) 400 mV/s of AC/AC capacitors in 1 M SBPBF4 in PC electrolyte.

The CV curve for KAC-5 displayed the fastest change from a rectangular to leaf form with an increase in the charge–discharge rate, whereas the leaf form was observed from 50 mV/s. Next, the CV curves for KAC-6 and KAC-8 were highly similar, whereas the leaf form was observed from 100 mV/s. The CV curve for KAC-7 stably maintained the rectangular form even at a high charge–discharge rate, whereas the leaf form was observed from 400 mV/s. It is worth noting that, at 5 mV/s, KAC-7 had the narrowest CV curve area, but at 400 mV/s, the broadest CV curve area was found; hence, despite the most outstanding levels of specific surface area and total pore volume, KAC-7 had the lowest specific capacitance at a low charge–discharge rate. At a high charge–discharge rate, however, the low ion-diffusion resistance led to the highest specific capacitance. In our previous study, the specific capacitance of EDLC was shown to be more dependent on the pore distribution required for ion storage than the specific surface area [32]. Thus, the smallest CV curve area of KAC-700 is thought to be due to low pore volume for better ion adsorption, despite the most outstanding pore characteristics. Nonetheless, the high proportion of mesopores of KAC-700 led to low ion-diffusion resistance to increase the scan rate, and to cause the highest specific capacitance.

Galvanostatic technique allows a highly useful analytic method for monitoring the specific capacitance of activated carbons. Figure 7 shows the specific capacitance of commercial activated carbons and KAC in relation to various current densities in a 0.1–10 A/g range. At relatively low current densities of 0.1–0.2 A/g, the specific capacitance decreased from the KAC with the highest proportion of micropores, as follows: CEP-21 > YP-50F > YP-FW > KAC-700. Although the micropore volume of KAC-700 was higher than YP-50F or YP-FW, the specific capacitance was lower at 0.1–0.2 A/g current densities. In our previous study, compared to specific surface area and micropore volume, pore size distribution was shown to have greater influence on the specific capacitance [33]. As the previous pore distribution curves showed, YP-50F, YP-FW, and KAC-700 had completely different pore-distribution curves despite sharing identical specific surface area and micropore volume, and YP-50F with the pore volume indicating closer space among ions is presumed to have the highest specific capacitance.

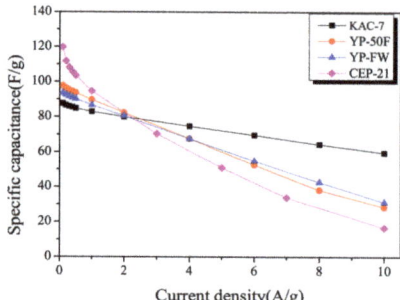

Figure 7. Specific capacitance curves of commercial activated carbons and kenaf-derived activated carbons with different current densities.

The specific capacitance across all activated carbons showed a linear decrease with an increase in current density. The slope of decrease in specific capacitance for each activated carbon was shown to be lower for higher mesopore proportions. Thus, at ≥ 2.5 A/g current densities, each activated carbon acquires an altered position in the order of specific capacitance; at 10 A/g current density, the specific capacitance was in reverse order as follows: KAC-700 > YP-FW > YP-50F > CEP-21. Notably, KAC-700 was shown to maintain 67% specific capacitance as the current density increased from 0.1 A/g to 10 A/g. Thus, to increase the energy density for EDLC, a microporous structure with a pore distribution comparable to that of the ion diameter is required, and to increase the power density, a mesoporous structure is required.

Electrochemical impedance spectroscopy is the most useful analytic method for monitoring the impedance of EDLC. The Nyquist plot for KAC and commercial activated carbons are presented in Figure 8. The Nyquist plot consists of a semi-circle and a 45° slope, each indicating the charge transfer resistance and the Warburg impedance. The semi-circle diameter for KAC-7 was shown to be greater than that across commercial activated carbons. Compared to commercial activated carbons based on the precursor of high-temperature physical activation or high-level conductivity, KAC-7 is likely to show low conductivity through the activation at a relatively low temperature [33], and the resulting increase in charge transfer resistance may account for the larger semi-circle diameter.

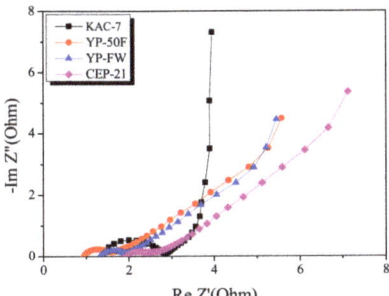

Figure 8. Nyquist plot of commercial activated carbons and kenaf-derived activated carbons.

In the Nyquist plot, the length of the 45° slope was in the following order: KAC-7 < YP-FW < YP-50F < CEP-21, with KAC-7 having the smallest slope. The result agreed with the rate of reduction of specific capacitance in the Galvanic technique result, based on the increase in current density. The Warburg impedance indicates the ion-diffusion resistance that decreases as the proportion of mesopores increases [32]. Thus, as KAC-7 has the highest mesopore volume, the reduction rate of the specific capacitance is presumed to be the lowest despite the high current density.

The results of the Galvanostatic technique and electrochemical impedance spectroscopy for the EDLC produced through the blending of KAC and commercial activated carbons are presented in Figure 9. After blending of KAC with CEP-21 and YP-50F in the ratio 8:2, the commercial activated carbons with a microporous structure, the consequent electrochemical behavior was monitored. In Figure 9a, the specific capacitance had a positive charge–discharge effect for commercial activated carbons through the blending with KAC. A trend of decreasing reduction rate of specific capacitance was noted after the blending of KAC with commercial activated carbons based on the high current density. At low current density, an improvement in specific capacitance was also observed.

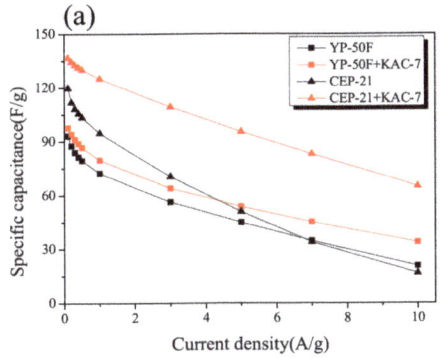 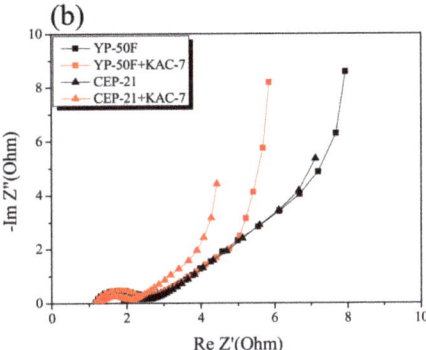

Figure 9. Specific capacitance and Nyquist plot of electrodes. (**a**) Specific capacitance of activated carbon and blended activated carbon electrode at different scan rates calculated from the Galvanostatic charge–discharge curve. (**b**) Nyquist plot for activated carbon and blended activated carbon electrode.

In Figure 9b, the Nyquist plot showed an improvement in impedance after the blending of KAC with commercial activated carbons. No significant change was observed for the semi-circle of commercial activated carbons even after the blending with KAC, but the Warburg impedance was greatly improved. The KAC was thus found to have a positive effect on greatly enhancing the original EDLC performance through the blending with commercial activated carbons with a microporous structure, although it may also serve independently as an active electrode material in EDLC for high power density.

3. Methods and Materials

3.1. Materials

Kenaf and phosphoric acid were obtained from the Jeonbuk Agricultural Research & Extension Services (Iksan, Korea) and Sigma-Aldrich (St Louis, MO, USA), respectively. Kenaf was immersed in phosphoric acid for 24 h, and stabilized at 150 °C. The stabilized kenaf was placed in a self-produced alumina tubular furnace for the activation at 500–800 °C in N_2 atmosphere for 30 min. At the end of activation, the furnace was cooled to room temperature. Subsequently, the KAC was washed with distilled water until pH7 was reached. The KAC was then pulverized to 10 μm particle size by using a ball mill (pulverisette 23, Fritsch GmbH, Idar-Oberstein, Germany) and dried in a 120 °C oven for 48 h. Subsequently, the KAC was compared with commercial activated carbons such as YP-50F, YP-FW (coconut based, steam acid activated; Kuraray, Tokyo, Japan), and CEP-21 (cokes based, KOH activated; power carbon technology, Gumi, Korea). The KAC was named KAC-activation temperature to show the experimental conditions.

3.2. Characteristics

The N_2/77K isothermal adsorption-desorption curve for activated carbons was drawn by using the BELSORP-max (MicrotracBEL Corp., Osaka, Japan). For the pore characteristics, the volumes of micropore and mesopore were calculated by using the isothermal adsorption curve and Brunauer–Emmett–Teller (BET), Barrett–Joyner–Halenda (BJH), Dubinin–Radushkevich (DR), and alpha-s plot (αs-plot) equations [34–37]. For the crystal structure of activated carbons, the X-ray diffraction (RINT2000 vertical goniometer) was used in 10–90° range at 2°/min injection rate, and the structure was analyzed via Raman spectroscopy (NTEGRA; NT-MDT Instruments, Moskow, Russia) in the 500–2000 cm^{-1} range.

3.3. Electrochemical Test

The electrode was produced by combining KAC, carbon black (Super-P, Timcal, Bodio, Switzerland), and polytetrafluoroethylene solution (PTFE 60 wt%, Sigma-Aldrich, St Louis, MO, USA) in an 85:10:5 ratio. After mixing the electrode materials in isopropyl alcohol, a manual roller was used for the mulling, and by applying the roll press at 120 °C, a sheet electrode of 200 μm thickness was produced. The KAC electrode was attached to an etching aluminum foil of 20 μm thickness with conductive glue, then dried in a 150 °C vacuum oven for 24 h. The KAC electrode punching set to φ12 mm diameter was used to form a cell with identical cathodes and anodes. The separation membrane was composed of a φ18 mm diameter cellulose (TF4035, NKK, Hyogo, Japan); the electrolyte was 1 M spirobipyrrolidinium tetrafluoroborate ($SBPBF_4$) in propylene carbonate (PC). The electrochemical properties of supercapacitors were measured at room temperature by using the Maccor 4300 battery tester (Maccor Inc., Tulsa, OK, USA) and VSP electrochemical workstation (Bio-Logic Science Instruments, Grenoble, France). For the galvanostatic charge/discharge (GCD) tests, a 0.1–10 A/g current density was applied at a voltage of 0.0–2.5 V with 20 cycles each of charge and discharge. From the 20-cycle discharge curve, the specific capacitance was calculated by using Equation (1):

$$Cs = \frac{\Delta T \times I}{\Delta V \times m}. \tag{1}$$

Here, I is the discharge current (A), ΔV is the range of constant voltage, m is the mass of the active material, and ΔT is the time taken for the discharge. The specific capacitance per unit mass (F/g) was calculated based solely on the mass of the active material without incorporating the binder and conductor weights. The cyclic voltammetry (CV) measurements were taken in the 0.0–2.5 V range at 5–400 mV/s scan rate. The electrochemical impedance spectroscopy (EIS) was conducted in the 300 kHz–10 MHz range at 10 mV amplitude.

4. Conclusions

In this study, a mesoporous kenaf-derived activated carbon was developed for the high-power density supercapacitor. Generally, KAC showed a mesoporous structure based on the kenaf decomposition, dehydration, and crosslinking via the use of phosphoric acid. The results in this study showed that kenaf is a suitable precursor in the production of activated carbon as it satisfies the ratio of the specific surface area and mesopore volume. The specific surface area and total pore volume were shown to be 1490–1942 m^2/g and 0.83–3.18 cm^3/g, respectively, for KAC. The specific surface area, total pore volume, and mesopore volume for KAC-7 were 1940 m^2/g, 3.18 cm^3/g, and 2.80 cm^3/g, respectively. The specific surface area and micropore volume were similar between KAC and commercial activated carbons, but the mesopore volume was 2–3 times higher for KAC than for commercial activated carbons. The high mesopore volume of KAC-7 lowered the ion-diffusion resistance, exhibiting the highest output characteristics. The specific capacitance of commercial activated carbons decreased at the rate of approximately 66–68% as the current density increased from 0.1 A/g to 10 A/g, whereas the blending of KAC-7 with commercial activated carbons led to an improvement not only in the specific capacitance, but also the ion-diffusion resistance. In conclusion, the KAC produced through phosphoric acid activation exhibited more outstanding output characteristics based on the mesoporous structure, and this implied that the mesopore control on the electrode material is the key determinant in improving the supercapacitor resistance and output in the future.

Author Contributions: Conceptualization, H.-M.L. and B.-J.K.; methodology, H.-M.L. and Y.-C.C.; software, J.-H.B., B.-H.L. and H.-M.L.; validation, H.-M.L. and B.-J.K.; formal analysis, H.-M.L. and B.-J.K.; investigation, J.-H.B. and B.-H.L.; resources, H.-M.L.; data curation, J.-H.B., B.-H.L., H.-M.L. and B.-J.K.; writing—original draft preparation, J.-H.B., B.-H.L. and H.-M.L.; writing—review and editing, B.-J.K.; visualization, Y.-C.C.; supervision, H.-M.L. and B.-J.K.; project administration, H.-M.L. and B.-J.K.; funding acquisition, H.-M.L. and B.-J.K. All authors have read and agreed to the published version of the manuscript.

Funding: This work was supported by the Technology Innovation Program (20016795, Development of manufacturing technology independence of advanced activated carbons and application for high performance supercapacitor) funded by the Ministry of Trade, Industry and Energy (MOTIE, Korea) and the Nano·Material Technology Development Program through the National Research Foundation of Korea (NRF) funded by Ministry of Science and ICT (2019M3A7B9071501, Development of manufacturing technology of polyolefin-based nanoporous activated carbon fibers for removal of ultrafine dust precursor and their fiber-shape control technology).

Institutional Review Board Statement: Not applicable.

Informed Consent Statement: Not applicable.

Data Availability Statement: The data presented in this study are available on request from the corresponding author.

Conflicts of Interest: The authors declare no conflict of interest.

References

1. Poizot, P.; Dolhem, F. Clean energy new deal for a sustainable world: From non-CO_2 generating energy sources to greener electrochemical storage devices. *Energy Environ. Sci.* **2011**, *4*, 2003–2019. [CrossRef]
2. Winter, M.; Brodd, R.J. What are batteries, fuel cell, and supercapacitor? *Chem. Rev.* **2004**, *104*, 4245–4270. [CrossRef] [PubMed]

3. Hu, X.; Deng, Z.; Suo, J.; Pan, Z. A high rate, high capacity and long life (LiMn$_2$O$_4$ + AC)/Li$_4$Ti$_5$O$_{12}$ hybrid battery–supercapacitor. *J. Power Sources* **2009**, *187*, 635–639. [CrossRef]
4. Cheng, J.; Lu, Z.; Zhao, X.; Chen, X.; Zhu, Y.; Chu, H. Electrochemical performance of porous carbons derived from needle coke with different textures for supercapacitor electrode materials. *Carbon Lett.* **2021**, *31*, 57–65. [CrossRef]
5. Burke, A.F. Batteries and ultracapacitors for electric, hybrid, and fuel cell vehicles. *Proc. IEEE* **2007**, *95*, 806–820. [CrossRef]
6. Allauoa, B.; Asnoune, K.; Mebarki, B. Energy management of PEM fuel cell/supercapacitor hybrid power sources for an electric vehicle. *Int. J. Hydrog. Energy* **2017**, *42*, 21158–21161. [CrossRef]
7. Cabrane, Z.; Ouassaid, M.; Maarouf, M. Management and control of storage photovoltaic energy using battery-supercapacitor combination. In Proceedings of the 2014 Second World Conference on Complex Systems (WCCS), Agadir, Morocco, 10–12 November 2014; IEEE: Manhattan, NY, USA, 2014; pp. 380–385.
8. Lahyani, A.; Venet, P.; Guermazi, A.; Troudi, A. Battery/supercapacitors combination in interruptible power supply. *IEEE Trans. Power Electron.* **2012**, *28*, 509–1522.
9. Rajagopal, R.; Kollimalla, S.K.; Mishra, M.K. Dynamic energy management of micro grids using battery super capacitor combined storage. In Proceedings of the 2012 Annual IEEE India Conference, Kochi, India, 28 January 2013; IEEE: Manhattan, NY, USA, 2013; pp. 1078–1083.
10. Bang, J.H.; Lee, H.M.; An, K.H.; Kim, B.J. A study on optimal pore development of modified commercial activated carbons for electrode materials of supercapacitors. *Appl. Surf. Sci.* **2017**, *415*, 61–66. [CrossRef]
11. Karthikeyan, K.; Amaresh, S.; Lee, S.N.; Sun, X.; Aravindan, V.; Lee, Y.G.; Lee, Y.S. Construction of high-energy-density supercapacitors from pine-cone-derived high surface-area carbons. *Chems. Sus. Chem.* **2014**, *7*, 1435–1442. [CrossRef] [PubMed]
12. Phiri, J.; Dou, J.; Vuorinenm, T.; GANE, P.A.C.; Maloney, T.C. Highly Porous Willow Wood-Derived Activated Carbon for High-Performance Supercapacitor Electrodes. *ACS Omega* **2019**, *4*, 18108–18117. [CrossRef] [PubMed]
13. Tian, X.; Ma, H.; Li, Z.; Yan, S.; Ma, L.; Yu, F.; Wang, G.; Guo, X.; Ma, Y.; Wong, C. Flute type micropores activated carbon from cotton stalk for high performance supercapacitors. *J. Power Sources* **2017**, *359*, 88–96. [CrossRef]
14. Yu, S.; Sano, H.; Zheng, G. Mesoporous carbon microspheres fabricated from KOH activation of sulfonated resorcinol-formaldehyde for "water-in-salt" electrolyte-based high-voltage (2.5 V) supercapacitors. *Carbon Lett.* **2022**, *32*, 285–294. [CrossRef]
15. He, X.; Ling, P.; Qiu, J.; Yu, M.; Zhang, X.; Yu, C.; Zheng, M. Efficient preparation of biomass-based mesoporous carbons for supercapacitors with both high energy density and high power density. *J. Power. Sources* **2013**, *240*, 109–113. [CrossRef]
16. Lee, K.S.; Seo, Y.J.; Jeong, H.T. Enhanced electrochemical performances of activated carbon (AC)-nickel-metal organic framework (SIFSIX-3-Ni) composite and ion-gel electrolyte based supercapacitor. *Carbon Lett.* **2021**, *31*, 635–642. [CrossRef]
17. Pham, D.T.; Lee, T.H.; Luong, D.H.; Yao, F.; Ghosh, A.; Le, V.T.; Kim, T.H.; Li, B.; Chang, J.; Lee, Y.H. Carbon nanotube-bridged graphene 3D building blocks for ultrafast compact supercapacitors. *ACS Nano* **2015**, *9*, 2018–2027. [CrossRef] [PubMed]
18. Basnayaka, P.A.; Ram, M.K.; Stefanakos, L.; Kumar, A. Graphene/polypyrrole nanocomposite as electrochemical supercapacitor electrode: Electrochemical impedance studies. *Grephene* **2013**, *2*, 81–87. [CrossRef]
19. Leng, C.; Zhao, Z.; Song, Y.; Sun, L.; Fan, Z.; Yang, Y.; Liu, X.; Wang, X.; Qiu, J. 3D carbon frameworks for ultrafast charge/discharge rate supercapacitors with high energy-power density. *Nano-Micro Lett.* **2021**, *13*, 8. [CrossRef]
20. Kim, J.H.; Lee, H.M.; Jung, S.C.; Chung, D.C.; Kim, B.J. Bamboo-based mesoporous activated carbon for high-power-density electric double-layer capacitors. *Nanomaterials* **2021**, *11*, 2750. [CrossRef] [PubMed]
21. Lee, B.H.; Lee, H.M.; Chung, D.C.; Kim, B.J. Effect of mesopore development on butane working capacity of biomass-derived activated carbon for automobile canister. *Nanomaterials* **2021**, *11*, 673. [CrossRef] [PubMed]
22. Andilolo, J.; Nikmatin, S.; Nugroho, N.; Alatas, H.; Wismogroho, A.S. Effect of kenaf short fiber loading on mechanical properties of biocomposites. *IOP Conf. Ser. Earth Environ. Sci.* **2017**, *65*, 012015. [CrossRef]
23. Sing, K.S.W.; Everett, D.H.; Haul, R.A.W.; Moscou, L.; Pierotti, R.A.; Rouquerol, J.; Siemieniewska, T. Reporting physisorption data for gas/solid systems with special reference to the determination of surface area and porosity. *Pure Appl. Chem.* **2009**, *57*, 603–619. [CrossRef]
24. Ioannidou, O.; Zabaniotou, A. Agricultural residues as precursors for activated carbon production—A review. *Renew. Sustain. Energy Rev.* **2007**, *11*, 1966–2005. [CrossRef]
25. Ruiz-Fernández, M.; Alexandre-Franco, M.; Fernández-González, C.; Gómez-Serrano, V. Development of activated carbon from vine shoots by physical and chemical activation methods. Some insight into activation mechanisms. *Adsorption* **2011**, *17*, 621–629. [CrossRef]
26. Lee, H.M.; Chung, D.C.; Jung, S.C.; An, K.H.; Park, S.J.; Kim, B.J. A study on pore development mechanism of activated carbons from polymeric precursor: Effects of carbonization temperature and nano crystallite formation. *Chem. Eng. J.* **2019**, *10*, 21. [CrossRef]
27. Lu, Y.H.; Zhang, S.L.; Yin, J.M.; Bai, C.C.; Zhang, J.H.; Li, Y.X.; Yang, Y.; Ge, Z.; Zhang, M.; Wei, L.; et al. Mesoporous activated carbon materials with ultrahigh mesopore volume and effective specific surface area for high performance supercapacitors. *Carbon* **2017**, *124*, 64–71. [CrossRef]
28. Jagtoyen, M.; Derbyshire, F. Activated carbons from yellow poplar and white oak by H$_3$PO$_4$ Activation. *Carbon* **1998**, *36*, 1085–1097. [CrossRef]

29. Baek, J.; Lee, H.M.; Roh, J.S.; Lee, H.S.; Kang, H.S.; Kim, B.J. Studies on preparation and applications of polymeric precursorbased activated hard carbons: I. Activation mechanism and microstructure analyses. *Microporous Mesoporous Mater.* **2016**, *219*, 258–264. [CrossRef]
30. Baek, J.; Shin, J.S.; Chung, D.C.; Kim, B.J. Studies on the correlation between nanostructure and pore development of polymeric precursor-based activated hard carbons: II. Transmission electron microscopy and Raman spectroscopy studies. *J. Ind. Eng. Chem.* **2017**, *54*, 324–331. [CrossRef]
31. Shimodaira, N.; Masui, A. Raman spectroscopic investigations of activated carbon materials. *J. Appl. Phys.* **2002**, *92*, 902–909. [CrossRef]
32. Lee, H.M.; An, K.H.; Park, S.J.; Kim, B.J. Mesopore-rich activated carbons for electrical double-layer capacitors by optimal activation condition. *Nanomaterials* **2019**, *9*, 608. [CrossRef] [PubMed]
33. Adinaveen, T.; John Kennedy, L.; Judith Vijaya, J.; Sekaran, G. Studies on structural, morphological, electrical and electrochemical properties of activated carbon prepared from sugarcane bagasse. *J. Ind. Eng. Chem.* **2013**, *19*, 1470–1476. [CrossRef]
34. Brauneur, S.; Emmet, P.; Telle, E. Adsorption of gases in multimolecular layer. *J. Am. Chem. Soc.* **1938**, *60*, 309–319. [CrossRef]
35. Barrett, E.P.; Joyner, L.G.; Halenda, P.P. The determination of pore volume and area distributions in porous substances. I. Computations from Nitrogen Isotherms. *J. Am. Chem. Soc.* **1951**, *73*, 373–380. [CrossRef]
36. Dubinin, M.M.; Radushkevich, L.V. Evaluation of microporous materials with a new isotherm. *Dokl. Akad. Nauk. SSSR* **1947**, *55*, 331–334.
37. Sing, K.S.W. Empirical method for analysis of adsorption isotherms. *Chem. Ind. Engl.* **1968**, *44*, 1520–1521.

Article

Prediction of Strong Transversal s(TE) Exciton–Polaritons in C_{60} Thin Crystalline Films

Vito Despoja [1,2,*] and Leonardo Marušić [3]

1. Institut za Fiziku, Bijenička 46, 10000 Zagreb, Croatia
2. Donostia International Physics Center (DIPC), P. Manuel de Lardizabal, 4, 20018 San Sebastián, Spain
3. Maritime Department, University of Zadar, M. Pavlinovića 1, 23000 Zadar, Croatia; lmarusic@unizd.hr
* Correspondence: vito@phy.hr

Abstract: If an exciton and a photon can change each other's properties, indicating that the regime of their strong bond is achieved, it usually happens in standard microcavity devices, where the large overlap between the 'confined' cavity photons and the 2D excitons enable the hybridization and the band gap opening in the parabolic photonic branch (as clear evidence of the strong exciton–photon coupling). Here, we show that the strong light–matter coupling can occur beyond the microcavity device setup, i.e., between the 'free' s(TE) photons and excitons. The s(TE) exciton–polariton is a polarization mode, which (contrary to the p(TM) mode) appears only as a coexistence of a photon and an exciton, i.e., it vanishes in the non-retarded limit ($c \to \infty$). We show that a thin fullerene C_{60} crystalline film (consisting of N C_{60} single layers) deposited on an Al_2O_3 dielectric surface supports strong evanescent s(TE)-polarized exciton–polariton. The calculated Rabi splitting is more than $\Omega = 500$ meV for $N = 10$, with a tendency to increase with N, indicating a very strong photonic character of the exciton–polariton.

Keywords: eksciton-polaritons; molecular crystals; optical conductivity; photonics

1. Introduction

The interaction between photons and polarization modes can result in the formation of hybrid photon polarization modes, called polaritons [1]. Very common platforms for studying strong light–matter interactions are the gapped systems, such as semiconductors [2,3] or molecules [4], placed in microcavity devices, where the cavity exciton–polaritons are formed. The quantum nature of a cavity exciton–polariton manifests in the form of the Bose–Einstein condensation, which has recently been experimentally detected [2,5,6]. The cavity exciton–polaritons are routinely observed in bulk [7–10] and quantum well systems [3,11], e.g., devised from GaAs [3]. Two-dimensional (2D) materials, such as semiconducting monolayers, thin heterostructures, and films, are even more attractive than their bulk counterparts, due to the reduced Coulomb screening and the corresponding large exciton binding energies [12–19] that enable the formation of well-defined exciton–polaritons even at room temperatures [20]. The first 2D exciton–polaritons were obtained in a monolayer of a transition metal dichalcogenide (TMD) MoS_2, where the Rabi splitting between the exciton and the cavity photon of $\sim$50 meV was observed [21]. Further photoluminescence studies showed clear anti-crossing behavior and splitting of the exciton–polariton in other 2D TMD cavity devices, e.g., in $MoSe_2$ [22], WS_2 [23], WSe_2 [24,25], and in the $MoSe_2$-WSe_2 heterostructure [26]. In addition, the real-space imaging of the exciton–polaritons has been done by means of near-field scanning optical microscopy for WSe_2 thin films [27]. Finally, a remarkable Rabi splitting of 440 meV was theoretically predicted in the monolayer hexagonal boron nitride cavity device [28], suggesting extraordinary strong light–matter interaction in 2D heterostructures.

The strongest exciton–photon coupling is achieved in the organic dye molecule thin films placed in a microcavity [29]. For example, Rabi splitting of $\Omega \leq 450$ meV [30–32],

0.7 eV [33], and even more than 1 eV [34,35] have been detected when various organic dye molecules were placed in a planar microcavity. The theoretical approach to such systems is mostly based on the two-level or boson–boson Hamiltonian model with an arbitrary coupling constant [4,36]. Using graphene [37] or perovskite [38]-layered heterostructures, one can obtain tunable microcavity devices of high performance, which can be applied as photonic detectors or emitters [38,39], but also as platforms for studying the exciton–photon coupling.

All these studies use the same concept: an exciton in a semiconducting nanostructure hybridizes with a photon 'confined' in a metallic microcavity. Such cavity photons and excitons are expected to interact stronger as the photon is more confined (i.e., the overlap between the exciton and photon is larger) and as the exciton oscillator strength [28] is larger. In this paper, we change the concept and explore the coupling between the 'free' photons and the excitons in the 2D nanostructures. The coupling between the 'free' photons and the polarization modes (such as plasmons, phonons, or excitons at surfaces or in 2D nanostructures) is a well-known, widely explored phenomenon [40–47]. The inherent property of all these eigenmodes (called plasmon polaritons, phonon polaritons, or exciton–polaritons) is their evanescent character, i.e., they are well-defined eigenmodes with the electromagnetic field (wave function) strongly localized at the interface or within the 2D nanostructure. Moreover, these modes usually have p or transverse-magnetic (TM) polarization, i.e., the electric field (and, therefore, the currents as well) has a component parallel to the direction of propagation. The most relevant point is that the p(TM) polarized plasmon polariton, phonon polariton, or exciton–polariton branches reduce to the plasmon, phonon, or exciton branches in the non-retarded ($c \to \infty$) limit. On the other hand, the electric field and the current of s or transverse-electric (TE) electromagnetic eigenmodes are perpendicular to the direction of their propagation. An especially attractive aspect of these polarization modes is that they **do not** exist in the non-retarded ($c \to \infty$) limit, i.e., they appear only as the coexistence of a photon and an exciton. The extent of the photon's participation in the s(TE) exciton–polariton is determined from the bending of the horizontal exciton branch (ω_{ex}) at the exciton–photon crossing ($\omega_{ex} = Q_{ex}c/\sqrt{\epsilon}$, where Q_{ex} is the photon wave vector at the exciton–photon crossing point), which we call the Rabi splitting Ω, to keep the terminology compatible with the cavity systems. Even though the s(TE) surface or 2D polaritons do exist [40] for some conditions, there is still no experimental evidence of such modes. However, the hybridization between the s(TE) Bloch surface waves (BSWs) (i.e., the photons confined between a truncated photonic crystal and a semi-infinite dielectric), and the excitons has been experimentally demonstrated in both inorganic (quantum well and TMD monolayer) and organic systems [48–51].

We show that very strong s(TE) exciton–polaritons may occur in layered van der Waals (vdW) heterostructures. The prototypical layered nanostructure we investigate in this paper is a thin film of the FCC fullerite (crystalline fullerene) cut along the (111) planes so that it formed several (N) molecular (C_{60}) layers. The crystalline C_{60} films were also deposited on a dielectric Al_2O_3 surface to make the simulation more realistic. The epitaxial growth of the C_{60} thin films of various thicknesses on various metallic or dielectric substrates under ambient conditions and in high vacuum was studied in references [52–58]. Some experimental studies even show that the crystalline growth of the C_{60} thin films on pentacene buffer layers is exclusively (111)-oriented [55]. Theoretical, molecular dynamic simulations of the C_{60} multilayer epitaxial growth and stability on various substrates were investigated in references [58–61]. These experimental/theoretical studies suggest that our model system is indeed highly realistic.

In this paper, the light–matter interaction was studied using our quantum electrodynamic Bethe–Salpeter equation approach (QE-BSE) developed in references [28,47]. This approach describes both excitons and photons by bosonic propagators σ and Γ, respectively, derived from the first principles. The C_{60} optical conductivity σ was calculated using *ab initio* G_0W_0-BSE method [47,62], and the free proton propagator Γ was derived by solving Maxwell's equation at the vacuum/dielectric interface [63,64]. The dielectric surface was described by the local dielectric function $\epsilon_M(\omega)$, also determined from the first principles. The exciton–photon coupling was achieved by dressing the free-photon propagator Γ with excitons at the random phase approximation (RPA) level. We studied the s(TE)-polarized exciton–polariton in the C_{60} thin crystalline film as a function of the number N of the C_{60} single layers. We obtained a very strong hybridization between the exciton in the C_{60} thin film and the s(TE) free photons. The hybridization increased with N, and for $N = 10$, we achieved the Rabi splitting Ω even larger than 1000 meV and 500 meV for self-standing and supported C_{60} films, respectively.

The paper is organized as follows. In Section 2, we present the geometry of the system and the derivation of the optical conductivity $\tilde{\sigma}$ of the C_{60} single layer using the G_0W_0-BSE approach with the solution of Dyson's equation for the electric field propagator $\mathcal{E} = \Gamma + \Gamma\tilde{\sigma}\mathcal{E}$. In Section 3, we present the spectra of the electromagnetic modes $S = \mathrm{Re}\mathcal{E}$, as well as the dispersion relations and intensities of the s(TE)-polarized exciton–polaritons in the C_{60} thin films of different thicknesses N, in vacuum, and at a Al_2O_3 surface. The conclusions are presented in Section 4.

2. Theoretical Formulation

We assume the the C_{60} molecules, upon deposition on the crystal surface, self-assemble in a regular FCC structure (the most stable bulk structure of crystalline fullerene) forming a (111) surface, as shown in Figure 1a. The FCC crystal lattice constant is taken to be $a_{3D} = 14$ Å [61], and the separation between the layers is fixed to be $\Delta = a_{3D}/\sqrt{3} = 8.1$ Å. Each crystal plane forms a 2D hexagonal Bravais lattice with the lattice constant $a_{2D} = a_{3D}/\sqrt{2} = 9.9$ Å, as shown in Figure 1b. The C_{60} films, occupying $z > 0$ half-space, are immersed in a dielectric medium described by a dielectric constant ϵ_0. The dielectric response of the substrate, occupying $z < 0$ half-space, is approximated by a local macroscopic dielectric function $\epsilon_M(\omega)$.

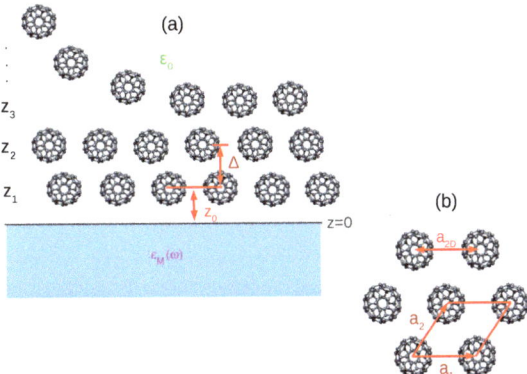

Figure 1. (a) C_{60} molecules upon deposition on the surface self-assemble in a regular FCC structure forming a (111) surface. The C_{60} layers, occupying $z > 0$ half-space, are immersed in a dielectric medium described by a dielectric constant ϵ_0. The dielectric response of the substrate, occupying $z < 0$ half-space, is approximated by a local macroscopic dielectric function $\epsilon_M(\omega)$. The FCC lattice constant is $a_{3D} = 14$ Å so that the separation between the layers is $\Delta = a_{3D}/\sqrt{3} = 8.1$ Å. (b) Each crystal plane forms a 2D-hexagonal Bravais lattice with lattice constant $a_{2D} = a_{3D}/\sqrt{2} = 9.9$ Å.

2.1. Calculation of Electric Field Propagator $\mathcal{E}$

The quantity we used to extract the information from the electromagnetic modes in the C_{60} films deposited on a dielectric surface was the electric field propagator $\mathcal{E}_{\mu\nu}$, which provides the electric field produced by an external oscillating point dipole $\mathbf{p}_0 e^{-i\omega t}$ placed at point $\mathbf{r}'$ as [63,64]

$$E_\mu(\mathbf{r},\omega) = \sum_{\nu=x,y,z} \mathcal{E}_{\mu\nu}(\mathbf{r},\mathbf{r}',\omega) p_\nu^0. \quad (1)$$

The propagator $\mathcal{E}$ is the solution of Dyson's equation [28,47,64,65]

$$\mathcal{E}_{\mu\nu}(\mathbf{r},\mathbf{r}',\omega) = \Gamma_{\mu\nu}(\mathbf{r},\mathbf{r}',\omega) + \sum_{\alpha,\beta=x,y,z}\int d\mathbf{r}_1 \int d\mathbf{r}_2\, \Gamma_{\mu\alpha}(\mathbf{r},\mathbf{r}_1,\omega)\sigma_{\alpha\beta}(\mathbf{r}_1,\mathbf{r}_2,\omega)\mathcal{E}_{\beta\nu}(\mathbf{r}_2,\mathbf{r}',\omega), \quad (2)$$

where the integration is performed over the entire space, σ is the nonlocal conductivity tensor of the deposited C_{60} thin film, and Γ is the electric field propagator in the absence of the C_{60} film, i.e., when $\sigma = 0$ [64,65]. The propagator Γ also includes the electromagnetic field scattering at the medium/substrate interface. If each molecule is approximated as a point polarizable dipole, then the optical conductivity of the C_{60} film can be written as

$$\sigma_{\mu\nu}(\mathbf{r},\mathbf{r}',\omega) = \sum_{i=1}^{N}\sum_{\mathbf{R}_\parallel} \sigma_i^{\mu\nu}(\omega)\delta(\boldsymbol{\rho}-\mathbf{R}_\parallel)\delta(z-z_i)\delta(\boldsymbol{\rho}'-\mathbf{R}_\parallel)\delta(z'-z_i), \quad (3)$$

where $\sigma_i^{\mu\nu}(\omega)$ is the optical conductivity tensor of a single molecule in the i-th molecular layer. This approximation is fully justified in the optical limit $2\pi c/\omega_{\text{light}} \gg R_M$, where R_M is the radius of a C_{60} molecule. Note that although all the molecules are equal, we distinguished between their conductivities in different layers $\sigma_i; i = 1, N$, due to the different influences of the substrates on a molecule in a different layer. The 2D Bravais lattice translation vectors spanning the molecular crystal are

$$\mathbf{R}_\parallel = n\mathbf{a}_1 + m\mathbf{a}_2; \quad n,m \in \mathbb{Z}, \quad (4)$$

where $\mathbf{a}_1$ and $\mathbf{a}_1$ are the primitive vectors, as illustrated in Figure 1b. The molecular layers occupy the planes

$$z_i = z_0, z_0 + \Delta, z_0 + 2\Delta, \ldots, z_0 + (N-1)\Delta,$$

where N is the number of molecular layers. Due to the planar translational invariance of the substrate, the Fourier transform of the propagator Γ is

$$\Gamma_{\mu\nu}(\mathbf{r},\mathbf{r}',\omega) = \int \frac{d\mathbf{Q}}{(2\pi)^2} \Gamma_{\mu\nu}(\mathbf{Q},\omega,z,z')e^{i\mathbf{Q}(\boldsymbol{\rho}-\boldsymbol{\rho}')}, \quad (5)$$

where $\mathbf{Q} = (Q_x, Q_x)$ are the two-dimensional wave vectors. The propagator $\mathcal{E}$ should also include the effects of the electromagnetic field Bragg scattering at the 2D crystal lattice, so that its Fourier transform is

$$\mathcal{E}_{\mu\nu}(\mathbf{r},\mathbf{r}',\omega) = \sum_{\mathbf{g}_\parallel}\int \frac{d\mathbf{Q}}{(2\pi)^2}\mathcal{E}_{\mathbf{g}_\parallel}^{\mu\nu}(\mathbf{Q},\omega,z,z')e^{i\mathbf{Q}\boldsymbol{\rho}}e^{-i(\mathbf{Q}+\mathbf{g}_\parallel)\boldsymbol{\rho}'}, \quad (6)$$

where

$$\mathbf{g}_\parallel = n\mathbf{b}_1 + m\mathbf{b}_2; \quad n,m \in \mathbb{Z} \quad (7)$$

are 2D reciprocal vectors, while $\mathbf{b}_1$ and $\mathbf{b}_1$ are primitive reciprocal vectors. After inserting (3), (5), and (6) in (2), it becomes an equation in the $(\mathbf{Q},\omega,z)$ space

$$\mathcal{E}_{\mathbf{g}_\parallel}^{\mu\nu}(\mathbf{Q},\omega,z,z') = \Gamma_{\mu\nu}(\mathbf{Q},\omega,z,z') +$$
$$\sum_{\alpha,\beta=x,y,z}\sum_i \sum_{\mathbf{g}_\parallel'} \Gamma_{\mu\alpha}(\mathbf{Q},\omega,z,z_i)\tilde{\sigma}_i^{\alpha\beta}(\omega)\mathcal{E}_{\mathbf{g}_\parallel+\mathbf{g}_\parallel'}^{\beta\nu}(\mathbf{Q}-\mathbf{g}_\parallel',\omega,z_i,z'), \quad (8)$$

where the surface optical conductivity is

$$\tilde{\sigma}_i^{\alpha\beta}(\omega) = \frac{1}{S_{\text{fcc}}} \sigma_i^{\alpha\beta}(\omega), \quad (9)$$

and $S_{\text{fcc}} = (\mathbf{a}_1 \times \mathbf{a}_2)\hat{\mathbf{z}} = a_{2D}^2 \sqrt{3}/2$ is the area of the 2D unit cell. If we neglect the electromagnetic field Bragg scattering, by introducing $\mathbf{g}_\parallel = \mathbf{g}'_\parallel = 0$ in Equation (8), and inserting $z = z_i$ and $z' = z_j$, the equation becomes the matrix tensor equation:

$$\mathcal{E}_{\mu\nu}(\mathbf{Q},\omega,z_i,z_j) = \Gamma_{\mu\nu}(\mathbf{Q},\omega,z_i,z_j) + \sum_{\alpha,\beta=x,y,z}\sum_k \Gamma_{\mu\alpha}(\mathbf{Q},\omega,z_i,z_k)\tilde{\sigma}_k^{\alpha\beta}(\omega)\mathcal{E}_{\beta\nu}(\mathbf{Q},\omega,z_k,z_j), \quad (10)$$

where $\mathcal{E}^{\mu\nu}(z_i, z_j)$ is the electric field propagator within ($i = j$) or between ($i \neq j$) the C_{60} layers. The electrical field propagator in the absence of the C_{60} film can be written as

$$\Gamma = \Gamma^0 + \Gamma^{sc}, \quad (11)$$

where the propagator of the 'free' electric field (or free photons propagator) is [63,64]

$$\Gamma^0(\mathbf{Q},\omega,z,z') = -\frac{4\pi i}{\epsilon_0\omega}\delta(z-z')\mathbf{z}\cdot\mathbf{z} - \frac{2\pi\omega}{\beta_0 c^2}e^{i\beta_0|z-z'|}\sum_{q=s,p}\mathbf{e}_q^0\cdot\mathbf{e}_q^0. \quad (12)$$

The propagator of the scattered electric field in the region $z, z' > 0$ is [63]

$$\Gamma^{sc}(\mathbf{Q},\omega,z,z') = -\frac{2\pi\omega}{\beta_0 c^2}e^{i\beta_0(z+z')}\sum_{q=s,p} r_q\, \mathbf{e}_q^+\cdot\mathbf{e}_q^-. \quad (13)$$

Here, the unit vectors of the s(TE)-polarized electromagnetic field are

$$\mathbf{e}_s^{0,\pm} = \mathbf{Q}_0 \times \mathbf{z}.$$

and the unit vectors of the p(TM) polarized electromagnetic field are

$$\mathbf{e}_p^{0,\pm} = \frac{c}{\omega\sqrt{\epsilon_0}}[\alpha_{0,\pm}\beta_0\mathbf{Q}_0 + Q\mathbf{z}],$$

where $\alpha_0 = -\text{sgn}(z-z')$, $\alpha_\pm = \mp 1$, and $\mathbf{Q}_0$ and $\mathbf{z}$ are the unit vectors in the $\mathbf{Q}$ and z directions, respectively. The reflection coefficients of the s(TE) and p(TM) polarized electromagnetic waves at the medium/substrate interface are

$$r_s = \frac{\beta_0 - \beta_M}{\beta_0 + \beta_M}$$

and

$$r_p = \frac{\beta_0\epsilon_M - \beta_M\epsilon_0}{\beta_0\epsilon_M + \beta_M\epsilon_0},$$

respectively, and the complex wave vectors in perpendicular (z) direction are

$$\beta_{0,M} = \sqrt{\frac{\omega^2}{c^2}\epsilon_{0,M}(\omega) - |\mathbf{Q}|^2}.$$

The β_0 and β_M determine the character of the electromagnetic modes at the medium/dielectric substrate interface. To simplify the interpretation, we assume that the dielectric medium is vacuum, i.e., $\epsilon_0 = 1$, and that the dielectric function of the substrate is constant $\epsilon_M(\omega) = \epsilon_M$, which is a plausible approximation for many wide band gap insulators. In the vacuum, for $\omega > Qc$, β_0 is a real number; therefore, the electromagnetic modes have a radiative character, and for $\omega < Qc$, β_0 is an imaginary number so that the electromagnetic modes have evanescent character. The two regions are separated by the so-called 'light-line'

$\omega = Qc$, as illustrated by the magenta line in Figure 2a. In analogy to that, the two regions in the substrate are separated by the $\omega = Qc/\sqrt{\epsilon_M}$ line, as illustrated by the green line in Figure 2a. Since $\epsilon_M > 1$, the slope of the light-line in the substrate is smaller than in the vacuum, so in the gap $Qc/\sqrt{\epsilon_M} < \omega < Qc$, the light propagates freely into the substrate but has an evanescent character in the vacuum region, as illustrated in Figure 2a. Therefore, the exciton–polariton mode $\omega_{\text{ex-pol}}$ is expected to appear in the fully evanescent region $\omega < Qc/\sqrt{\epsilon_M}$, since in that region it cannot be irradiated into the surrounding media. The evanescent character of the electric field produced by the exciton–polariton in the C_{60} film is illustrated in Figure 2b.

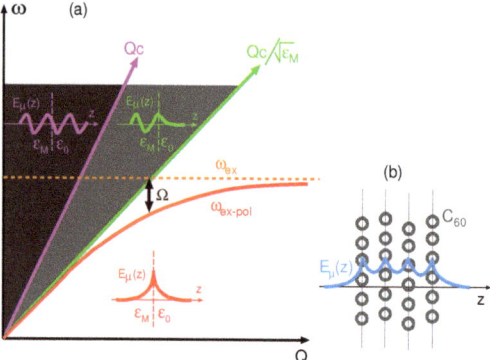

Figure 2. (**a**) The character of the electromagnetic modes at the dielectric/vacuum (ϵ_M/ϵ_0) interface. In the region $\omega > Qc$, the electromagnetic modes are entirely radiative (both in vacuum and in the dielectric), in the region $Qc/\sqrt{\epsilon_M} < \omega < Qc$, they are radiative in the dielectric and evanescent in vacuum, and in the $\omega < Qc/\sqrt{\epsilon_M}$ region, they have fully evanescent character. In the latter region, the photon and molecular exciton (ω_{ex}) hybridize, and an exciton–polariton ($\omega_{\text{ex-pol}}$) occurs. The measure of the coupling strength between the exciton ω_{ex} and the photon is given by Rabi splitting Ω. (**b**) The evanescent electric field $E_\mu(z)$ produced by an exciton–polariton in the C_{60} film.

We chose the electromagnetic modes to propagate in the $\mathbf{Q}_0 = \mathbf{y}$ directions. In this case, the Dyson Equation (10) decouples into the independent matrix and the matrix tensor equations for the s(TE) and p(TM) polarizations, respectively. Here, we explore the s(TE)-polarized electromagnetic modes, which satisfy the matrix equation:

$$\mathcal{E}_{xx}(|\mathbf{Q}|\mathbf{y}, \omega, z_i, z_j) = \Gamma_{xx}(|\mathbf{Q}|\mathbf{y}, \omega, z_i, z_j) + \sum_k \Gamma_{xx}(|\mathbf{Q}|\mathbf{y}, \omega, z_i, z_k) \tilde{\sigma}_k^{xx}(\omega) \mathcal{E}_{xx}(|\mathbf{Q}|\mathbf{y}, \omega, z_k, z_j), \quad (14)$$

where after using (11)–(13)

$$\Gamma_{xx}(|\mathbf{Q}|\mathbf{y}, \omega, z_i, z_j) = -\frac{2\pi\omega}{\beta_0 c^2} \left[e^{i\beta_0 |z_i - z_j|} + r_s e^{i\beta_0 (z_i + z_j)} \right]. \quad (15)$$

The first term in (15) represents the incident electromagnetic field, while the second term represents the one reflected at the vacuum/substrate boundary. In the electrostatic or non-retarded limit $c \to \infty$ and

$$\lim_{c \to \infty} \Gamma_{xx}(|\mathbf{Q}|\mathbf{y}, \omega, z_i, z_j) = 0, \quad (16)$$

so that all the properties presented below are a direct consequence of the binding between the s(TE)-polarized light (photons) and the molecular excitons, and they vanish, i.e., do not exist in the electrostatic limit.

2.2. Calculation of the Optical Conductivity of a Single Molecule

First, we explain the calculation of the molecular conductivity $\sigma_i^{\mu\nu}(\omega)$ in a 'self-standing molecule' ($z_0 \to \infty$), and then we extend that to the case when the molecule is close to a dielectric surface, i.e., when z_0 is finite, and chosen to be a characteristic vdW distance. The basic ingredients we need to calculate the molecular conductivity $\sigma_{\mu\nu}(\omega)$ are the molecular orbitals $\phi_n(\mathbf{r})$ and the energies E_n, which can be obtained by solving the Kohn–Sham equation self-consistently. We assume that the molecules are periodically repeating so that they form a simple cubic (sc) Bravais lattice with the unit cell a and volume $\Omega_{sc} = a^3$. The unit cell a is chosen so that there is no intermolecular overlap. This allows the molecular states $|\phi_n\rangle$ to be calculated at the Γ point only. It should be emphasized that the sc lattice and the unit cell a are not related to the previously described FCC lattice with the unit cell 'a_{3D}'. The purpose of the sc lattice is only to determine the molecular states $|\phi_n\rangle$ at the Γ point using the plane-wave DFT code. From now on, the conductivity of the 3D molecular crystal will be denoted as $\sigma_{\mu\nu}^{3D}(\omega)$.

The nonlocal optical conductivity tensor in the 3D molecular crystal is [62,64–66]

$$\sigma_{\mu\nu}^{3D}(\mathbf{r},\mathbf{r}',\omega) = \frac{2i}{\omega} \sum_{nm} \sum_{n'm'} \mathcal{K}_{n \to n'}^{m \leftarrow m'}(\omega) j_{nm}^{\mu}(\mathbf{r})[j_{n'm'}^{\nu}(\mathbf{r}')]^*, \quad (17)$$

where

$$j_{nm}^{\alpha}(\mathbf{r}) = \frac{e\hbar}{2im}\{\phi_n^*(\mathbf{r})\partial_\alpha \phi_m(\mathbf{r}) - [\partial_\alpha \phi_n^*(\mathbf{r})]\phi_m(\mathbf{r})\} \quad (18)$$

represents the current produced by the transition between the molecular states $|\phi_n\rangle \to |\phi_m\rangle$. Considering that the Bloch wave functions at the Γ point ϕ_n are periodic functions, tensor (17) can be expanded in the Fourier series

$$\sigma_{\mu\nu}^{3D}(\mathbf{r},\mathbf{r}',\omega) = \frac{1}{\Omega_{sc}} \sum_{\mathbf{GG}'} e^{i\mathbf{Gr}} e^{-i\mathbf{G}'\mathbf{r}'} \sigma_{\mathbf{GG}'}^{\mu\nu}(\omega), \quad (19)$$

where the Fourier coefficients are

$$\sigma_{\mathbf{GG}'}^{\mu\nu,3D}(\omega) = \frac{i}{\omega}\frac{2}{\Omega_{sc}} \sum_{nm} \sum_{n'm'} j_{nm}^{\mu}(\mathbf{G}) \mathcal{K}_{n \to n'}^{m \leftarrow m'}(\omega)[j_{n'm'}^{\nu}(\mathbf{G}')]^*, \quad (20)$$

and the current vertices are

$$j_{nm}^{\alpha}(\mathbf{G}) = \int_{\Omega_{sc}} d\mathbf{r} \, e^{-i\mathbf{Gr}} j_{nm}^{\alpha}(\mathbf{r}). \quad (21)$$

The four-point polarizability $\mathcal{K}$ can be obtained by solving the Bethe–Salpeter (BS) equation [28,47]

$$\mathcal{K}_{n \to n'}^{m \leftarrow m'}(\omega) = \mathcal{L}_{n \to n'}^{m \leftarrow m'}(\omega) + \sum_{n_1 m_1} \sum_{n_2 m_2} \mathcal{L}_{n \to n_1}^{m \leftarrow m_1}(\omega) \, \Xi_{m_1 \to n_2}^{m_1 \leftarrow m_2} \, \mathcal{L}_{n_2 \to n'}^{m_2 \leftarrow m'}(\omega), \quad (22)$$

where the time-ordered electron–hole propagator is defined as

$$\mathcal{L}_{n \to n'}^{m \leftarrow m'}(\omega) = \delta_{nn'} \delta_{mm'} \int_{-\infty}^{\infty} \frac{d\omega'}{2\pi i} G_n(\omega') G_m(\omega + \omega'). \quad (23)$$

The propagator (or Green's function) of an electron or a hole in a molecular state $|\phi_n\rangle$ is

$$G_n(\omega) = \frac{1}{\omega - E_n + E_n^{XC} - \Sigma_n^X - \Sigma_n^{C,0}(\omega)}, \quad (24)$$

where the exchange of self-energy is

$$\Sigma_n^X = -\frac{1}{\Omega_{sc}} \sum_m f_m \sum_{\mathbf{GG}'} \rho_{nm}^*(\mathbf{G}) V_{\mathbf{GG}'}^C(\mathbf{Q}) \rho_{nm}(\mathbf{G}'). \quad (25)$$

The correlation self-energy in the G_0W_0 approximation is

$$\Sigma_n^{C,0}(\omega) = \frac{1}{\Omega_{sc}} \sum_m \sum_{\mathbf{GG'}} \rho_{nm}^*(\mathbf{G})\rho_{nm}(\mathbf{G'})\left\{(1-f_m)\,\Gamma_{\mathbf{GG'}}^0(\omega - E_m) - f_m\,\Gamma_{\mathbf{GG'}}^0(E_m - \omega)\right\}, \qquad (26)$$

where the correlation propagator is defined as

$$\Gamma_{\mathbf{GG'}}^0(\omega) = \int_0^\infty d\omega'\, \frac{S_{\mathbf{GG'}}^0(\omega')}{\omega - \omega' + i\delta}, \qquad (27)$$

and $f_n = \theta(E_F - E_n)$ is the Fermi–Dirac distribution at $T = 0$. The spectrum of the electronic excitation in a self-standing molecule is

$$S_{\mathbf{GG'}}^0(\omega) = -\frac{1}{\pi} \mathrm{Im} W_{\mathbf{GG'}}^0(\omega). \qquad (28)$$

To avoid double-counting, we excluded the DFT exchange–correlation contribution E_n^{XC} from the KS energy E_n in Equation (24). In the quasi-particle (QP) approximation, the electrons and holes have energies E_n^{QP}, which are the real poles of Green's function (24), i.e., they satisfy the equation

$$E_n - E_n^{XC} + \Sigma_n^X + \mathrm{Re}\Sigma_n^{C,0}(E_n^{QP}) = E_n^{QP}. \qquad (29)$$

The electron/hole Green functions can now be approximated as

$$G_n^{QP}(\omega) = \frac{1 - f_n^{QP}}{\omega - E_n^{QP} + i\delta} + \frac{f_n^{QP}}{\omega - E_n^{QP} - i\delta}, \qquad (30)$$

and after they are used in (23), the time-ordered electron–hole propagator becomes

$$\mathcal{L}_{n\to n'}^{m\leftarrow m'}(\omega) = \frac{f_n^{QP} - f_m^{QP}}{\omega + E_n^{QP} - E_m^{QP} + i\delta\,\mathrm{sgn}(E_m^{QP} - E_n^{QP})}\delta_{nn'}\delta_{mm'}, \qquad (31)$$

where $\delta = 0^+$. The 'time-ordered' screened Coulomb interaction is the solution of Dyson's equation

$$W_{\mathbf{GG'}}^0(\omega) = V_{\mathbf{GG'}}^C + \sum_{\mathbf{G_1 G_2}} V_{\mathbf{GG_1}}^C \chi_{\mathbf{G_1 G_2}}^0(\omega) W_{\mathbf{G_2 G'}}^0(\omega), \qquad (32)$$

where the matrix of the 'time-ordered' irreducible polarizability is

$$\chi_{\mathbf{GG'}}^0(\omega) = \frac{2}{\Omega_{sc}} \sum_{nm} \frac{(f_n - f_m)\rho_{nm}(\mathbf{G})\,\rho_{nm}^*(\mathbf{G'})}{\hbar\omega + E_n - E_m + i\delta\,\mathrm{sgn}(E_m - E_n)}, \qquad (33)$$

and the charge vertices are

$$\rho_{nm}(\mathbf{G}) = \int_{\Omega_{sc}} d\mathbf{r}\, \phi_n^*(\mathbf{r}) e^{-i\mathbf{Gr}} \phi_m(\mathbf{r}). \qquad (34)$$

Since we calculate the optical conductivity of a single isolated benzene molecule, we have to exclude the effect on its polarizability due to the interaction with the surrounding molecules in the sc lattice. This is accomplished by using the truncated Coulomb interaction [67]

$$V_C(\mathbf{r} - \mathbf{r'}) = \frac{\Theta(|\mathbf{r} - \mathbf{r'}| - R_C)}{|\mathbf{r} - \mathbf{r'}|}, \qquad (35)$$

where Θ is the Heaviside step function, and R_C is the range of the Coulomb interactions, i.e., the radial cutoff. The Coulomb interaction matrix to be used in (32) is then

$$V^C_{GG'} = \frac{1}{\Omega}\int_\Omega d\mathbf{r}\int_\Omega d\mathbf{r}' e^{-i\mathbf{Gr}} V_C(\mathbf{r},\mathbf{r}') e^{i\mathbf{G}'\mathbf{r}'} = \frac{4\pi}{|\mathbf{G}|^2}[1-\cos|\mathbf{G}|R_C]\delta_{GG'}. \quad (36)$$

The Bethe–Salpeter kernel is

$$\Xi^{m\leftarrow m'}_{n\to n'} = \Xi^{H,m\leftarrow m'}_{n\to n'} - \frac{1}{2}\Xi^{F,m\leftarrow m'}_{n\to n'} \quad (37)$$

where the BS–Hartree kernel is

$$\Xi^{H,m\leftarrow m'}_{n\to n'} = \frac{1}{\Omega_{sc}}\sum_{\mathbf{G}_1\mathbf{G}_2}\rho^*_{nm}(\mathbf{G}_1)V^C_{\mathbf{G}_1\mathbf{G}_2}\rho_{n'm'}(\mathbf{G}_2), \quad (38)$$

and the BS–Fock kernel is

$$\Xi^{F,m\leftarrow m'}_{n\to n'} = \frac{1}{\Omega_{sc}}\sum_{\mathbf{G}_1\mathbf{G}_2}\rho^*_{nn'}(\mathbf{G}_1)W^0_{\mathbf{G}_1\mathbf{G}_2}(\omega=0)\rho_{mm'}(\mathbf{G}_2). \quad (39)$$

Here, the index '0' in W, S, Γ, and Σ^C emphasizes that we consider the screened interaction in a self-standing molecule. Finally, considering that the interaction V_C prevents the correlations between the conductivities in the adjacent cells, the conductivity of an isolated molecule is equal to the conductivity per unit cell

$$\sigma_{\mu\nu}(\omega) = \int_{\Omega_{sc}} d\mathbf{r}\int_{\Omega_{sc}} d\mathbf{r}'\sigma^{3D}_{\mu\nu}(\mathbf{r},\mathbf{r}',\omega). \quad (40)$$

After using (19) in (40), the optical conductivity of a single molecule becomes

$$\sigma_{\mu\nu}(\omega) = \Omega_{sc}\sigma^{\mu\nu,3D}_{\mathbf{G}=0\mathbf{G}'=0}(\omega). \quad (41)$$

After combining Equations (9), (20), and (41), we determine the explicit expression for the surface optical conductivity

$$\tilde{\sigma}_{\mu\nu}(\omega) = \frac{2i}{\omega S_{fcc}}\sum_{nm}\sum_{n'm'} j^\mu_{nm}(\mathbf{G}=0)\mathcal{K}^{m\leftarrow m'}_{n\to n'}(\omega)[j^\nu_{n'm'}(\mathbf{G}'=0)]^*, \quad (42)$$

which can be used in Dyson's equation for the electric field propagator (10). It is important to note that the dimension of the conductivity (42) is exactly the quantum of conductance $G_0 = \frac{2\pi e^2}{\hbar}$, already the standardized unit for describing the optical conductivity in 2D crystals [28,47,65]. Accordingly, the $\tilde{\sigma}_{\mu\nu}(\omega)$ represents the optical conductivity of one (e.g., i-th) molecular layer.

2.3. Optical Conductivity in a Molecule Physisorbed at a Dielectric Surface

We assume that a fullerene molecule, centered at $z = z_i$, is physisorbed at the supporting crystal occupying the $z < 0$ half-space, as illustrated in Figure 3.

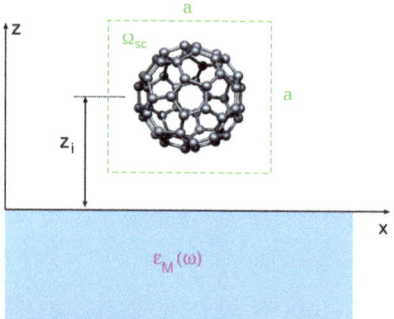

Figure 3. The fullerene molecule C_{60} centered at $z = z_i$ is physisorbed at the supporting crystal occupying the half-space $z < 0$, with the dielectric properties approximated by the macroscopic dielectric function $\epsilon_M(\omega)$.

We further assume that the bonding between the molecule and the supporting crystal has the vdW character, which implies a small electronic overlap between the molecule and the substrate and, therefore, a small impact on the short-range electronic correlations to the molecular QP and optical properties. More precisely, the processes involving the electronic hopping between the molecule and the supporting crystal are neglected. However, we shall retain the processes of scattering an electron or a hole or an excited electron–hole pair by the potential induced at the crystal surface, ΔV.

If we have two valence electrons at points $\mathbf{r}$ and $\mathbf{r}'$ in a self-standing molecule, they interact via the bare Coulomb truncated potential V_C (35), but they also polarize the molecule itself so that the total interaction between them is given by the potential W^0, obtained as the solution of Equation (32). After the polarizable surface is brought close to the molecule, the electrons at $\mathbf{r}$ and $\mathbf{r}'$ polarize the surface as well so that the interaction between them, neglecting the polarization of the molecule, becomes

$$V_S(\mathbf{r}, \mathbf{r}', \omega) = V_C(\mathbf{r}, \mathbf{r}') + \Delta V_S(\mathbf{r}, \mathbf{r}', \omega), \tag{43}$$

where $\Delta V_S(\mathbf{r}, \mathbf{r}', \omega)$ is the induced dynamic Coulomb potential coming from the excitations of the electronic modes or phonons at the surface. The total interaction between the electrons (including the polarization of the molecule) is then the solution of Dyson's equation

$$W^S_{\mathbf{G}\mathbf{G}'}(\omega) = V^S_{\mathbf{G}\mathbf{G}'}(\omega) + \sum_{\mathbf{G}_1 \mathbf{G}_2} V^S_{\mathbf{G}\mathbf{G}_1}(\omega) \chi^0_{\mathbf{G}_1 \mathbf{G}_2}(\omega) W^S_{\mathbf{G}_2 \mathbf{G}'}(\omega), \tag{44}$$

where

$$\Delta V^S_{\mathbf{G}\mathbf{G}'}(\omega) = \frac{1}{\Omega_{sc}} \int_{\Omega_{sc}} d\mathbf{r} \int_{\Omega_{sc}} d\mathbf{r}' e^{-i\mathbf{G}\mathbf{r}} \Delta V_S(\mathbf{r}, \mathbf{r}', \omega) e^{i\mathbf{G}'\mathbf{r}'}. \tag{45}$$

Here, the integration is constrained within the unit cell Ω_{sc} centered at $\mathbf{r}_0 = (0, 0, z_i)$ (to avoid interaction with the neighboring molecules, via ΔV_S), as shown in Figure 3. The induced interaction in the region $z > 0$ can be written as [62]

$$\Delta V(\mathbf{r}, \mathbf{r}', \omega) = \int \frac{d\mathbf{Q}}{(2\pi)^2} v_Q D(\mathbf{Q}, \omega) e^{i\mathbf{Q}(\boldsymbol{\rho} - \boldsymbol{\rho}')} e^{-Q(z+z')}, \tag{46}$$

where $v_Q = \frac{2\pi}{|\mathbf{Q}|}$. Since the supporting crystal dielectric response is approximated by the local 3D macroscopic dielectric function $\epsilon_M(\omega)$, the surface excitation propagator can be approximated as [68]

$$D(\mathbf{Q}, \omega) \approx D(\omega) = \frac{1 - \epsilon_M(\omega)}{1 + \epsilon_M(\omega)}. \tag{47}$$

After (46) and (47) are used in (45), we have

$$\Delta V_{\mathbf{GG'}}^{S}(\omega) = \frac{D(\omega)}{\Omega_{sc}} \int \frac{d\mathbf{Q}}{(2\pi)^2} v_Q e^{-2Qz_i} F_{\mathbf{G}}(\mathbf{Q}) F_{\mathbf{G'}}^*(\mathbf{Q}), \qquad (48)$$

where the form factors are defined as

$$F_{\mathbf{G}}(\mathbf{Q}) = 8(-1)^{n_z} \frac{\sin[(Q_x - G_x)\frac{a}{2}] \sin[(Q_y - G_y)\frac{a}{2}] \sinh[Q\frac{a}{2}]}{(Q_x - G_x)(Q_y - G_y)(Q + iG_z)}.$$

The reciprocal vectors of the sc lattice are $\mathbf{G} = (G_x, G_y, G_z)$, where $G_x = \frac{2\pi n_x}{a}$, $G_y = \frac{2\pi n_y}{a}$, $G_z = \frac{2\pi n_z}{a}$ and $n_x, n_y, n_z \in \mathbb{Z}$. After we determine the 'bare' potential (which includes the polarization of the surface) V_S and the 'total' potential (which includes the polarization of the surface and of the molecule) W_S, the calculation of the QP and optical properties of a molecule near the dielectric surface is equal to the procedure described in Section 2.2, except for that in the BS–Hartree kernel (38), we need to replace

$$V_{\mathbf{G}_1\mathbf{G}_2}^{C} \to V_{\mathbf{G}_1\mathbf{G}_2}^{S}(\omega),$$

in the BS–Fock kernel (39)

$$W_{\mathbf{G}_1\mathbf{G}_2}^{0}(\omega = 0) \to W_{\mathbf{G}_1\mathbf{G}_2}^{S}(\omega = 0),$$

and the spectrum of the electronic excitations (28) used to calculate the correlation self-energy (26) is

$$S_{\mathbf{GG'}}^{0}(\omega) \to S_{\mathbf{GG'}}^{S}(\omega) = -\frac{1}{\pi} Im W_{\mathbf{GG'}}^{S}(\omega). \qquad (49)$$

2.4. Computational Details

The fullerene KS orbitals $\phi_n(\mathbf{r})$ and energies E_n were calculated using the plane-wave self-consistent field DFT code (PWSCF) within the QUANTUM ESPRESSO (QE) package [69]. The core-electron interaction was approximated by the norm-conserving pseudopotentials [70,71] so that the number of occupied valance states was 120. The exchange-correlation (XC) potential was approximated by the Perdew–Burke–Ernzerhof (PBE) generalized gradient approximation (GGA) functional [72]. The plane-wave cut-off energy was 60 Ry. The molecules were arranged in the simple cubic Bravais lattice of the unit cell $a = 18$ Å with one molecule per unit cell. Since there was no intermolecular overlap, the ground state electronic density was calculated at the Γ point only. The geometries were fully relaxed, with all forces $\lesssim 0.02$ eV/Å. The RPA 'time-ordered' screened Coulomb interactions $W^{0,S}$ (Equations (32) and (44)) were calculated using the energy cut-off 2 Ry (~27 eV), and the band summations '(n, m)' in the irreducible polarizability (33) were performed over 240 molecular states. The exchange self-energy (25) was calculated using the energy cut-off 8 Ry (~109 eV) and the correlation self-energy (26) was determined (according to the cut-off in W_0) using the energy cut-off 2 Ry (~27 eV); the band summation 'm' was performed using 240 molecular orbitals. The BS–Hartree kernel (38) and the 'bare' BS–Fock kernel (the Equation (39), derived using the bare interaction V_C), were calculated using the energy cut-off 8 Ry (~109 eV); the induced Fock kernel (the Equation (39), derived using the induced interaction $W_{0,S} - V_C$), was calculated using the energy cut-off 2 Ry (~27 eV). During the evaluation of the BSE–HF kernels, we used 42 occupied (HOMO − 41, ..., HOMO) and 42 unoccupied (LUMO, ..., LUMO + 41) molecular states, i.e., the dimensions of the BSE–HF kernel matrix was $2 \times 42 \times 42 = 3528$. To achieve the accurate (experimental) exciton energy, the calculation was performed beyond the Tamm–Dancoff approximation. The damping parameters δ used in (31) and in (33) were 50 meV and 200 meV, respectively. We assume that the dielectric medium was vacuum (i.e., $\epsilon_0 = 1$), and that the substrate was the aluminum-oxide Al_2O_3, described by the macroscopic dielectric function

$$\epsilon_M(\omega) = 1/\epsilon^{-1}_{G=0G'=0}(\mathbf{q} \approx 0, \omega), \quad (50)$$

where the dielectric matrix is $\hat{\epsilon} = \hat{I} - \hat{V}\hat{\chi}^0$. The irreducible polarizability χ^0 is determined using Equation (33) for $\Omega_{sc} \to \Omega$, $n \to (n, \mathbf{k})$ and $m \to (m, \mathbf{k} + \mathbf{q})$. Here $\mathbf{k}$, $\mathbf{q}$, and $\mathbf{G}$ are the 3D wave vector, the transfer wave vector, and the reciprocal lattice vector, respectively, corresponding to the bulk Al_2O_3 crystal. The bare Coulomb interaction is $V_{GG'}(\mathbf{q}) = \frac{4\pi}{|\mathbf{q}+\mathbf{G}|^2} \delta_{GG'}$. The ground state electronic density of the bulk Al_2O_3 is calculated using $9 \times 9 \times 3$ K-mesh, the plane-wave cut-off energy is 50 Ry, and the Bravais lattices are hexagonal (12 Al and 18 O atoms in the unit cell) with the lattice constants $a_{Al_2O_3} = 4.76$ Å and $c_{Al_2O_3} = 12.99$ Å. The Al_2O_3 irreducible polarizability χ^0 is calculated using the $21 \times 21 \times 7$ k-point mesh and the band summations (n, m) are performed over 120 bands. The damping parameter is $\delta = 100$ meV and the temperature is $T = 10$ meV. For the optically small wave vectors $\mathbf{q} \approx 0$, the crystal local field effects are negligible, i.e., the crystal local field effects cut-off energy is set to zero. Using this approach, the Reϵ_M is almost constant (Re$\epsilon_M \approx 3$) for low frequencies ($\omega < 3$ eV), i.e., in the IR and visible range, while Imϵ_M is equal to zero up to the band gap energy ($E_g \sim 6$ eV). Therefore, Al_2O_3 is a good choice for the substrate in the visible and near-UV frequency range, since its electronic excitations are above that range, and its IR active SO phonons (at $\omega_{SO} < 100$ meV) [73] are far below the C_{60} excitons, which means that in the frequency range of interest, there is no dissipation of the electromagnetic energy in the substrate (it is transparent). In addition to that, the dielectric function is mostly constant, but in this calculation, we used the fully dynamical and complex $\epsilon_M(\omega)$. The integration in (48) was performed over the two-dimensional wave vectors $\mathbf{Q} = (Q_x, Q_x)$ using a 121×121 rectangular mesh and the cut-off wave vector $Q_C = 0.5$ a.u. For the radial cut-off in the truncated interaction (36), we used $R_C = a/2 = 9$ Å.

3. Results and Discussion

Strong exciton–photon hybridization usually occurs due to the large overlap between excitons of large oscillatory strengths and confined photons. This is experimentally achieved by placing nanostructures or molecules of large oscillatory strength in a metallic cavity. Our goal here was to explore the electromagnetic eigenmodes, which are a mixture of 'free' s(TE) photons and excitons, i.e., a 'free' traveling photon (not confined between metallic walls) captured by an exciton and 'trapped' in a molecular nanostructure. This concept makes it possible to clearly estimate the contribution of the 'free' photon in the hybrid exciton–photon mode.

First, we determined the QP and optical properties of a self-standing layer of the C_{60} molecules, which corresponded with the properties of the gas-phase molecule, since at the G_0W_0-BSE stage of the calculations, the molecules were not coupled. The calculated G_0W_0 HOMO-LUMO gap was $E_g = 4.66$ eV, which is in good agreement with the experimental value 5 eV [74–77]. When a molecular layer is deposited on an Al_2O_3 dielectric surface, the band gap reduces to $E_g = 4.26$ eV. Here, the separation was chosen to be $z_0 = 6.5$ Å which corresponds to the characteristic vdW atom–atom separation of 3 Å (Note that z_0 is defined as the distance between the substrate and the center of the molecule, as denoted in Figure 1, so the C_{60} molecule radius (3.57 Å) has to be added to the vdW atom–atom distance). For comparison, we determined the HOMO-LUMO gap for the molecule at a silver (Ag) surface also described in terms of the *ab initio* macroscopic dielectric function. For the same separation ($z_0 = 6.5$ Å), the gap was $E_g = 3.81$ eV; compared with the insulator surface, the reduction is twice as strong. The image theory estimation of the HOMO-LUMO gap at the metallic surface is $E_g \sim 3.56$ eV which, as expected, overestimates the G_0W_0 result. Figure 4a shows the calculated G_0W_0-BSE optical conductivities $\tilde{\sigma}_x(\omega)$ in the self-standing molecular layer (black solid) and in the molecular layer deposited at the Al_2O_3

dielectric surface (cyan dashed). For comparison, the experimental optical absorption in the fullerene C_{60} [78] is presented by the red circles. This experimental study, as well as some others [79,80], show three broad excitation bands at $\omega_{ex1} \sim 3.9$ eV, $\omega_{ex2} \sim 4.9$ eV and $\omega_{ex3} \sim 6$ eV, which is in very good agreement with our results. It should be emphasized that, since the band gap is $E_g = 4.66$ eV, only the excitation band ω_{ex1} can be considered as an exciton by definition, and its binding energy is $E_g - \hbar\omega_{ex1} \sim 0.76$ eV. In the literature, the strong maximum ω_{ex3} is usually referred to as the π plasmon [81,82]. All three excitons are the result of the electronic transitions within the C_{60} π-complex [83]. When the molecule is deposited at the Al_2O_3 surface, the excitation band barely changes at all, due to the well-known cancellation effect: the substrate weakens the interaction between the excited electron and the hole, which reduces the exciton binding energy and, therefore, cancels the gap reduction. This phenomenon was studied in detail in references [62,66,84]. Since the influence of the dielectric surface on the molecular optical conductivity is weak, we shall further approximate $\tilde{\sigma}_i^x(\omega) \approx \tilde{\sigma}^x(\omega)$, where $\tilde{\sigma}^x(\omega)$ is the optical conductivity in the self-standing molecular single layer, i.e., for $z_0 \to \infty$.

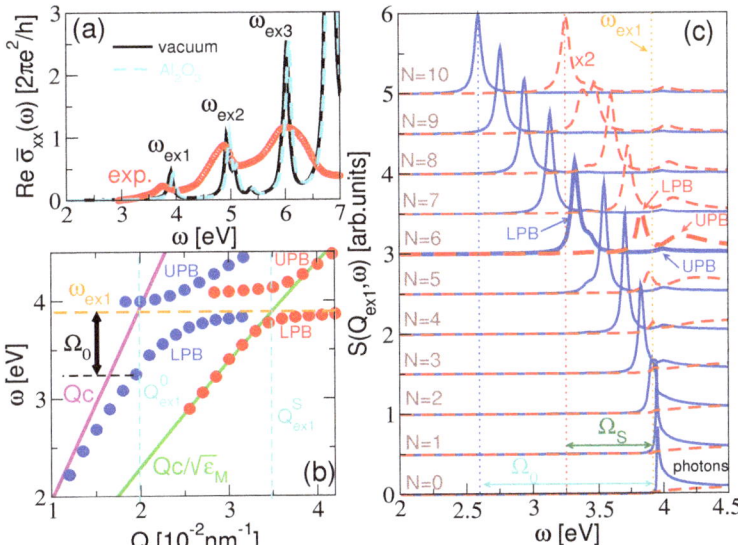

Figure 4. (a) The optical conductivities $\tilde{\sigma}_{xx}(\omega)$ in the C_{60} single layer in vacuum (black solid), in the C_{60} single layer at the Al_2O_3 dielectric surface (cyan dashed) (where $z_0 = 6.5$ Å), and the experimental optical absorption in the gas phase fullerene C_{60} (red circles). (b) The lower and upper exciton–polariton branches, LPB and UPB, respectively, in the self-standing C_{60} film (blue dots) and the C_{60} film deposited at the Al_2O_3 surface (red dots). The LPB corresponds to the dispersion relation of the exciton-polariton $\omega_{ex1\text{-pol}}(Q)$ appearing in Equation (53). The number of C_{60} layers is $N = 6$. (c) The spectra of the s(TE)-polarized electromagnetic modes $S(Q_{ex1}, \omega)$ in the self-standing C_{60} films for $Q_{ex1}^0 = 0.02$ nm^{-1} (blue solid) and in the C_{60} films at the Al_2O_3 surface for $Q_{ex1} = 0.035$ nm^{-1} (red dashed). The number of the single layers is $N = 0, 1, 2, 3, \ldots, 10$, where the case $N = 0$ corresponds to the spectrum of the free photons in vacuum or at the vacuum/Al_2O_3 interface (the photon continuum). All spectra for the supported films are multiplied by factor 2.

This means that the impact of the dielectric substrate on the electromagnetic modes in the C_{60} films is reduced to the propagator Γ^{sc}, which appears in Dyson's Equation (14). The spectra of the s(TE)-polarized electromagnetic modes in the C_{60} films will be analyzed using the real part of the propagator $\mathcal{E}$ in the topmost molecular layer

$$S(Q, \omega) = Re\mathcal{E}_{xx}(|\mathbf{Q}|\mathbf{y}, \omega, z_N, z_N). \tag{51}$$

As already mentioned, the spacing parameters were chosen to be $z_0 = 6.5$ Å and $\Delta = 8.1$ Å. In the near field spectroscopy experiment, the incident photon of the wavelength

$$\lambda_{\text{ex1}} = \frac{2\pi c}{\omega_{\text{ex1}} \sqrt{\epsilon(\omega_{\text{ex1}})}} \tag{52}$$

couples to polarization modes in some sub-wavelength nanostructure (e.g., AFM tip) and is scattered (diffracted) to all $\lambda > \lambda_{\text{ex1}}$ and $\lambda < \lambda_{\text{ex1}}$ so that it could excite all radiative electromagnetic modes, as well as all evanescent ones at a fixed frequency ω_{ex1}. The Rabi splitting (the measure of the free-photon participation in the exciton–polariton mode) is defined as the difference between the exciton ω_{ex1} and the exciton–polariton $\omega_{\text{ex1-pol}}$

$$\hbar\Omega = \hbar\omega_{\text{ex1}} - \hbar\omega_{\text{ex1-pol}}(Q_{ex1}), \tag{53}$$

where $Q_{ex1} = 2\pi/\lambda_{\text{ex1}}$ is the wave vector at which the exciton ω_{ex1} and the photon cross.

Figure 4b shows the dispersion relations of the exciton–polaritons $\omega_{\text{ex1-pol}}(Q)$ in the self-standing C_{60} film (blue dots), and in the C_{60} film deposited at a Al_2O_3 surface (red dots). The number of C_{60} single layers in the film was chosen to be $N = 6$. The energies of the exciton–polaritons $\omega_{\text{ex1-pol}}(Q)$ correspond to the maxima in the spectral function $S(Q,\omega)$, appearing below the exciton energy ω_{ex1} [see Figure 4c]. For the self-standing film, $\epsilon(\omega) = 1$, so that the crossing wave vector, according to the Equation (52), was $Q^0_{\text{ex1}} = 0.02$ nm^{-1}, and the Rabi splitting, according to the Equation (53), was $\hbar\Omega_0 = 650$ meV. The red dots in Figure 4b show that the presence of the substrate reduces the bending of the exciton–polariton dispersion relation significantly, therefore reducing the corresponding Rabi splitting. In this case, $\epsilon(\omega) = \epsilon_M(\omega)$, which gives the crossing wave vector $Q^S_{\text{ex1}} = 0.035$ nm^{-1} and, thus, the Rabi splitting $\hbar\Omega_S = 103$ meV.

Figure 4c shows the spectra of the s(TE)-polarized electromagnetic modes $S(Q_{ex1},\omega)$ in the self-standing C_{60} films for $Q^0_{\text{ex1}} = 0.02$ nm^{-1} (blue solid) and in the C_{60} films deposited at the Al_2O_3 surface for $Q^S_{\text{ex1}} = 0.035$ nm^{-1} (red dashed). The number of layers is set to be $N = 0, 1, 2, 3, \ldots, 10$, where the case $N = 0$ corresponds to the spectra of the free photons in vacuum or at the vacuum/Al_2O_3 interface (photons continuum). Due to the lower intensity, all spectra for the supported films were multiplied by factor 2. We can clearly see the exciton–polariton peaks separating from the photon continuum as the number of layers increases. In the self-standing films, the exciton–polariton is already well separated for $N = 2$, while in the supported films, it occurs for $N = 4$. For $N = 10$, the giant light–matter coupling was achieved providing large Rabi splitting $\Omega_0 = 1334$ meV and $\Omega_S = 670$ meV in the self-standing and the supported films, respectively. Here we limited our study to the exciton–polaritons in the quasi-2D nanostructures so that the maximum number of the molecular layers was limited to $N = 10$, still satisfying the sub-wavelength limit $\lambda_{\text{ex1}} > (N-1)\Delta$. For a further increase of the number of layers N, the Rabi splitting continued to increase, but the experimental limitations in the realization of such a system raise the question of the plausibility of such a result.

We need to emphasize that the anti-crossing behavior and the Rabi splitting into the lower polariton band (LPB) and the upper polariton band (UPB) is a well-defined concept only when describing the interaction between the well-defined eigenmodes, such as excitons and cavity photons [33] or BSWs [48]. In our case, the only well-defined modes (boson) were the excitons ω_{ex1}-ω_{ex3}. On the other hand, the photons appeared as a continuum of eigenmodes above the light lines Qc or $Qc/\sqrt{\epsilon_M}$. This means that it only made sense to observe the deformation of the exciton dispersion ω_{ex1}, characterized by the exciton bending parameter Ω. However, in Figure 5, it seems that the edge of the photon continuum Qc or $Qc/\sqrt{\epsilon_M}$ behaves similar to a well-defined eigenmode, so in Figure 4b,c we also introduced and denoted the LPB and UPB, exactly as they appear in Figure 5b,e. The LPB corresponds to the exciton–polariton dispersion relation $\omega_{\text{ex1-pol}}(Q)$ appearing in (53).

To present a more extensive picture of the electromagnetic modes in the C_{60} films, including the hybridizations of the photon and the higher excitons ω_{ex2} and ω_{ex3}, Figure 5a–c show the spectral intensity of the s(TE)-polarized electromagnetic modes $S(Q, \omega)$ in the self-standing C_{60} films for $N = 3, 6$, and 9, respectively. Figure 5d–f show the same for the C_{60} films deposited at the Al_2O_3 surface. It is immediately obvious that the most intensive electromagnetic modes occur in the evanescent regions $\omega < Qc$ and $\omega < Qc/\epsilon_M(\omega)$, for the self-standing and supported films, respectively. These modes produce an electric field, which spreads within the film and exponentially decays outside of the film [as illustrated in Figure 2b], which is one of the inherent properties of the 2D exciton–polaritons. Moreover, as the field is more confined, it is stronger, so that the polariton modes confined at a sub-wavelength scale are very interesting for many applications. The free photon continua can be seen as the low-intensity patterns in the regions $\omega > Qc$, and $\omega > Qc/\epsilon_M(\omega)$ in the self-standing and supported films, respectively, with their intensity weakening with the number of layers N. If the electromagnetic eigenmode, for example in the C_{60} film at the Al_2O_3, would occur in the region $\omega > Qc/\epsilon_M(\omega)$, it would be irradiated (decay radiatively) into the Al_2O_3 crystal, so that these modes appear only as weak resonances. Such weak resonances can be seen as weak horizontal patterns (parallel with the excitons ω_{ex1}-ω_{ex3}) entering the radiative region.

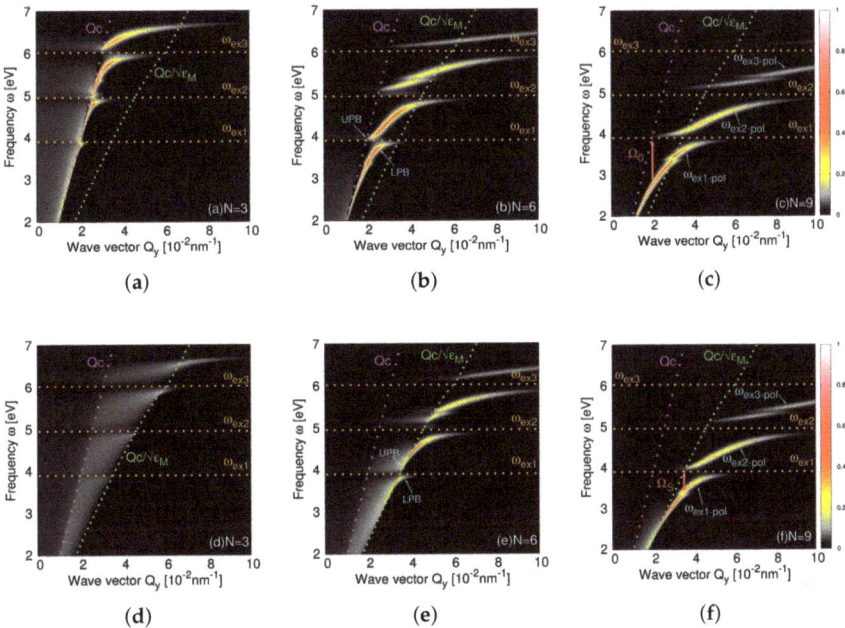

Figure 5. The spectral intensity of the s(TE)-polarized electromagnetic modes $S(Q, \omega)$ in the self-standing C_{60} films for (**a**) $N = 3$, (**b**) $N = 6$, and (**c**) $N = 9$. Figures (**d**–**f**) show the same for the C_{60} films deposited at the Al_2O_3 surface. In both cases, the strong electromagnetic modes ($\omega_{ex1\text{-pol}}$, $\omega_{ex2\text{-pol}}$, and $\omega_{ex3\text{-pol}}$) occur in the evanescent regions $\omega < Qc$ and $\omega < Qc/\epsilon_M(\omega)$.

In Figure 5a, we can see the beginning of the hybridization between the free photon and the three C_{60} excitons (ω_{ex1}-ω_{ex3}), which visibly intersect the photon line $\omega = Qc$. In Figure 5b,c, the photon line is already significantly deformed between the excitons and pushed far into the evanescent region, which means that the photon Qc and the three excitons ω_{ex1}, ω_{ex2}, and ω_{ex3} are strongly coupled and converted into three exciton–polaritons $\omega_{ex1\text{-pol}}$, $\omega_{ex2\text{-pol}}$, and $\omega_{ex3\text{-pol}}$. In Figure 5d–f, the exciton–polaritons are less intense and pushed into the evanescent region $\omega < Qc/\epsilon_M(\omega)$, the use of the substrate obviously weakens the intensity of the exciton–polaritons. However, in Figure 5e,f, we can see the formation of the three strong exciton–polariton modes $\omega_{\text{ex1-pol}}$–$\omega_{\text{ex3-pol}}$, for larger numbers of layers $N = 6$ and 9, respectively. Finally, this confirms that the strong binding between the transverse s(TE) photon and the molecular excitons is a realistic physical phenomenon that can be achieved experimentally. The use of a metal substrate is not recommended because a metallic surface will strongly quench the exciton–polariton modes. To support this argument, we performed the same calculation using a silver (Ag) substrate and obtained much weaker exciton–polariton modes.

Finally, we noticed that the hybridization between the exciton and the photon increases weakly if just the exciton oscillatory strength (e.g., S) increases, but strongly if the number of spatially separated molecular layers N increases. For example, the exciton–photon binding is much stronger in the N-separated layers of the oscillatory strength S than in one layer of the oscillatory strength N× S. This suggests that the photon-exciton coupling can be increased simply by increasing the number of layers N in a multilayered van der Waals heterostructure.

4. Conclusions

We showed that the 2D-layered heterostructures, consisting of a larger number of exciton-active single layers or 2D crystals (e.g., $N > 5$), can support evanescent s(TE)-polarized exciton–polaritons with strong photonic character. We obtained giant Rabi splitting of more than $\Omega_0 = 1000$ meV and $\Omega_S = 500$ meV in the self-standing and supported C_{60} films for $N = 10$, respectively. This investigation has fundamental and practical contributions. We predict the existence of the evanescent s(TE) polarization modes with significant photonic character, which vanish exactly without the photon admixture (for $c \to \infty$). Unlike the well-known p(TM) polarization modes (e.g., the plasmon polariton for $c \to \infty$ collapses into a plasmon), this is a new fundamental contribution. We also demonstrated that exciton–photon coupling can be manipulated simply by changing the number of single layers (N) in a vdW-layered heterostructure. Moreover, due to the fact that the vdW heterostructure with a thickness of just a few molecular (or atomic) layers supports the confined photons, it can be easily implemented in the photonic integrated circuits or photonics chips. The disadvantage of these types of photonic modes is that they do not couple directly to the free photons (external light). However, once excited, these modes can be easily manipulated (since they are trapped in the nanostructure). For example, by patterning the vdW nanoribbons on a dielectric wafer (patterning the photonics circuits), the direction of the photon propagation can easily be modified. Moreover, the exciton–polaritons can be easily switched (at the contact) into evanescent modes in another nanostructure. Finally, these layered vdW heterostructures can be applied in photonic devices, such as light sources (LED), telecommunications (as waveguides or optical cables), or chemical and biological sensing.

Author Contributions: Conceptualization, V.D.; methodology, V.D.; software, V.D. and L.M.; validation, V.D. and L.M.; formal analysis, V.D.; investigation, V.D. and L.M.; resources, V.D. and L.M.; data curation, V.D.; writing—original draft preparation, V.D.; writing—review and editing, V.D. and L.M.; visualization, V.D.; supervision, V.D.; project administration, V.D.; funding acquisition, V.D. All authors have read and agreed to the published version of the manuscript.

Funding: This research was funded by the Croatian Science Foundation (grant no. IP-2020-02-5556) and the European Regional Development Fund for the "QuantiXLie Centre of Excellence" (grant KK.01.1.1.01.0004).

Institutional Review Board Statement: Not applicable.

Informed Consent Statement: Not applicable.

Data Availability Statement: Not applicable.

Acknowledgments: The authors acknowledge support from the Donostia International Physic Center (DIPC) computing center, which provided the computational resources.

Conflicts of Interest: The authors declare no conflict of interest.

References

1. Basov, D.N.; Asenjo-Garcia, A.; Schuck, P.J.; Xiaoyang, Z.; Rubio, A. Polariton panorama. *Nanophotonics* **2021**, *10*, 549–577. [CrossRef]
2. Deng, H.; Haug, H.; Yamamoto, Y. Exciton-polariton Bose-Einstein condensation. *Rev. Mod. Phys.* **2010**, *82*, 1489. [CrossRef]
3. Weisbuch, C.; Nishioka, M.; Ishikawa, A.; Arakawa, Y. Observation of the coupled exciton-photon mode splitting in a semiconductor quantum microcavity. *Phys. Rev. Lett.* **1992**, *69*, 3314. [CrossRef]
4. Ribeiro, R.F.; Martínez-Martínez, L.A.; Du, M.; Campos-Gonzalez-Angulo, J.; Yuen-Zhou, J. Polariton chemistry: Controlling molecular dynamics with optical cavities. *Chem. Sci.* **2021**, *9*, 6325–6329. [CrossRef]
5. Kasprzak, J.; Richard, M.; Kundermann, S.; Baas, A.; Jeambrun, P.; Keeling, J.M.J.; Marchetti, F.M.; Szymańska, M.H.; André, R.; Staehli, J.L.; et al. Bose–Einstein condensation of exciton polaritons. *Nature* **2006**, *443*, 409–414. [CrossRef] [PubMed]
6. Balili, R.; Hartwell, V.; Snoke, D.; Pfeiffer, L.; West, K. Bose-Einstein Condensation of Microcavity Polaritons in a Trap. *Science* **2007**, *316*, 1007–1010. [CrossRef]
7. Christopoulos, S.; Von Högersthal, G.B.H.; Grundy, A.J.D.; Lagoudakis, P.G.; Ka-vokin, A.V.; Baumberg, J.J.; Christmann, G.; Butté, R.; Feltin, E.; Carlin, J.F.; et al. Room-Temperature Polariton Lasing in Semiconductor Microcavities. *Phys. Rev. Lett.* **2007**, *98*, 126405. [CrossRef]
8. Baumberg, J.J.; Kavokin, A.V.; Christopoulos, S.; Grundy, A.J.D.; Butté, R.; Christ-mann, G.; Solnyshkov, D.D.; Malpuech, G.; von Högersthal, G.B.H.; Feltin, E.; et al. Spontaneous Polarization Buildup in a Room-Temperature Polariton Laser. *Phys. Rev. Lett.* **2008**, *101*, 136409. [CrossRef]
9. Zhang, S.; Chen, J.; Shi, J.; Fu, L.; Du, W.; Sui, X.; Mi, Y.; Jia, Z.; Liu, F.; Shi, J.; et al. Trapped Exciton–Polariton Condensate by Spatial Confinement in a Perovskite Microcavity. *ACS Photonics* **2020**, *7*, 327–337. [CrossRef]
10. Bao, W.; Liu, X.; Xue, F.; Zheng, F.; Tao, R.; Wang, S.; Xia, Y.; Zhao, M.; Kim, J.; Yang, S.; et al. Observation of Rydberg exciton polaritons and their condensate in a perovskite cavity. *Proc. Natl. Acad. Sci. USA* **2019**, *116*, 20274–20279. [CrossRef]
11. Horikiri, T.; Byrnes, T.; Kusudo, K.; Ishida, N.; Matsuo, Y.; Shikano, Y.; Löffler, A.; Höfling, S.; Forchel, A.; Yamamoto, Y. Highly excited exciton-polariton condensates. *Phys. Rev. B* **2017**, *95*, 245122. [CrossRef]
12. Ramasubramaniam, A. Large excitonic effects in monolayers of molybdenum and tungsten dichalcogenides. *Phys. Rev. B* **2012**, *86*, 115409. [CrossRef]
13. Li, Y.; Chernikov, A.; Zhang, X.; Rigosi, A.; Hill, H.M.; van der Zande, A.M.; Chenet, D.A.; Shih, E.M.; Hone, J.; Heinz, T.F. Measurement of the optical dielectric function of monolayer transition-metal dichalcogenides: MoS_2, $MoSe_2$, WS_2, and WSe_2. *Phys. Rev. B* **2014**, *90*, 205422. [CrossRef]
14. Qiu, D.Y.; Jornada, F.H.; Louie, S.G. Optical Spectrum of MoS_2: Many-Body Effects and Diversity of Exciton States. *Phys. Rev. Lett.* **2013**, *111*, 216805. [CrossRef] [PubMed]
15. Lin, Y.; Ling, X.; Yu, L.; Huang, S.; Hsu, A.L.; Lee, Y.H.; Kong, J.; Dresselhaus, M.S.; Palacios, T. Dielectric Screening of Excitons and Trions in Single-Layer MoS_2. *Nano Lett.* **2014**, *14*, 5569–5576. [CrossRef] [PubMed]
16. Yan, J.; Jacobsen, K.W.; Thygesen, K.S. Optical properties of bulk semiconductors and graphene/boron nitride: The Bethe-Salpeter equation with derivative discontinuity-corrected density functional energies. *Phys. Rev. B* **2012**, *86*, 045208. [CrossRef]
17. Ferreira, F.; Ribeiro, R.M. Improvements in the GW and Bethe-Salpeter-equation calculations on phosphorene. *Phys. Rev. B* **2017**, *96*, 115431. [CrossRef]
18. Villegas, C.E.P.; Rodin, A.S.; Carvalho, A.C.; Rocha, A.R. Two-dimensional exciton properties in monolayer semiconducting phosphorus allotropes. *Phys. Chem. Chem. Phys.* **2016**, *18*, 27829–27836. [CrossRef]
19. Wang, X.; Jones, A.M.; Seyler, K.L.; Tran, V.; Jia, Y.; Zhao, H.; Wang, H.; Yang, L.; Xu, X.; Xia, F. Highly anisotropic and robust excitons in monolayer black phosphorus. *Nat. Nanotechnol.* **2015**, *10*, 517–521. [CrossRef]
20. Low, T.; Chaves, A.; Caldwell, J.D.; Kumar, A.; Fang, N.X.; Avouris, P.; Heinz, T.F.; Guinea, F.; Martin-Moreno, L.; Koppens, F. Polaritons in layered two-dimensional materials. *Nat. Mater.* **2017**, *16*, 182–194. [CrossRef]
21. Liu, X.; Galfsky, T.; Sun, Z.; Xia, F.; Lin, E.C.; Lee, Y.H.; Kéna-Cohen, S.; Menon, V.M. Strong light–matter coupling in two-dimensional atomic crystals. *Nat. Photonics* **2014**, *9*, 30–34. [CrossRef]
22. Dufferwiel, S.; Schwarz, S.; Withers, F.; Trichet, A.A.P.; Li, F.; Sich, M.; Del Pozo-Zamudio, O.; Clark, C.; Nalitov, A.; Solnyshkov, D.D.; et al. Exciton–polaritons in van der Waals heterostructures embedded in tunable microcavities. *Nat. Commun.* **2015**, *6*, 8579. [CrossRef] [PubMed]
23. Flatten, L.C.; He, Z.; Coles, D.M.; Trichet, A.A.P.; Powell, A.W.; Taylor, R.A.; Warner, J.H.; Smith, J.M. Room-temperature exciton-polaritons with two-dimensional WS_2. *Sci. Rep.* **2016**, *6*, 33134. [CrossRef] [PubMed]
24. Dufferwiel, S.; Lyons, T.P.; Solnyshkov, D.D.; Trichet, A.A.P.; Catanzaro, A.; Withers, F.; Malpuech, G.; Smith, J.M.; Novoselov, K.S.; Skolnick, M.S.; et al. Valley coherent exciton-polaritons in a monolayer semiconductor. *Nat. Commun.* **2018**, *9*, 4797. [CrossRef]
25. Gu, J.; Walther, V.; Waldecker, L.; Rhodes, D.; Raja, A.; Hone, J.C.; Heinz, T.F.; Ké-na-Cohen, S.; Pohl, T.; Menon, V.M. Enhanced nonlinear interaction of polaritons via excitonic Rydberg states in monolayer WSe_2. *Nat. Commun.* **2021**, *12*, 2269. [CrossRef]

26. Förg, M.; Colombier, L.; Patel, R.K.; Lindlau, J.; Mohite, A.D.; Yamaguchi, H.; Glazov, M.M.; Hunger, D.; Högele, A. Cavity-control of interlayer excitons in van der Waals heterostructures. *Nat. Commun.* **2019**, *10*, 3697. [CrossRef]
27. Fei, Z.; Scott, M.E.; Gosztola, D.J.; Foley IV, J.; Yan, J.; Mandrus, D.G.; Wen, H.; Zhou, P.; Zhang, D.W.; Sun, Y.; et al. Nano-optical imaging of WSe$_2$ waveguide modes revealing light-exciton interactions. *Phys. Rev. B* **2016**, *94*, 081402(R). [CrossRef]
28. Novko, D.; Despoja, V. Cavity exciton polaritons in two-dimensional semiconductors from first principles. *Phys. Rev. Res.* **2021**, *3*, L032056. [CrossRef]
29. Lidzey, D.G.; Bradley, D.D.C.; Skolnick, M.S.; Virgili, T.; Walker, S.; Whittaker, D.M. Strong exciton-photon coupling in an organic semiconductor microcavity. *Nature* **1998**, *395*, 53–55. [CrossRef]
30. Hobson, P.A.; Barnes, W.L.; Lidzey, D.G.; Gehring, G.A.; Whittaker, D.M.; Skolnick, M.S.; Walker, S. Strong exciton–photon coupling in a low-Q all-metal mirror microcavity. *Appl. Phys. Lett.* **2002**, *81*, 3519–3521. [CrossRef]
31. Takada, N.; Kamata, T.; Bradley, D.D.C. Polariton emission from polysilane-based organic microcavities. *Appl. Phys. Lett.* **2003**, *82*, 1812–1814.
32. Kena-Cohen, S.; Forrest, S.R. Giant Davydov splitting of the lower polariton branch in a polycrystalline tetracene microcavity. *Phys. Rev. B* **2008**, *77*, 073205. [CrossRef]
33. Schwartz, T.; Hutchison, J.A.; Genet, C.; Ebbesen, T.W. Reversible Switching of Ultrastrong Light-Molecule Coupling. *Phys. Rev. Lett.* **2011**, *106*, 196405. [CrossRef] [PubMed]
34. Gambino, S.; Mazzeo, M.; Genco, A.; Di Stefano, O.; Savasta, S.; Patane, S.; Ballarini, D.; Mangione, F.; Lerario, G.; Sanvitto, D.; et al. Exploring Light-Matter Interaction Phenomena under Ultrastrong Coupling Regime. *ACS Photonics* **2014**, *1*, 1042–1048. [CrossRef]
35. Kena-Cohen, S.; Maier, S.A.; Bradley, D.D.C. Ultrastrongly Coupled Exciton–Polaritons in Metal-Clad Organic Semiconductor Microcavities. *Adv. Opt. Mater.* **2013**, *1*, 827–833. [CrossRef]
36. Martínez-Martínez, L.A.; Ribeiro, R.F.; Campos-Gonzá lez-Angulo, J.; Yuen-Zhou, J. Can Ultrastrong Coupling Change Ground-State Chemical Reactions? *ACS Photonics* **2018**, *5*, 167–176. [CrossRef]
37. Dudek, D.; Kowerdziej, R.; Pianelli, A.; Parka, J. Graphene-based tunable hyperbolic microcavity. *Sci. Rep.* **2021**, *11*, 74. [CrossRef] [PubMed]
38. Tsai, W.L.; Chen, C.Y.; Wen, Y.T.; Yang, L.; Cheng, Y.L.; Lin, H.W. Band Tunable Microcavity Perovskite Artificial Human Photoreceptors. *Adv. Mater.* **2019**, *31*, 1900231. [CrossRef]
39. Xia, K.; Sardi, F.; Sauerzapf, C.; Kornher, T.; Becker, H.W.; Kis, Z.; Kovacs, L.; Der-tli, D.; Foglszinger, J.; Kolesov, R.; et al. Tunable microcavities coupled to rare-earth quantum emitters. *Optica* **2022**, *9*, 445–450. [CrossRef]
40. Vinogradov, E.A. Vibrational polaritons in semiconductor films on metal surfaces. *Phys. Rep.* **1992**, *217*, 159–223. [CrossRef]
41. Pitarke, J.M.; Silkin, V.M.; Chulkov, E.V.; Echenique, P.M. Theory of surface plasmons and surface-plasmon polaritons. *Rep. Prog. Phys.* **2007**, *70*, 1. [CrossRef]
42. Zhang, H.; Abhiraman, B.; Zhang, Q.; Miao, J.; Jo, K.; Roccasecca, S.; Knight, M.W.; Davoyan, A.R.; Jariwala, D. Hybrid Exciton-Plasmon-Polaritons in van der Waals Semiconductor. *Nat. Commun.* **2020**, *11*, 3552. [CrossRef] [PubMed]
43. Kumar, P.; Lynch, J.; Song, B.; Ling, H.; Barrera, F.; Kisslinger, K.; Zhang, H.B.; Anantharaman, S.; Digani, J.; Zhu, H.H.; et al. Light-matter coupling in large-area van der Waals superlattices. *Nat. Nanotechnol.* **2022**, *17*, 182–189. [CrossRef] [PubMed]
44. Anantharaman, S.B.; Stevens, C.E.; Lynch, J.; Song, B.; Hou, J.; Zhang, H.; Jo, K.; Kumar, P.; Blancon, J.C.; Mohite, A.D.; et al. Self-Hybridized Polaritonic Emission from Layered Perovskites. *Nano Lett.* **2021**, *21*, 6245–6252. [CrossRef]
45. Anantharaman, S.B.; Jo, K.; Jariwala, D. Exciton–Photonics: From Fundamental Science to Applications. *ACS Nano* **2021**, *15*, 12628–12654. [CrossRef]
46. Jariwala, D.R.; Davoyan, A.; Tagliabue, G.C.; Sherrott, M.; Wong, J.A. Atwater, H. Near-Unity Absorption in van der Waals Semiconductors for Ultrathin Optoelectronics. *Nano Lett.* **2016**, *16*, 5482–5487. [CrossRef]
47. Novko, D.; Lyon, K.; Mowbray, D.J.; Despoja, V. Ab initio study of electromagnetic modes in two-dimensional semiconductors: Application to doped phosphorene. *Phys. Rev. B* **2021**, *104*, 115421. [CrossRef]
48. Pirotta, S.; Patrini, M.; Liscidini, M.; Galli, M.; Dacarro, G.; Canazza, G.; Guizzetti, G.; Comoretto, D.; Bajoni, D. Strong coupling between excitons in organic semiconductors and Bloch surface waves. *Appl. Phys. Lett.* **2014**, *104*, 051111. [CrossRef]
49. Lerario, G.; Ballarini, D.; Fieramosca, A.; Cannavale, A.; Genco, A.; Mangione, F.; Gambino, S.; Dominici, L.; De Giorgi, M.; Gigli, G.; et al. Hig-speed flow of interacting organic polaritons. *Light. Sci. Appl.* **2017**, *6*, e16212. [CrossRef]
50. Barachati, F.; Fieramosca, A.; Hafezian, S.; Gu, J.; Chakraborty, B.; Ballarini, D.; Martinu, L.; Menon, V.; Sanvitto, D.; Kena-Cohen, S. Interacting polariton fluids in a monolayer of tungsten disulfide. *Nat. Nanotechnol.* **2018**, *13*, 906–909. [CrossRef]
51. Hou, S.; Khatoniar, M.; Ding, K.; Qu, Y.; Napolov, A.; Menon, V.M.; Forrest, S.R. Ultralong-Range Energy Transport in a Disordered Organic Semiconductor at Room Temperature Via Coherent Exciton-Polariton Propagation. *Adv. Mater.* **2020**, *32*, 2002127. [CrossRef] [PubMed]
52. Amelines-Sarria, O.; dos Santos Claro, P.C.; Schilardi, P.L.; Blum, B.; Rubert, A.; Benitez, G.; Basiuk, V.A.; González Orive, A.; Hernández Creus, A.; Díaz, C.; et al. Electronic and magnetic properties of C60 thin films under ambient conditions: A multitechnique study. *Org. Electron.* **2011**, *12*, 1483–1492. [CrossRef]
53. Elschner, C.A.; Levin, A.; Wilde, L.; Grenzer, J.; Schroer, C.; Leo, K.; Riede, M. Determining the C60 molecular arrangement in thin films by means of X-ray diffraction. *J. Appl. Cryst.* **2011**, *44*, 983–990. [CrossRef]
54. Elsenbeck, D.K.; Das, S.; Velarde, L. Substrate influence on the interlayer electron–phonon couplings in fullerene films probed with doubly-resonant SFG spectroscopy. *Phys. Chem. Chem. Phys.* **2017**, *19*, 18519–18528. [CrossRef] [PubMed]

55. Huttner, A.; Breuer, T.; Witte, G. Controlling Interface Morphology and Layer Crystallinity in Organic Heterostructures: Microscopic View on C_{60} Island Formation on Pentacene Buffer Layers. *ACS Appl. Mater. Interfaces* **2019**, *11*, 35177–35184. [CrossRef] [PubMed]
56. Seydel, E.; Hoffmann-Vogel, R.; Marz, M. Epitaxial growth of C_{60} on highly oriented pyrolytic graphite surfaces studied at low temperatures. *Nanotechnology* **2019**, *30*, 025703. [CrossRef]
57. Oreshkina, A.I.; Bakhtizin, R.Z.; Sadowski, J.T.; Sakurai, T. Epitaxial Growth of C_{60} Thin Films on the Bi(0001)/Si(111) Surface. *Bull. Russ. Acad. Sci. Phys.* **2009**, *73*, 883. [CrossRef]
58. Bommel, S.; Kleppmann, N.; Weber, C.; Spranger, H.; Schafer, P.; Novak, J.; Roth, S.V.; Schreiber, F.; Klapp, S.H.L.; Kowarik, S. Unravelling the multilayer growth of the fullerene C60 in real time. *Nat. Commun.* **2014**, *5*, 5388. [CrossRef]
59. Acevedo, Y.M.; Cantrell, R.A.; Berard, P.G.; Koch, D.L.; Clancy, P. Multiscale Simulation and Modeling of Multilayer Heteroepitactic Growth of C_{60} on Pentacene. *Langmuir* **2016**, *32*, 3045–3056. [CrossRef]
60. Janke, W.; Speck, T. Modeling of epitaxial film growth of C60 revisited. *Phys. Rev. B* **2020**, *101*, 125427. [CrossRef]
61. Chen, L.; Wang, X.; Kumar, S. Thermal Transport in Fullerene Derivatives Using Molecular Dynamics Simulations. *Sci. Rep.* **2015**, *5*, 12763. [CrossRef] [PubMed]
62. Despoja, V.; Lončarić, I.; Mowbray, D.J.; Marušić, L. Quasiparticle spectra and excitons of organic molecules deposited on substrates: G0W0-BSE approach applied to benzene on graphene and metallic substrates. *Phys. Rev. B* **2013**, *88*, 235437. [CrossRef]
63. Tomaš, M.S. Green function for multilayers: Light scattering in planar cavities. *Phys. Rev. A* **1995**, 2545. [CrossRef] [PubMed]
64. Despoja, V.; Šunjić, M.; Marušić, L. Propagators and spectra of surface polaritons in metallic slabs: Effects of quantum-mechanical nonlocality. *Phys. Rev. B* **2009**, *80*, 075410. [CrossRef]
65. Novko, D.; Šunjić, M.; Despoja, V. Optical absorption and conductivity in quasi-two-dimensional crystals from first principles: Application to graphene. *Phys. Rev. B* **2016**, *93*, 125413. [CrossRef]
66. Despoja, V.; Mowbray, D.J. Using surface plasmonics to turn on fullerene's dark excitons. *Phys. Rev. B* **2014**, *89*, 195433. [CrossRef]
67. Rozzi, C.A.; Varsano, D.; Marini, A.; Gross, E.K.U.; Rubio, A. Exact Coulomb cutoff technique for supercell calculations. *Phys. Rev. B* **2006**, *73*, 205119. [CrossRef]
68. Despoja, V.; Jakovac, J.; Golenić, N.; Marušić, L. Bias-controlled plasmon switching in lithium-doped graphene on dielectric model Al2O3 substrate. *npj 2D Mater. Appl.* **2020**, *4*, 19. [CrossRef]
69. Giannozzi, P.; Baroni, S.; Bonini, N.; Calandra, M.; Car, R.; Cavazzoni, C.; Ceresoli, D.; Chiarotti, G.L.; Cococcioni, M.; Dabo, I.; et al. QUANTUM ESPRESSO: A modular and open-source software project for quantum simulations of materials. *J. Phys. Conden. Matter* **2009**, *21*, 395502. [CrossRef]
70. Troullier, N.; Martins, J.L. Efficient pseudopotentials for plane-wave calculations. *Phys. Rev. B* **1991**, *43*, 1993. [CrossRef]
71. Hamann, D.R. Optimized norm-conserving Vanderbilt pseudopotentials. *Phys. Rev. B* **2013**, *88*, 085117. [CrossRef]
72. Perdew, J.P.; Burke, K.; Ernzerhof, M. Generalized Gradient Approximation Made Simple. *Phys. Rev. Lett.* **1996**, *77*, 3865. [CrossRef] [PubMed]
73. Ong, Z.Y.; Fischetti, M.V. Charged impurity scattering in top-gated graphene nanostructures. *Phys. Rev. B* **2012**, *86*, 165422. [CrossRef]
74. Weaver, J.H.; Martins, J.L.; Komeda, T.; Chen, Y.; Ohno, T.R.; Kroll, G.H.; Troullier, N.; Haufler, R.E.; Smalley, R.E. Electronic structure of solid C60: Experiment and theory. *Phys. Rev. Lett.* **1991**, *66*, 1741. [CrossRef] [PubMed]
75. Lof, R.W.; van Veenendaal, M.A.; Koopmans, B.; Jonkman, H.T.; Sawatzky, G.A. Band gap, excitons, and Coulomb interaction in solid C60. *Phys. Rev. Lett.* **1992**, *68*, 3924. [CrossRef]
76. Sau, J.D.; Neaton, J.B.; Choi, H.J.; Louie, S.G.; Cohen, M.L. Electronic energy levels of weakly coupled nanostructures: C-60-metal interfaces. *Phys. Rev. Lett.* **2008**, *101*, 026804. [CrossRef]
77. Blase, X.; Attaccalite, C.; Olevano, V. First-principles GW calculations for fullerenes, porphyrins, phtalocyanine, and other molecules of interest for organic photovoltaic applications. *Phys. Rev. B* **2011**, *83*, 115103. [CrossRef]
78. Berkowitz, J. Sum rules and the photoabsorption cross sections of C_{60}. *J. Chem. Phys.* **1999**, *111*, 1446–1453. [CrossRef]
79. Lobanov, B.V.; Murzashev, A.I. Optical Absorption of Fullerene C_{60} Within the Concept of a Strongly Correlated State. *Russ. Phys. J.* **2016**, *59*, 856–861. [CrossRef]
80. Tiago, M.L.; Kent, P.R.C.; Hood, R.Q.; Reboredo, F.A. Neutral and charged excitations in carbon fullerenes from first-principles many-body theories. *J. Chem. Phys.* **2008**, *129*, 084311. [CrossRef]
81. Shnitov, V.V.; Mikoushkin, V.M.S.; Gordeev, Y. Fullerite C_{60} as electron-beam resist for 'dry' nanolithography. *Microelectron. Eng.* **2003**, *69*, 429–434. [CrossRef]
82. Shnitov, V.V.; Mikoushkin, V.M. Analysis of Fullerite C_{60} Electron Induced Modification in Terms of Effective Destruction Cross-section. *Fullerenes Nanotub. Carbon Nanostruct.* **2012**, *20*, 324–327. [CrossRef]
83. Nikolaev, A.V. Expansion of the pi-Molecular Orbitals of the C_{60} Fullerene in Spherical Harmonics. *Int. J. Quantum Chem.* **2011**, *111*, 2478–2481. [CrossRef]
84. Despoja, V.; Marušić, L. Use of surface plasmons for manipulation of organic molecule quasiparticles and optical properties. *J. Phys. Condens. Matter* **2014**, *26*, 485012. [CrossRef] [PubMed]

Article

Constructing a Carbon-Encapsulated Carbon Composite Material with Hierarchically Porous Architectures for Efficient Capacitive Storage in Organic Supercapacitors

Rene Mary Amirtha [1,†], Hao-Huan Hsu [1,†], Mohamed M. Abdelaal [1,2], Ammaiyappan Anbunathan [1], Saad G. Mohamed [2], Chun-Chen Yang [1,3,4] and Tai-Feng Hung [1,*]

1. Battery Research Center of Green Energy, Ming Chi University of Technology, 84 Gungjuan Rd., Taishan Dist., New Taipei City 24301, Taiwan; reneamirthasdj@gmail.com (R.M.A.); u08137122@mail2.mcut.edu.tw (H.-H.H.); mohamedbec@yahoo.com (M.M.A.); ammaiyappananbunathan1@gmail.com (A.A.); ccyang@mail.mcut.edu.tw (C.-C.Y.)
2. Tabbin Institute for Metallurgical Studies (TIMS), Tabbin, Helwan 109, Cairo 11421, Egypt; sgmmohamed@gmail.com
3. Department of Chemical Engineering, Ming Chi University of Technology, 84 Gungjuan Rd., Taishan Dist., New Taipei City 24301, Taiwan
4. Department of Chemical and Materials Engineering, Chang Gung University, 259 Wenhua 1st Rd., Guishan Dist., Taoyuan 33302, Taiwan
* Correspondence: taifeng@mail.mcut.edu.tw; Tel.: +886-2-2908-9899 (ext. 4957)
† These authors contributed equally to this work.

Abstract: Hierarchical porous activated carbon (HPAC) materials with fascinating porous features are favored for their function as active materials for supercapacitors. However, achieving high mass-loading of the HPAC electrodes remains challenging. Inspired by the concepts of carbon/carbon (C/C) composites and hydrogels, a novel hydrogel-derived HPAC (H-HPAC) encapsulated H-HPAC (H@H) composite material was successfully synthesized in this study. In comparison with the original H-HPAC, it is noticed that the specific surface area and pore parameters of the resulting H@H are observably decreased, while the proportions of nitrogen species are dramatically enhanced. The free-standing and flexible H@H electrodes with a mass-loading of 7.5 mg/cm^2 are further prepared for electrochemical measurements. The experiments revealed remarkable reversible capacitance (118.6 F/g at 1 mA/cm^2), rate capability (73.9 F/g at 10 mA/cm^2), and cycling stability (76.6% of retention after 30,000 cycles at 5 mA) are delivered by the coin-type symmetric cells. The cycling stability is even better than that of the H-HPAC electrode. Consequently, the findings of the present study suggest that the nature of the HPAC surface is a significant factor affecting the corresponding capacitive performances.

Keywords: supercapacitors; hierarchical porous activated carbon; hydrogel; composite materials; clean energy technology

1. Introduction

Electrochemical-based energy storage devices such as metal–ion batteries/capacitors and supercapacitors are recognized as alternative choices for electricity storage owing to their high flexibility, remarkable reversibility, and simple maintenance as compared to other electric storage technologies [1,2]. Recently, preparing electrodes with high-mass loading has attracted much attention because the active material ratio in devices should be increased as much as possible to provide high total capacitances and gravimetric or volumetric energy densities [3–5]. However, challenges associated with this target still remain, especially in employing the hierarchical porous activated carbon (HPAC) as the active material for supercapacitors. This can be attributed to its huge specific surface area (typically more than 1000 m^2/g), leading to the limited mass loading (normally 1 to

2 mg/cm^2 by a doctor blade method) [6]. Therefore, developing an HPAC that retains the distinctive textural properties and increases the mass loading of the resulting electrode is of interest and highly desirable [3,4,7–9].

Carbon/carbon (C/C) composites are demonstrated to possess a variety of characteristics, i.e., high specific strength, remarkable electrical and thermal conductivities, and excellent dimensional stability [10]. Given the diverse properties, they are beneficial in the field of biomedical, automobile industries, and aeronautics. To realize the C/C composites, it is reported that polymer infiltration pyrolysis and chemical vapor infiltration were commonly adapted [11,12]. As a result, compact and dense C/C composites were obtained, particularly from the repeatedly manufacturing processes. Such a configuration would be beneficial to enhance the mass loading, but not favorable in terms of electrolyte penetration and ionic transportation.

To maximize the electrolyte storage and ionic conductivity within a high mass loading electrode, the hierarchical porosities including micro-, meso-, and macro-pores are crucial. For example, the charges are primarily adsorbed/desorbed inside the micropores. As for the latter two, they can contribute to (i) an electrolyte reservoir, (ii) enlarging the ionic diffusion rate, and (iii) facilitating the migration of large ions/molecules [13–16]. Undoubtedly, utilizing a template and activation are straightforward approaches for synthesizing the HPAC [14,17]. Even so, it will be highly appreciated if the greener templates and activators were chosen owing to resolve environmental issues and promote cost-effectiveness issues [9,18–21].

In our recent study, hydrogel-derived HPAC (H-HPAC) synthesized by pyrolysis of polyvinylpyrrolidone hydrogel under an argon atmosphere at 900 °C was successfully obtained [9]. The merits of H-HPAC can be attributed to (i) numerous water molecules encapsulated within PVP hydrogel efficiently serving as green templates, and (ii) the simultaneous function of K_2CO_3 as an initiator for hydrogel formation and an activator to enable rich porous conformations. Accordingly, the resultant H-HPAC revealed fascinating structural features and distinguished capacitive performances for electrochemical storage applications. Inspired by the concepts of C/C composites and hydrogels, an alternative H-HPAC encapsulated H-HPAC (H@H) composite material was proposed in the present study. After systemically investigating the physicochemical and morphological properties, the H@H electrodes with a mass loading of 7.5 mg/cm^2 were prepared to evaluate the electrochemical performances of supercapacitors that were assembled with an organic electrolyte. Moreover, various factors such as the physicochemical and textural properties that affect the corresponding electrochemical performances were also explored. On the basis of the results and viewpoints reported here, it is reasonably anticipated that such a strategy also has general applicability to other C/C composites.

2. Results and Discussion

2.1. Characterizations of Hydrogel-Derived Hierarchical Porous Activated Carbon (H-HPAC)-Encapsulated H-HPAC (H@H) Composite Material

To explore the physicochemical and morphological properties of H@H composite material, it was systematically investigated by PXRD, Raman, SEM, TEM, BET, EA, and XPS, with the corresponding characteristics compared with those of H-HPAC. Figure 1 depicts the normalized PXRD pattern of H@H to show its crystalline structure. As can be seen, only two broad peaks, assigned to the (002) and (100) planes of carbon (JCPDS No.: 41-1487), were reflected, which was consistent with the original H-HPAC and other activated carbon materials [9,14,17,20,22]. In line with the possible formation mechanism for PVP hydrogel proposed in our previous study, the cross-linking reactions among the polymer chains were initiated by the coordination between potassium cations (K^+) and oxygen anions (O^-) [9]. When the K_2CO_3 solution was added to the PVP/H-HPAC solution, it is reasonably postulated that the K^+ would also interact with the H-HPAC because 4.7% of the oxygen present in the original H-HPAC was verified by elemental analysis [9]. If so, the K^+ coordination among the H-HPAC would be covered by the PVP hydrogel. Even so, no peaks associated with unreacted K_2CO_3 were observed from the PXRD pattern, implying that the purity of H@H was not affected by the presence of H-HPAC after thoroughly rinsing with DI water.

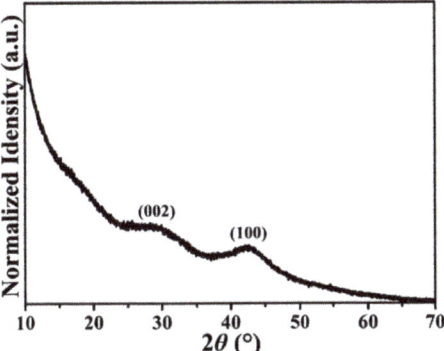

Figure 1. Normalized PXRD pattern of H@H composite material.

Raman characterization is another approach that can directly examine the crystallinity of carbonaceous materials. The normalized Raman spectrum illustrated in Figure 2 shows two distinct peaks representing the D (~1327 cm^{-1}) and G (~1593 cm^{-1}) bands. Besides, it is meaningful to discuss the intensity ratio between D and G bands (I_D/I_G) because the degree of defects within the carbonaceous materials can be further evaluated. The value calculated from the H@H was 1.16, the same as the H-HPAC and close to that of the A-PVP-NC (1.18) [21]. The high I_D/I_G ratio suggests that many defects and/or highly disordered degrees exist, as is generally observed in the carbonaceous materials with numerous functional groups [23,24]. Consequently, we could ascribe this result to the presence of heteroatoms (i.e., N and O) and lower crystallinity, as demonstrated in the original H-HPAC [9]. On the other hand, the corresponding Raman spectrum was sequentially deconvoluted into four peaks (labeled peaks (1)–(4)) since the integrated area ratio of sp^3 to sp^2 (A_{sp}^3/A_{sp}^2) has been shown to provide helpful information on the nature of carbon, e.g., a low A_{sp}^3/A_{sp}^2 ratio indicates that a large amount of carbon exists as the sp^2 type [25,26]. Among them, peaks (2) and (4) are associated with sp^2-type carbon, whereas the others are related to sp^3-type carbon [27]. The integrated area ratio of sp^3 to sp^2 (A_{sp}^3 / A_{sp}^2) was calculated to be 0.28, which was identical to that of H-HPAC [9]. This result signifies that the H@H still retained a high proportion of sp^2-type carbons, even with intrinsically lower crystallinity.

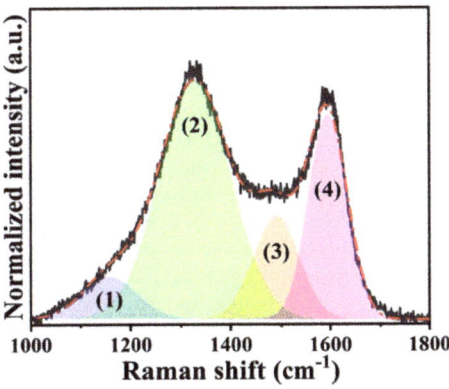

Figure 2. Fitted Raman spectrum of H@H composite material.

To examine the morphological features of H@H, micrographs were captured using SEM and TEM. The hierarchically porous architectures constructed by interconnected carbonaceous frameworks were clearly visible from the low-magnification SEM micrographs in Figure 3a,b. It is worth mentioning that rough surfaces with numerous voids were found, as indicated by white circles in Figure 3b. The diverse porous configurations are reasonably attributed to the water molecules encapsulating within PVP/H-HPAC hydrogel being evaporated and the activation process by interacting the carbonized residues with K_2CO_3 under 900 °C (i.e., $K_2CO_3 + 2C \rightarrow 2K + 3CO$, $K_2CO_3 \rightarrow K_2O + CO_2$, $C + CO_2 \rightarrow 2CO$) [18,19]. Based on the morphologies found in SEM, it is expected that similar characteristics were also exhibited, as shown in Figure S2 and Figure 3c at different TEM magnifications. Moreover, it is seen that the short-range disorders, such as carbon lattices, highlighted by white circles were displayed in Figure 3d, which might be correlated with skeleton collapse after high-temperature pyrolysis [28].

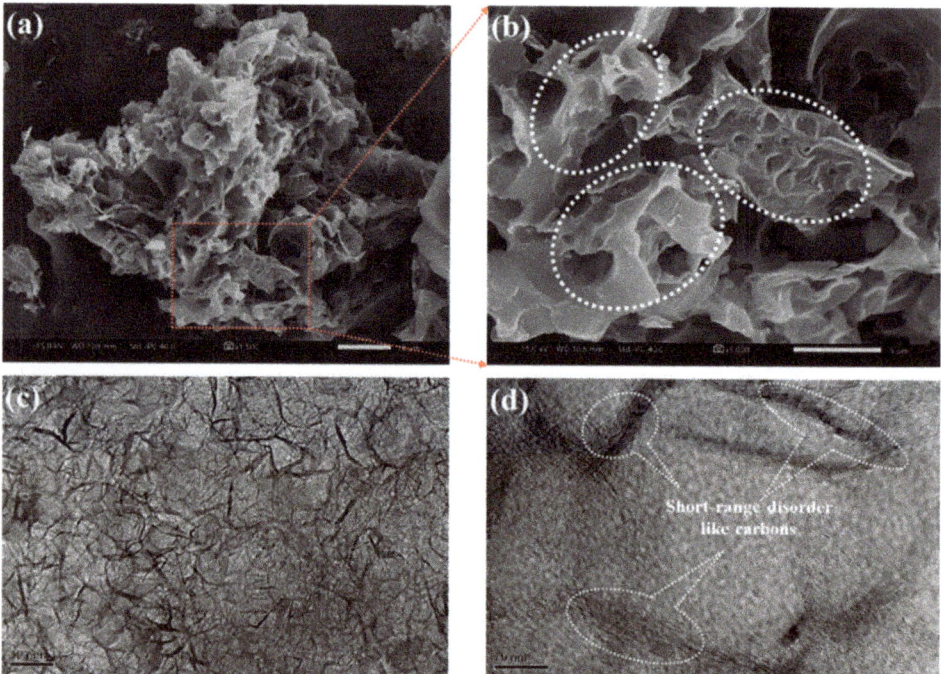

Figure 3. (**a**,**b**) SEM and (**c**,**d**) TEM micrographs of the H@H composite material. Scale bar: (**a**) 10 μm, (**b**) 5 μm, (**c**) 50 nm, and (**d**) 10 nm.

Given the positive results found in SEM and TEM, it is believed that the textural characteristics of H@H would not be significantly affected. To accurately classify pores and determine the specific surface area (SSA), the nitrogen adsorption–desorption measurement was conducted, and the corresponding isotherm is shown in Figure 4. As plotted, not only a high volume of nitrogen gases were adsorbed and desorbed at low relative pressure (i.e., Type I isotherm), but also the predominant pore diameter was less than 2 nm (inset of Figure 4), confirming the microporous feature for H@H [9,21,29]. However, all values diminished except for the pore size distributions of ultramicropores as compared with H-HPAC (see Table 1). For instance, the SSA value determined by the Brunauer–Emmett–Teller (BET) method decreased by about 35%. Accordingly, the SSA values contributed by micropores and mesopores were decrease to 1246 m^2/g and 45 m^2/g, respectively. In particularly, the latter was reduced by even approximately 84%. When preparing the

PVP/H-HPAC composite hydrogel, these pores within the H-HPAC would be filled with the viscous PVP solution so that the PVP blocked the original pores after drying of the composite hydrogel at 120 °C. In addition to the issue mentioned above, the decrease in the textural parameters might also be actuated by the possible interaction between H-HPAC and K^+. Such a phenomenon is rationally postulated to affect the cross-linking degree of the PVP/H-HPAC composite hydrogel, as shown by the XRD characterization, leading to fewer sites for activation.

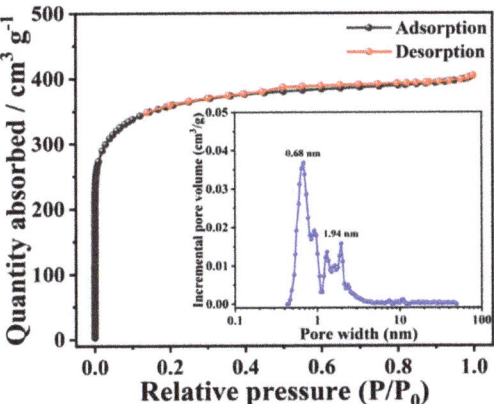

Figure 4. Nitrogen adsorption–desorption isotherm of H@H composite material collected by an accelerated surface area and porosimetry system at 77 K. Inset shows the pore size distribution curve determined by the 2D-NLDFT model.

Table 1. Textural properties of H@H composite material and H-HPAC.

Properties Samples	SSA (m^2/g)	V_t [1] (cm^3/g)	V_{ultra} [2] (cm^3/g)	V_{micro} [3] (cm^3/g)	V_{meso} [4] (cm^3/g)
H@H	1316	0.62	0.21	0.51	0.11
H-HPAC [5]	2012	1.16	0.11	0.69	0.47

[1] V_t: total pore (single-point) volume obtained from the amount of adsorbed nitrogen at $P/P_0 = 0.995$. [2] V_{ultra}: volume of ultramicropores (pores < 0.7 nm). [3] V_{micro}: volume of micropores (pores < 2.0 nm). [4] V_{meso}: volume of mesopores (difference between V_t and V_{micro}). [5] The values were obtained from Ref. [9].

The compositional information and chemical environments of H@H were identified through EA and XPS, respectively. As quantified by the former, the proportions of carbon, nitrogen, and oxygen in the as-synthesized H@H were 76.2%, 0.58%, and 4.5%, respectively. In comparison with the original H-HPAC, the carbon content was decreased (76.2% vs. 95.1%), but the nitrogen species was enhanced (0.58% vs. 0.23%), while the oxygen species was similar (4.5% vs. 4.7%). It would be attributed the variation in carbon to the cross-linking degree of PVP/H-HPAC composite hydrogel. As for the increase in the nitrogen species, the following possible reasons could be given. It is reported that the nitrogen species included in the carbon precursor/char were preferentially removed during chemical activation with K-based salts [20]. However, as previously mentioned in XRD characterization, K^+ would also interact with the original H-HPAC due to the presence of oxygen, reducing the concentration of K^+ that was coordinated to the PVP as compared to the preparation of the original H-HPAC. According to Ref. [20] and our experimental results [9,21], the higher nitrogen percentage in the H@H could be attributed to the lower interaction between the K^+ and carbonaceous residues.

Figure 5 provides the high-resolution XPS spectra that were analyzed using the Gaussian–Lorentzian fitting method. From the EA result, the presence of nitrogen atoms within the H@H was already demonstrated. Therefore, the C-N bonding in the C 1s spectrum did not particularly point out for better reading. As revealed in Figure 5a, the C 1s spectrum was deconvoluted into four peaks: (1) C=C bond at 284.8 eV, (2) C-O bond at 285.9 eV, (3) C=O bond at 287.8 eV, and (4) O=C-O bond at 290.2 eV [30,31]. It is known that the first peak was assigned to the sp^2-type carbon, while the rest corresponded to the contribution of sp^3-type carbon [27]. As for the O 1s spectrum (Figure 5b), three peaks fitted at 531.5 eV, 533.1 eV, and 535.0 eV have appeared, representing (1) O=C-O, (2) C=O, and (3) C-O bonds, respectively [32]. Even though lower nitrogen content was shown in the EA results, the N 1s peak in the binding energy between 396 eV and 402 eV still can be detected (Figure 5c) [9,21,33].

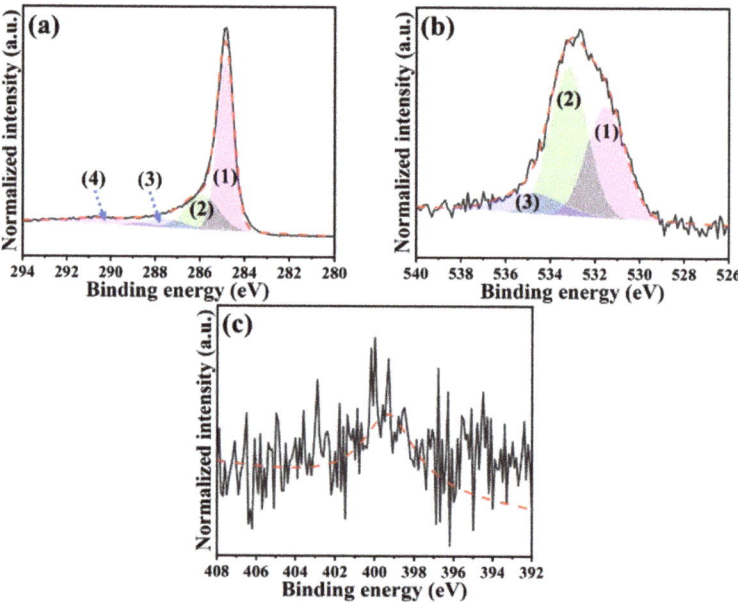

Figure 5. High-resolution XPS spectra of H@H composite material: (**a**) C 1s ((1) for C=C, (2) for C−O, (3) for C=O, and (4) for O=C−O bonds), (**b**) O 1s ((1) O=C−O, (2) C=O, and (3) C−O bonds), and (**c**) N 1s.

Based on the results discussed in this section, the as-prepared H@H produced from thermal pyrolysis of the PVP/H-HPAC composite hydrogel combines various benefits, such as good purity, hierarchical porous characteristics, and high proportions of sp^2-type carbons, as well as nitrogen species. Although the structural parameters were significantly altered, it is of interest to consider the influence of physicochemical and textural features of the H@H on the corresponding electrochemical performance as organic supercapacitors.

2.2. Electrochemical Performances of H@H in Symmetric Supercapacitor

To evaluate the capacitive efficiencies of the H@H electrode, coin-type symmetric cells were fabricated to conduct the cyclic voltammetry (CV) and galvanostatic charge-discharge (GCD) measurements in the voltage window between 0 V and 2.7 V by different scanning rates and current densities. Although the SSA values and relative textural parameters were less than the H-HPAC, as discussed previously, the typical curves in nearly rectangular shapes with good symmetries were reflected from the H@H (Figure 6a), despite gradually increasing the scanning rate to 10 mV/s. During CV cycling, the integral area from the cyclic voltammogram is associated with the charges adsorbed and desorbed among the active materials. Figure 6b compares the integral area of H@H and H-HPAC acquired in terms of the forward and backward scanning. As can be seen, the values linearly increased with the scanning rate. In addition, the values for the H@H electrode were superior those for the H-HPAC electrode; this was correlated with the different mass loading (7.5 mg/cm^2 vs. 5.1 mg/cm^2). On the other hand, the voltage delay (ΔV) is regarded as an important indicator providing similar information to the IR drop. To discuss this discrepancy, we consider the voltage that reached zero for a current density at 1 mV/s as a reference. The ΔV values were then calculated by comparing the difference between the voltage recorded from each scanning rate and the reference; the corresponding data are visible in Figure S3. Under the voltage range of 0 V to 2.7 V and a scanning rate of 1 mV/s, the reference data for each free-standing electrode were 45 mV (HPAC electrode: 230 μm [7]), 22 mV (H@H electrode: 160 μm) and 12 mV (H-HPAC electrode: 100 μm [9]), respectively. With increase in scanning rate to 10 mV/s, it was found that the ΔV value was 128 mV, i.e., a 64% increase as compared to that of the H-HPAC (78 mV) [9]. This could be ascribed to the thickness of the H@H electrode, which was ~60% more than the H-HPAC electrode, prolonging the pathway for electron transportation.

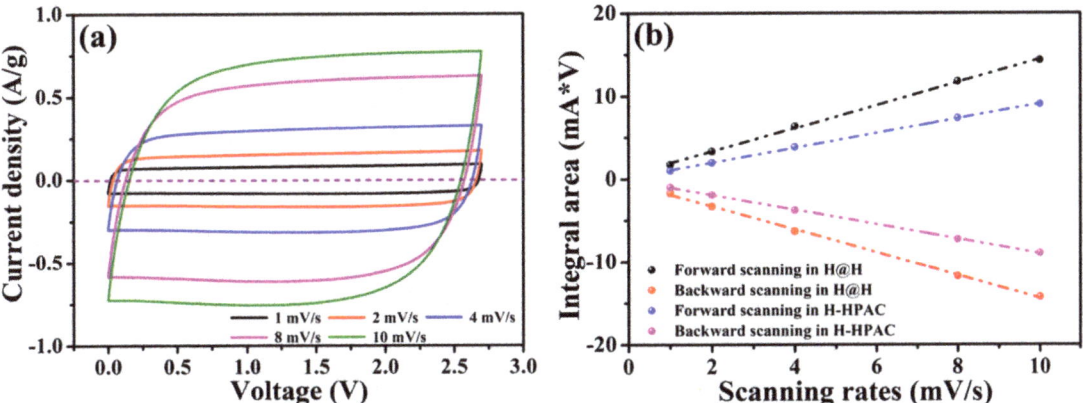

Figure 6. (**a**) Cyclic voltammograms collected in the voltage range between 0 and 2.7 V at scanning rates from 1 to 10 mV/s and (**b**) dependence of the integral area on scanning rate, with the values acquired from (**a**) and Ref. [9] for H@H and H-HPAC, respectively.

Figure 7a presents the GCD profiles measured using the same voltage window as in the CV test but with current densities from 1 mA/cm^2 to 10 mA/cm^2. Contributing to the ideal electric double-layer behavior and high reversibility, as shown in Figure 6a, the linear and symmetric charge–discharge behavior at each current density was observed. The specific capacitance discharged from the H@H electrode at the 100th cycle was 118.6 F/g at 1 mA/cm^2 with 99% Coulombic efficiency. This result was slightly higher than that outputted from the H-HPAC electrode (117.5 F/g [9]), implying that the specific capacitance was not appreciably affected by the thickness when applying a small current density. Additionally, the stable discharge capacitances of 110.2, 98.8, 81.6, and 73.9 F/g are compared in Figure 7b. The tendency for capacitance decay was the same as the H-HPAC, but the values were observably declined while the current densities were above 4 mA/cm^2, which could also be attributed to the change in thickness. However, 96.4% of the recovery in capacitance after 100 cycles was obtained when the current density was returned to 1 mA/cm^2. Figure 7c shows the EIS spectra that were recorded before and after rate-capability testing. It is recognized that charge transfer resistance (R_{ct}) includes ionic and electronic resistances. The former is the resistance to the mobility of ionic electrolytes inside the textual pores of the electrode, while the latter comprises the intrinsic resistance of the electrode material and the contact resistance between the active layer and the current collector [34]. For the rate-capability testing, the same electrode conditions (i.e., composition, working area, and thickness) were used, based on the hypothesis that the electronic resistance should be no significant differences. Following repeated charging–discharging processes, the R_{ct} value increased from 18.5 Ohm to 29.1 Ohm. As reported, the diameter of solvated ions for TEA$^+$ and BF$_4^-$ are 1.35 nm and 1.40 nm, respectively [35]. Hence, the bulky solvated ions would accumulate within the pores of H@H, further blocking the ionic transport and causing an increase in R_{ct} as well as capacitance decay [36].

The energy and power densities calculated from the data presented in Figure 7b and the equations shown in Section 3.4 are plotted in Figure 7d. As indicated, the values ranged from 30.0 Wh/kg (@ 1 mA/cm^2) to 18.7 Wh/kg (@ 10 mA/cm^2) for the former, and from 88.1 W/kg (@ 1 mA/cm^2) to 881.7 W/kg (@ 10 mA/cm^2) for the latter. To compare with the H-HPAC electrode, ~74% of the power density was outputted by the H@H electrode for all current densities. On the basis of the same electrolyte and similar voltage window, the H@H electrode exhibits reasonable energy and power densities, comparable with the results reported previously (Figure 7d) [9,15,37–43]. Considering the cycling stability, the accelerated experiment conducted in the voltage ranged from 1.35 to 2.7 V (i.e., 50% of the state of discharge) and the current of 5 mA was used; the corresponding result is displayed in Figure 7e. The initial capacitance discharge to 1.35 V was ~0.5 F. After 30,000 cycles, about 76.6% of capacitance retention and ≥99.5% of Coulombic efficiency were found, respectively. Although the R_{ct} value increased from 32.8 Ohm to 45.1 Ohm (see Figure S4), the H@H electrode provided a better lifespan than that of the H-HPAC electrode (capacitance retention: 76% after 10,000 cycles) [9]. It is reported that increasing the mass loading or electrode thickness leads to a decrease in capacitance and the rate capability of the electrode materials, which is related to the decreased accessible surface area, increased electrical resistance, prolonged ion transport channels, and poor electrolyte wetting [4]. Besides, the variety of heteroatom dopants and their corresponding amounts, as well as the porosity characteristics, within HPAC were also significant influences [17,44]. The comparison of the electrochemical performance of H@H and that reported for HPAC in symmetric supercapacitors using 1 M TEABF$_4$/PC electrolyte is listed in Table S1. The variations in the electrochemical performance of H@H can be attributed to the following. First, even though the SSA value and pore parameters were lower than those for H-HPAC, the increased thickness of the H@H electrode would compensate for their active sites of capacitive storage, because the specific capacitance generated from the H@H electrode at 1 mA/cm^2 was slightly enhanced. Second, the number of nitrogen species doped in the H@H was increased by up to 152% in comparison with the H-HPAC, so the GCD profiles of the first and last five cycles in Figure 7f showed high symmetry, meaning that the overall resistance was not significant, even when increasing the thickness by 60% and after 30,000 cycles.

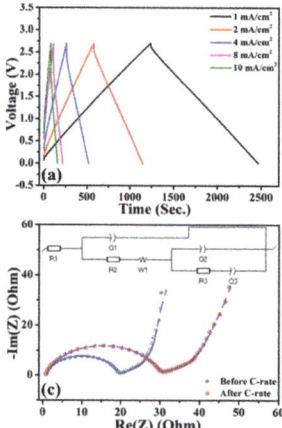
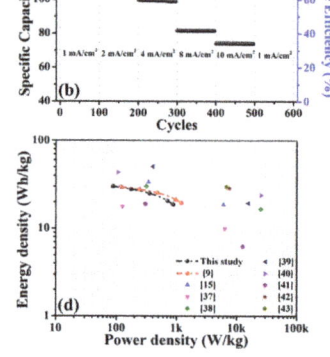

Figure 7. *Cont.*

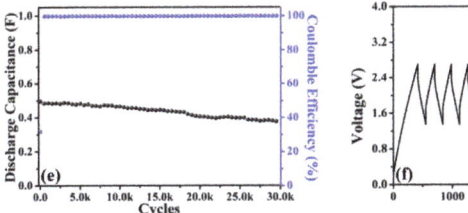
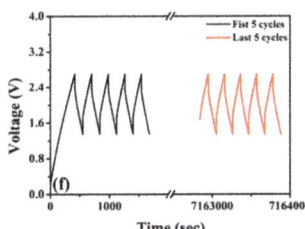

Figure 7. (a) Galvanostatic charge–discharge profiles, (b) rate capabilities, (c) electrochemical impedance spectra, (d) Ragone plots, (e) discharge capacitance as a function of cycle number of the H@H electrode in a coin-type symmetric cell, and (f) galvanostatic charge–discharge profiles of the first and last five cycles received from (e). The inset of (c) illustrates the equivalent circuit model used for the parameter fitting [45].

3. Materials and Methods

3.1. Chemicals

All reagents, including polyvinylpyrrolidone (PVP, $(C_6H_9NO)_n$, average MW: 1,300,000, Sigma-Aldrich, St. Louis, MO, USA), potassium carbonate (anhydrous, K_2CO_3, 99%, Alfa Aesar, Heysham, UK), carbon black (Super P®, Timcal Ltd., Bodio, Switzerland), vapor-grown carbon nanofibers (VGCFs, 7 µm in length and 0.11 µm in diameter, Yonyu Applied Technology Material Co., Ltd., Tainan, Taiwan), and colloidal polytetrafluoroethylene (PTFE) dispersion (D1-E, Daikin Industries Ltd., Osaka, Japan), were adopted without further purification. Deionized (DI) water produced from a Milli-Q Integral water purification system (Millipore Ltd., Burlington, MA, USA) was utilized throughout the experiments.

3.2. Preparation of Hydrogel-Derived Hierarchical Porous Activated Carbon (H-HPAC) Encapsulated H-HPAC (H@H) Composite Material

To construct the H@H composite material, PVP powders and K_2CO_3 were well-dissolved in DI water individually. Here, the mass ratio between K_2CO_3 and PVP was controlled at 2 as no hydrogel is formed when the ratio was less than 2, as demonstrated in our previous study [9]. The concentrations of PVP and K_2CO_3 solutions were 14.3 wt.% and 40.7 wt.%, respectively. The volume ratio between PVP solution and K_2CO_3 solution was 2. Then, 0.5 g of H-HPAC was carefully added to the PVP solution, whereas the resultant was vigorously stirred to ensure homogenous mixing. Following the addition of K_2CO_3 solution dropwise to the PVP/H-HPAC solution, the black elastomer-like sample was completely obtained within 5 min (see Figure S1). The resulting composite hydrogel was dried in an oven at 120 °C for 12 h to completely evaporate the water molecules that were encapsulated within the matrix. The residues were then thermally pyrolyzed in a tube furnace at 900 °C for 2 h under an argon atmosphere with a flow rate of 200 mL/min, so the newly formed H-HPAC converted from the PVP would encapsulate the original H-HPAC. After repeated rinses with DI water, drying, and grinding procedures, the loose H@H powders can be obtained.

3.3. Characterizations

The crystalline structure of the as-prepared H@H composite material was identified using a powder X-ray diffractometer (XRD, D2 PHASER, Bruker AXS Inc., Karlsruhe, Germany) with a Cu target (λ = 1.541 Å) that was excited at 30 kV and 10 mA. The corresponding PXRD pattern was recorded in the range of 2θ from 10° to 70° at a scanning rate of 0.5 s/step. The Raman spectrum was collected between 1000 cm^{-1} and 1800 cm^{-1} by a confocal Raman microscope (inVia, Renishaw, UK) equipped with a 633 nm laser source. For morphological observations, the micrographs were acquired from the scanning electron microscope (SEM, JSM-IT200, JEOL Ltd., Tokyo, Japan) and a transmission electron microscope (TEM, JEM-2100, JEOL Ltd., Tokyo, Japan). To examine the textural properties, the N$_2$ adsorption–desorption isotherm was measured at 77 K on a surface area and porosity analyzer (ASAP 2020 V3.00, Micromeritics Instrument Corporation, Norcross, GA, USA) after degassing under vacuum at 160 °C for 8 h. An elemental analyzer (FLASH 2000, Thermo Fisher Scientific Inc., Waltham, MA, USA) was applied for determining the percentages of carbon, nitrogen, and oxygen in the H@H composite material. The chemical environments were analyzed with X-ray photoelectron spectroscopy (XPS, PHI 5000 VersaProbe III, ULVAC-PHI, Inc., Kanagawa, Japan) with a beam size of 100 µm under Al K$_\alpha$ radiation (λ = 8.3406 Å). Their corresponding high-resolution spectra were further deconvoluted by the Gaussian–Lorentzian fitting method using an XPSPEAK 4.1 software.

3.4. Electrochemical Measurements

The electrochemical tests throughout this study were conducted in the symmetric two-electrode configuration at ambient conditions. To prepare the free-standing H@H electrodes, the ingredients (80 wt.% of H@H, 5 wt.% of Super P®, 5 wt.% of VGCFs, and 10 wt.% of PTFE) were mechanically blended and repeatedly calendared. The as-prepared H@H electrodes with a thickness of 160 ± 7 µm and a mass loading of 7.5 ± 0.8 mg/cm^2 were obtained after drying at 130°C. To assemble the coin-type cells, 1 M TEABF$_4$/PC and cellulose-based membrane (TF4535, NKK, Kochi, Japan) were used as the organic electrolyte and separator, respectively. The cyclic voltammograms (CVs) and electrochemical impedance spectroscopy measurements were recorded using a multichannel electrochemical workstation (VSP-3e, Bio-Logic, Seyssinet-Pariset, France). The electrochemical impedance spectra (EIS) were recorded at open circuit potential (OCP) from 100 kHz to 0.01 Hz with an AC potential amplitude of 5 mV. The galvanostatic charge–discharge (GCD) profiles and the cycling stabilities were evaluated through a computer-controlled system (CT-4008T-5V50mA, Neware Technology Limited, Shenzhen, China). To determine the specific capacitance (C_s, F/g) of the H@H electrode in the symmetric supercapacitor, the value can be calculated from the GCD profiles by $C_s = 2\ It/mV$, where I is the applied current (A), t is the recorded discharge time (s), m is the mass of active material at one electrode (g), and V is the voltage window (volts). As for the energy density (E, Wh/kg) and power density (P, W/kg), they can be further acquired based on the equations $E = C_s V^2/(2 \times 4 \times 3.6)$ and $P = 3600\ E/t$, respectively [9,15].

4. Conclusions

In summary, this study presents an alternative concept for the construction of the hydrogel-derived HPAC (H-HPAC) encapsulated H-HPAC (H@H) composite material through the thermal pyrolysis of a PVP/H-HPAC hydrogel under an argon atmosphere at 900 °C. Compared to the original H-HPAC, the as-prepared H@H retains good purity, lower crystallinity, and high proportions of sp^2-type carbons. However, H@H has a lower specific surface area and decreased pore parameters, but a substantial increase in the percentage of nitrogen species. Even with the notable change in the textural features, the symmetric supercapacitor assembled by the H@H electrode with a mass loading of 7.5 mg/cm^2 and organic electrolyte still exhibits good reversible capacitance, comparable rate capability, and excellent cyclability. The results presented in this study support the H@H as a promising electrode material for other electrochemical energy storage fields, such as metal–ion capacitors.

Supplementary Materials: The following supporting information can be downloaded at: https://www.mdpi.com/article/10.3390/ijms23126774/s1.

Author Contributions: Conceptualization, T.-F.H.; methodology, T.-F.H., R.M.A. and H.-H.H.; validation, T.-F.H., R.M.A., H.-H.H., M.M.A. and S.G.M.; investigation, T.-F.H., R.M.A., H.-H.H., M.M.A. and A.A.; resources, T.-F.H. and C.-C.Y.; writing—original draft preparation, T.-F.H. and R.M.A.; writing—review and editing, T.-F.H., R.M.A., M.M.A., S.G.M. and C.-C.Y.; visualization, T.-F.H., R.M.A., H.-H.H., M.M.A. and A.A.; supervision, T.-F.H.; funding acquisition, T.-F.H. All authors have read and agreed to the published version of the manuscript.

Funding: This research was funded by the Ministry of Science and Technology (MOST) of Taiwan (grant number: MOST 110-2222-E-131-001-MY3) and Ming Chi University of Technology (grant number: VK003-6100-110).

Institutional Review Board Statement: Not applicable.

Informed Consent Statement: Not applicable.

Data Availability Statement: Data sharing is not applicable to this article.

Conflicts of Interest: The authors declare no conflict of interest.

References

1. Yu, L.; Chen, G.Z. Supercapatteries as high-performance electrochemical energy storage devices. *Electrochem. Energ. Rev.* **2020**, *3*, 271–285. [CrossRef]
2. Iqbal, M.Z.; Faisal, M.M.; Ali, S.R. Integration of supercapacitors and batteries towards high-performance hybrid energy storage devices. *Int. J. Energy Res.* **2021**, *45*, 1449–1479. [CrossRef]
3. Vijayakumar, M.; Santhosh, R.; Adduru, J.; Rao, T.N.; Karthik, M. Activated carbon fibres as high-performance supercapacitor electrodes with commercial level mass loading. *Carbon* **2018**, *140*, 465–476. [CrossRef]
4. Dong, Y.; Zhu, J.; Li, Q.; Zhang, S.; Song, H.; Jia, D. Carbon materials for high mass-loading supercapacitors: Filling the gap between new materials and practical applications. *J. Mater. Chem. A* **2020**, *8*, 21930–21946. [CrossRef]
5. He, W.; Chen, K.; Pathak, R.; Hummel, M.; Reza, K.M.; Ghimire, N.; Pokharel, J.; Lu, S.; Gu, Z.; Qiao, Q.; et al. High-mass-loading Sn-based anode boosted by pseudocapacitance for long-life sodium-ion batteries. *Chem. Eng. J.* **2021**, *414*, 128638. [CrossRef]
6. Lin, Y.T.; Chang-Jian, C.W.; Hsieh, T.H.; Huang, J.H.; Weng, H.C.; Hsiao, Y.S.; Syu, W.L.; Chen, C.P. High-performance Li-ion capacitor constructed from biomass-derived porous carbon and high-rate $Li_4Ti_5O_{12}$. *Appl. Surf. Sci.* **2021**, *543*, 148717. [CrossRef]
7. Hung, T.F.; Hsieh, T.H.; Tseng, F.S.; Wang, L.Y.; Yang, C.C.; Yang, C.C. High-mass loading hierarchically porous activated carbon electrode for pouch-type supercapacitors with propylene carbonate-based electrolyte. *Nanomaterials* **2021**, *11*, 785. [CrossRef]
8. Guo, W.; Yu, C.; Li, S.; Qiu, J. Toward commercial-level mass-loading electrodes for supercapacitors: Opportunities, challenges and perspectives. *Energy Environ. Sci.* **2021**, *14*, 576–601. [CrossRef]
9. Abdelaal, M.M.; Hung, T.C.; Mohamed, S.G.; Yang, C.C.; Hung, T.F. Two birds with one stone: Hydrogel-derived hierarchical porous activated carbon toward the capacitive performance for symmetric supercapacitors and lithium-ion capacitors. *ACS Sustain. Chem. Eng.* **2022**, *10*, 4717–4727. [CrossRef]
10. Zhang, S.; Ma, Y.; Suresh, L.; Hao, A.; Bick, M.; Tan, S.C.; Chen, J. Carbon nanotube reinforced strong carbon matrix composites. *ACS Nano* **2020**, *14*, 9282–9319. [CrossRef]
11. Han, Y.; Li, S.; Chen, F.; Zhao, T. Multi-scale alignment construction for strong and conductive carbon nanotube/carbon composites. *Mater. Today Commun.* **2016**, *6*, 56–68. [CrossRef]
12. Lee, J.; Kim, T.; Jung, Y.; Jung, K.; Park, J.; Lee, D.M.; Jeong, H.S.; Hwang, J.Y.; Park, C.R.; Lee, K.H.; et al. High-strength carbon nanotube/carbon composite fibers via chemical vapor infiltration. *Nanoscale* **2016**, *8*, 18972–18979. [CrossRef] [PubMed]
13. Liu, T.; Zhang, F.; Song, Y.; Li, Y. Revitalizing carbon supercapacitor electrodes with hierarchical porous structures. *J. Mater. Chem. A* **2017**, *5*, 17705–17733. [CrossRef]
14. Wu, L.; Li, Y.; Fu, Z.; Su, B.L. Hierarchically structured porous materials: Synthesis strategies and applications in energy storage. *Natl. Sci. Rev.* **2020**, *7*, 1667–1701. [CrossRef]
15. Wu, J.; Xia, M.; Zhang, X.; Chen, Y.; Sun, F.; Wang, X.; Yang, H.; Chen, H. Hierarchical porous carbon derived from wood tar using crab as the template: Performance on supercapacitor. *J. Power Source* **2020**, *455*, 227982. [CrossRef]
16. Yin, J.; Zhang, W.; Alhebshi, N.A.; Alshareef, H.N. Synthesis strategies of porous carbon for supercapacitor applications. *Small Methods* **2020**, *4*, 1900853. [CrossRef]
17. Díez, N.; Fuertes, A.B.; Sevilla, M. Molten salt strategies towards carbon materials for energy storage and conversion. *Energy Storage Mater.* **2021**, *38*, 50–69. [CrossRef]
18. Díez, N.; Ferrero, G.A.; Fuertes, A.B.; Sevilla, M. Sustainable salt template-assisted chemical activation for the production of porous carbons with enhanced power handling ability in supercapacitors. *Batter. Supercaps* **2019**, *2*, 701–711. [CrossRef]
19. Díez, N.; Ferrero, G.A.; Sevilla, M.; Fuertes, A.B. A sustainable approach to hierarchically porous carbons from tannic acid and their utilization in supercapacitive energy storage systems. *J. Mater. Chem. A* **2019**, *7*, 14280–14290. [CrossRef]
20. Sevilla, M.; Díez, N.; Fuertes, A.B. More sustainable chemical activation strategies for the production of porous carbons. *ChemSusChem* **2021**, *14*, 94–117. [CrossRef]
21. Abdelaal, M.M.; Hung, T.C.; Mohamed, S.G.; Yang, C.C.; Huang, H.P.; Hung, T.F. A comparative study of the influence of nitrogen content and structural characteristics of NiS/nitrogen-doped carbon nanocomposites on capacitive performances in alkaline medium. *Nanomaterials* **2021**, *11*, 1867. [CrossRef] [PubMed]
22. Plachy, T.; Kutalkova, E.; Skoda, D.; Holcapkova, P. Transformation of cellulose via two-step carbonization to conducting carbonaceous particles and their outstanding electrorheological performance. *Int. J. Mol. Sci.* **2022**, *23*, 5477. [CrossRef] [PubMed]
23. Vatankhah, A.R.; Hosseini, M.A.; Malekiec, S. The characterization of gamma-irradiated carbon-nanostructured materials carried out using a multi-analytical approach including Raman spectroscopy. *Appl. Surf. Sci.* **2019**, *488*, 671–680. [CrossRef]

24. Hsu, C.H.; Chung, C.H.; Hsieh, T.H.; Lin, H.P. Green and highly-efficient microwave synthesis route for sulfur/carbon composite for Li-S battery. *Int. J. Mol. Sci.* **2022**, *23*, 39. [CrossRef] [PubMed]
25. Duan, W.; Zhu, Z.; Li, H.; Hu, Z.; Zhang, K.; Cheng, F.; Chen, J. Na$_3$V$_2$(PO$_4$)$_3$@C core-shell nanocomposites for rechargeable sodium-ion batteries. *J. Mater. Chem. A* **2014**, *2*, 8668–8675. [CrossRef]
26. Zolkin, A.; Semerikova, A.; Chepkasov, S.; Khomyakov, M. Characteristics of the Raman spectra of diamond-like carbon films. Influence of methods of synthesis. *Mater. Today Proc.* **2017**, *4*, 11480–11485. [CrossRef]
27. Hung, T.F.; Cheng, W.J.; Chang, W.S.; Yang, C.C.; Shen, C.C.; Kuo, Y.L. Ascorbic acid-assisted synthesis of mesoporous sodium vanadium phosphate nanoparticles with highly sp^2-coordinated carbon coatings as efficient cathode materials for rechargeable sodium-ion batteries. *Chem. Eur. J.* **2016**, *22*, 10620–10626. [CrossRef] [PubMed]
28. Lv, Y.; Ding, L.; Wu, X.; Guo, N.; Guo, J.; Hou, S.; Tong, F.; Jia, D.; Zhang, H. Coal-based 3D hierarchical porous carbon aerogels for high performance and super-long life supercapacitors. *Sci. Rep.* **2020**, *10*, 7022. [CrossRef]
29. Kim, J.H.; Jung, S.C.; Lee, H.M.; Kim, B.J. Comparison of pore structures of cellulose-based activated carbon fibers and their applications for electrode materials. *Int. J. Mol. Sci.* **2022**, *23*, 3680. [CrossRef]
30. Kubicka, M.; Bakierska, M.; Chudzik, K.; Świętosławski, M.; Molenda, M. Nitrogen-doped carbon aerogels derived from starch biomass with improved electrochemical properties for Li-ion batteries. *Int. J. Mol. Sci.* **2021**, *22*, 9918. [CrossRef]
31. Xiao, Z.; Zhao, L.; Yu, Z.; Zhang, M.; Li, S.; Zhang, R.; Ayub, M.; Ma, X.; Ning, G.; Xu, C. Multilayered graphene endowing superior dispersibility for excellent low temperature performance in lithium-ion capacitor as both anode and cathode. *Chem. Eng. J.* **2022**, *429*, 132358. [CrossRef]
32. Zhang, L.; Tu, L.; Liang, Y.; Chen, Q.; Li, Z.; Li, C.; Wang, Z.; Li, W. Coconut-based activated carbon fibers for efficient adsorption of various organic dyes. *RSC Adv.* **2018**, *8*, 42280–42291. [CrossRef] [PubMed]
33. Lazar, P.; Mach, R.; Otyepka, M. Spectroscopic fingerprints of graphitic, pyrrolic, pyridinic, and chemisorbed nitrogen in N-doped graphene. *J. Phys. Chem. C* **2019**, *123*, 10695–10702. [CrossRef]
34. Yang, I.; Kim, S.G.; Kwon, S.H.; Kim, M.S.; Jung, J.C. Relationships between pore size and charge transfer resistance of carbon aerogels for organic electric double-layer capacitor electrodes. *Electrochim. Acta* **2017**, *223*, 21–30. [CrossRef]
35. Kim, Y.J.; Masutzawa, Y.; Ozaki, S.; Endo, M.; Dresselhaus, M.S. PVDC-based carbon material by chemical activation and its application to nonaqueous EDLC. *J. Electrochem. Soc.* **2004**, *151*, E199. [CrossRef]
36. Lin, R.; Taberna, P.L.; Chmiola, J.; Guay, D.; Gogotsi, Y.; Simon, P. Microelectrode study of pore size, ion size, and solvent effects on the charge/discharge behavior of microporous carbons for electrical double-layer capacitors. *J. Electrochem. Soc.* **2009**, *156*, A7. [CrossRef]
37. Zhou, L.; Cao, H.; Zhu, S.; Hou, L.; Yuan, C. Hierarchical micro-/mesoporous N- and O-enriched carbon derived from disposable cashmere: A competitive cost-effective material for high-performance electrochemical capacitors. *Green Chem.* **2015**, *17*, 2373–2382. [CrossRef]
38. Bai, Q.; Li, H.; Zhang, L.; Li, C.; Shen, Y.; Uyama, H. Flexible solid-state supercapacitors derived from biomass konjac/polyacrylonitrile-based nitrogen-doped porous carbon. *ACS Appl. Mater. Interfaces* **2020**, *12*, 55913–55925. [CrossRef]
39. Sun, F.; Wu, H.; Liu, X.; Liu, F.; Zhou, H.; Gao, J.; Lu, Y. Nitrogen-rich carbon spheres made by a continuous spraying process for high-performance supercapacitors. *Nano Res.* **2016**, *9*, 3209–3221. [CrossRef]
40. Zhu, Y.; Guo, Y.; Wang, C.; Qiao, Z.; Chen, M. Fabrication of conductive carbonaceous spherical architecture from pitch by spray drying. *Chem. Eng. Sci.* **2015**, *135*, 109–116. [CrossRef]
41. Yang, C.; Nguyen, Q.D.; Chen, T.; Helal, A.S.; Li, J.; Chang, J. Functional group-dependent supercapacitive and aging properties of activated carbon electrodes in organic electrolyte. *ACS Sustain. Chem. Eng.* **2018**, *6*, 1208–1214. [CrossRef]
42. Chang, P.; Wang, C.; Kinumoto, T.; Tsumura, T.; Chen, M.; Toyoda, M. Frame-filling C/C composite for high-performance EDLCs with high withstanding voltage. *Carbon* **2018**, *131*, 184–192. [CrossRef]
43. Zhang, Y.; Zhu, Y.; Jiao, M.; Zhang, J.; Chen, M.; Wang, C. Synthesis of size-controllable lignin-based nanospheres and its application in electrical double layer capacitors. *ChemistrySelect* **2020**, *5*, 8265–8273. [CrossRef]
44. Ghosh, S.; Barg, S.; Jeong, S.M.; Ostrikov, K. Heteroatom-doped and oxygen-functionalized nanocarbons for high-performance supercapacitors. *Adv. Energy Mater.* **2020**, *10*, 2001239. [CrossRef]
45. Samartzis, N.; Athanasiou, M.; Dracopoulos, V.; Yannopoulos, S.N.; Ioannides, T. Laser-assisted transformation of a phenol-based resin to high quality graphene-like powder for supercapacitor applications. *Chem. Eng. J.* **2022**, *430*, 133179. [CrossRef]

Article

Function of Graphene Oxide as the "Nanoquencher" for Hg^{2+} Detection Using an Exonuclease I-Assisted Biosensor

Ting Sun [1], Xian Li [1], Xiaochuan Jin [1], Ziyi Wu [1], Xiachao Chen [2] and Jieqiong Qiu [1,*]

1. College of Life Sciences and Medicine, Zhejiang Sci-Tech University, Hangzhou 310018, China; z1178390272@163.com (T.S.); adadallen@163.com (X.L.); jxc20012022@163.com (X.J.); ziyiwu1999@163.com (Z.W.)
2. School of Material Sciences & Engineering, Zhejiang Sci-Tech University, Hangzhou 310018, China; chenxiachao@zstu.edu.cn
* Correspondence: qiujieqiong@zstu.edu.cn or qiujieqiongqjq@163.com; Tel.: +86-0571-86843192

Abstract: Graphene oxide is well known for its excellent fluorescence quenching ability. In this study, positively charged graphene oxide (pGO25000) was developed as a fluorescence quencher that is water-soluble and synthesized by grafting polyetherimide onto graphene oxide nanosheets by a carbodiimide reaction. Compared to graphene oxide, the fluorescence quenching ability of pGO25000 is significantly improved by the increase in the affinity between pGO25000 and the DNA strand, which is introduced by the additional electrostatic interaction. The FAM-labeled single-stranded DNA probe can be almost completely quenched at concentrations of pGO25000 as low as 0.1 μg/mL. A simple and novel FAM-labeled single-stranded DNA sensor was designed for Hg^{2+} detection to take advantage of exonuclease I-triggered single-stranded DNA hydrolysis, and pGO25000 acted as a fluorescence quencher. The FAM-labeled single-stranded DNA probe is present as a hairpin structure by the formation of T–Hg^{2+}–T when Hg^{2+} is present, and no fluorescence is observed. It is digested by exonuclease I without Hg^{2+}, and fluorescence is recovered. The fluorescence intensity of the proposed biosensor was positively correlated with the Hg^{2+} concentration in the range of 0–250 nM (R^2 = 0.9955), with a seasonable limit of detection (3σ) cal. 3.93 nM. It was successfully applied to real samples of pond water for Hg^{2+} detection, obtaining a recovery rate from 99.6% to 101.1%.

Keywords: positively charged graphene oxide (pGO); exonuclease I; fluorescence quencher; hairpin structure; T–Hg^{2+}–T

1. Introduction

Water-soluble mercury(II) ion (Hg^{2+}), as one of the most familiar environmental pollutants, is a toxic heavy metal that can exist in metallic, inorganic, and organic forms, especially in freshwater and marine ecosystems [1]. After prolonged exposure, it is extremely toxic to the brain, kidney, and other organs of organisms at very low mercury(II) concentrations [2]. The accumulation of heavy metals can occur in animal and human bodies via the food chain and damage the reproductive, gastrointestinal, and cardiovascular systems. Based on the guidelines of the United States Environmental Protection Agency (EPA), the Ministry of Health of the People's Republic of China (MOH), and the World Health Organization (WTO), the maximum mercury(II) concentration in drinking water should be as low as 10 nM [3].

Currently, many traditional techniques have been developed for Hg^{2+} analysis and detection, including inductively coupled plasma–optical emission spectrometry (ICP-OES) [4], chemical vapor generation–inductively coupled plasma–optical emission spectrometry (CVG-ICP-OES) [5], inductively coupled plasma–mass spectrometry (ICP-MS) [6], cold vapor atomic absorption spectroscopy (CVAAS) [7], atomic fluorescent spectroscopy [8], and electrochemical methods. Hg^{2+} can be detected at pM concentrations by most of the abovementioned methods. However, high-cost, complex sample preparation and professional operation are also needed. Therefore, it is essential to explore rapid, specific, sensitive, cost-efficient, convenient, and real-time biosensors instead of traditional approaches for monitoring heavy metal ions.

Fluorescence-based methods have been widely used as potential techniques for Hg^{2+} detection. In view of the sensitivity improvement for Hg^{2+} detection, an increasing number of researchers have paid attention to signal amplification in DNA-based strategies by exonucleases, including exonuclease I and III (Exo I and III) [9–13]. For example, Exo I is a 3′–5′ exonuclease that can cleave single-stranded DNA (ssDNA) without sequence-dependence [14,15]. Hg^{2+} can promote the formation of a DNA duplex (dsDNA) via T–Hg^{2+}–T formation, which is not allowed to be digested by Exo I. However, Exo I can hydrolyze ssDNA. Thus, there is potential for applying Exo I in Hg^{2+} detection on the basis of DNA-based signal amplification strategies.

However, traditional fluorescent DNA probes, such as Taqman probes, molecular beacons (MBs), and scorpions, cannot meet the requirements [16,17] because no free 3′-OH is present. For this reason, a label-free assay has been developed based on the fluorescence "turn-on" caused by dye intercalation into special DNA structures [18–20]. However, such label-free dyes have a non-negligible fluorescent background.

To solve these problems, guanine bases [16] and nanostructures [21] have been used as quenchers for DNA probes. Various nanostructures, such as gold nanoparticles (AuNPs) [22,23], single-walled carbon nanotubes (SWCNTs) [24], graphene oxide (GO) [25], fullerene (C_{60}) [26,27], multiwalled carbon nanotubes (MWCNTs) [9], and positive carbon dots (P-CDs) [28], have been successfully used as nanoquenchers for mercury(II) ion detection. As a two-dimensional (2D) material, GO exhibits high-efficiency fluorescence quenching, good water dispersibility, low cost, and various surface modifications. Therefore, it is frequently used in biosensors [29]. Previous studies have confirmed that ssDNA is labelled with a fluorescent dye, which can be quenched by GO due to fluorescence resonance energy transfer (FRET) [12,30]. This result is attributed to the hydrogen bond and π–π stacking caused by nucleobases and GO, which make FRET more efficient. The fluorescence quenching efficiency of GO is dependent on the GO quantity used. A high concentration of GO can limit its application in Hg^{2+} detection, e.g., in cells. Therefore, it is necessary to improve the fluorescence quenching ability and efficiency of GO. In addition, the fluorescence quenching efficiency of GO can be increased by partially reducing graphene oxide due to the increase in π–π stacking interaction [25].

In this work, positively charged graphene oxide (pGO25000) was synthesized by grafting polyetherimide (PEI) onto GO nanosheets by a carbodiimide reaction. The first use of positively charged pGO25000 as an efficient fluorescence quencher was demonstrated. Compared to GO, the fluorescence quenching efficiency of pGO25000 can be enhanced by the positively charged surface that allows attraction of the negatively charged DNA strands via electrostatic interaction. Based on the special property of pGO25000, a FAM–ssDNA probe was designed for the highly selective and ultrasensitive detection of mercury(II) ions with the assistance of the Exo I enzyme under mild conditions.

2. Results and Discussion

2.1. Strategy for Ultrasensitive Detection of Hg^{2+}

A FAM–ssDNA probe was designed for the highly sensitive and highly selective detection of Hg^{2+}, with pGO25000 being a fluorescence nanoquencher and Exo I being a special enzyme for hydrolyzing ssDNA in the 3′→5′ direction, as shown in Scheme 1. GO, as a fluorescence quencher, can quench fluorescent dye via FRET when GO and fluorescent dye are sufficiently close to each other. Considering the binding affinity of ssDNA to GO, which results from π–π stacking and hydrogen bonding between ssDNA and GO, the dye-labelled ssDNA probes were designed for Hg^{2+} detection based on the GO fluorescence quenching ability. Positively charged GO (PEI-GO) has been reported as a fluorescence quencher of anionic dyes (i.e., Merocyanine 540) via electrostatic interactions [31]. However, dye-labelled DNA probes have never been reported as fluorescence quenchers. In this work, pGO25000 was synthesized by grafting PEI (M.W. = 25,000) onto GO nanosheets, which can selectively bind to ssDNA/dsDNA at very low concentrations. Thus, the fluorescence quenching efficiency was enhanced based on the additional electrostatic attraction between the phosphate group and positively charged PEI. The fluorescence was almost quenched for the FAM–ssDNA probe by pGO25000; however, FAM–dsDNA can be also quenched by pGO25000. To solve this problem, enzyme-based technology was applied in the FAM–ssDNA/pGO25000 system. Exo I is a sequence-independent 3′–5′ exonuclease that cleaves ssDNA. It has been reported that the digestion of Exo I is limited by binding to the targets to form the DNA duplex and G-quadruplex structures [11,14,15,32]. In this study, pGO25000 was synthesized as an efficient nanoquencher of fluorescence for the proposed strategy to detect Hg^{2+}. After adding Exo I, the special enzyme efficiently digested the FAM–ssDNA in the direction of 3′ to 5′, and the fluorescence was restored. However, if dsDNA was present because of the formation of the T–Hg^{2+}–T construct after adding Hg^{2+}, which could suppress the activity of Exo I, no fluorescence was restored. Therefore, the fluorescence "turn-on" indicated that no Hg^{2+} was present in the analytical sample, and vice versa. It is expected that this strategy could provide a novel method to detect Hg^{2+} with great sensitivity and high selectivity.

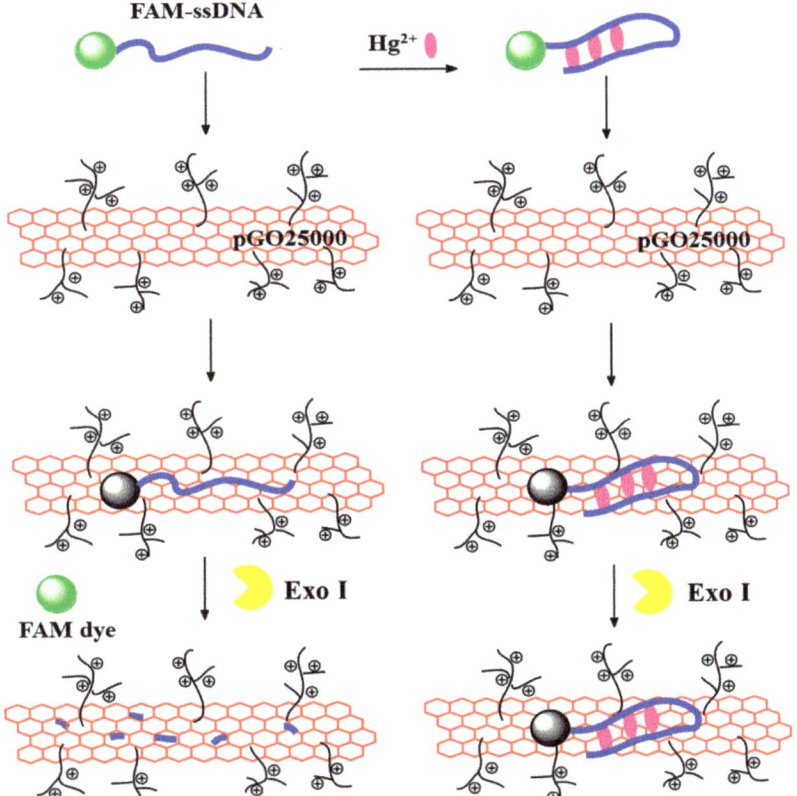

Scheme 1. Schematic diagram of the designed fluorescence "turn-off" strategy for Hg^{2+} detection with the assistance of Exo I nuclease using a FAM–ssDNA probe.

2.2. Characterization of pGO25000

GO has good water solubility due to the abundance of hydrophilic groups (hydroxyl, carboxylic, epoxy) that have been introduced onto the surface of GO after a series of chemical modification processes. Because pGO25000 was prepared by grafting PEI onto GO nanosheets, pGO25000 also has good dissolvability. To study the surface charge of GO and pGO25000, zeta potential analysis was performed at concentrations of 1 mg/mL GO and pGO25000 solution. The GO solution showed a negative zeta potential level of −37.6 Mv, as shown in Figure 1, while pGO25000 had a positive zeta potential level of 25 mV because the PEI linkers completely changed the pGO25000 surface charge; thus, positively charged GO (pGO25000) was obtained [33]. To investigate the structural change in the condensation reaction of pGO25000 synthesis, Raman spectroscopy, FT-IR analysis, and high-resolution XPS were performed. Figure 2A shows the Raman spectra of GO and pGO25000, and two bands located at approximately 1320 cm^{-1} and 1596 cm^{-1} can be attributed to the D and G bands of graphitic materials, respectively. It is well known that the defect level of graphene sheets can be evaluated by the peak intensity ratio of the D band to the G band (I_D/I_G), and a higher I_D/I_G commonly indicates an increase in the degree of disorder [34]. pGO25000 gave a higher I_D/I_G ratio of 2.30 compared with GO (1.85), which can be attributed to the condensation reaction by incorporation of PEI, reducing the oxygen functional group and increasing the sp^3 carbon form [35–37]. Figure 2B shows the FT-IR spectra of GO and pGO25000. Peaks located at ~1720 cm^{-1}, ~1620 cm^{-1}, ~1400 cm^{-1}, and ~1090 cm^{-1} can be assigned to the stretching vibrations of the C=O, C–C, C–OH, and C–O (epoxy) groups [35,38,39]. Of note, the FT-IR spectrum of pGO25000 showed that PEI was successfully grafted onto the GO surface. Compared to GO, the N–C=O peak at 1650 cm^{-1} appeared with the disappearance of the C=O peak at 1720 cm^{-1} in pGO25000. Meanwhile, the C–O (epoxy) peak was replaced by the C–N peak (1384 cm^{-1}) on pGO25000. The N–C=O and C–N groups were produced by the amine reacting with the COOH and C–O (epoxy) groups. The band at 1580 cm^{-1} appeared first, which corresponded to the C=N stretch by Schiff's base reaction [40,41].

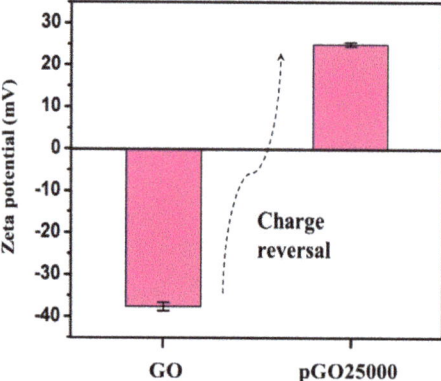

Figure 1. Charge analysis of GO and pGO25000.

As shown in Figure 3A, there was almost no N1s signal in the spectrum of GO, whereas the spectrum of pGO25000 presented a clear N1s peak. After calibration of the binding energy position with C1s (284.4 eV) in XPS spectra, the five main peaks of carbon bonding in the C1s XPS spectra of GO with binding energies at 283.7, 284.4, 286.1, 286.9, and 288.3 eV (Figure S1A) were attributed to the C=C, C–C, C–O (hydroxyl and epoxy), C=O, and C(O)O bonds, respectively [25,42]. After reacting with PEI, the signal at 285.3 eV (C–N bond) appeared along with the disappearance of the C(O)O bond signal, which indicated that the condensation reaction between the amino group and carboxyl group was completed. The peak at 286.0 eV (C–O) was dramatically decreased due to the epoxy reacting with PEI (Figure 3B). The N1s spectrum had fitted curves at 400.4, 399.1, and 398.2 eV (Figure 3C), corresponding to the binding energies of nitrogen atoms in NH_3^+, CONH, and PEI [40,43,44]. Compared to the O1s spectrum of GO (Figure S1B), pGO25000 was deconvoluted into four peaks (Figure 3D), three of which were similar to those of GO, i.e., C=O (530.5 eV), C–OH (531.7 eV), and C–O (532.4 eV) [43]; a new peak with a bonding energy of 530.9 eV appeared, corresponding to the CONH bond. These results indicated that pGO25000 was successfully obtained by GO reacting with PEI.

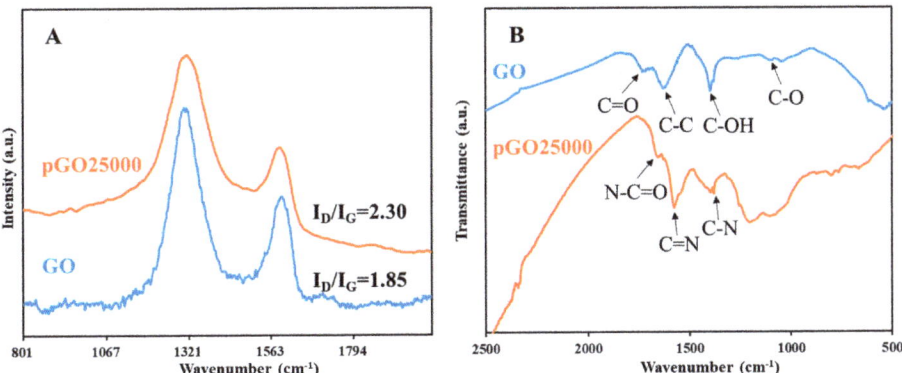

Figure 2. Characterization of GO and pGO25000 by (**A**) Raman spectroscopy and (**B**) FT-IR.

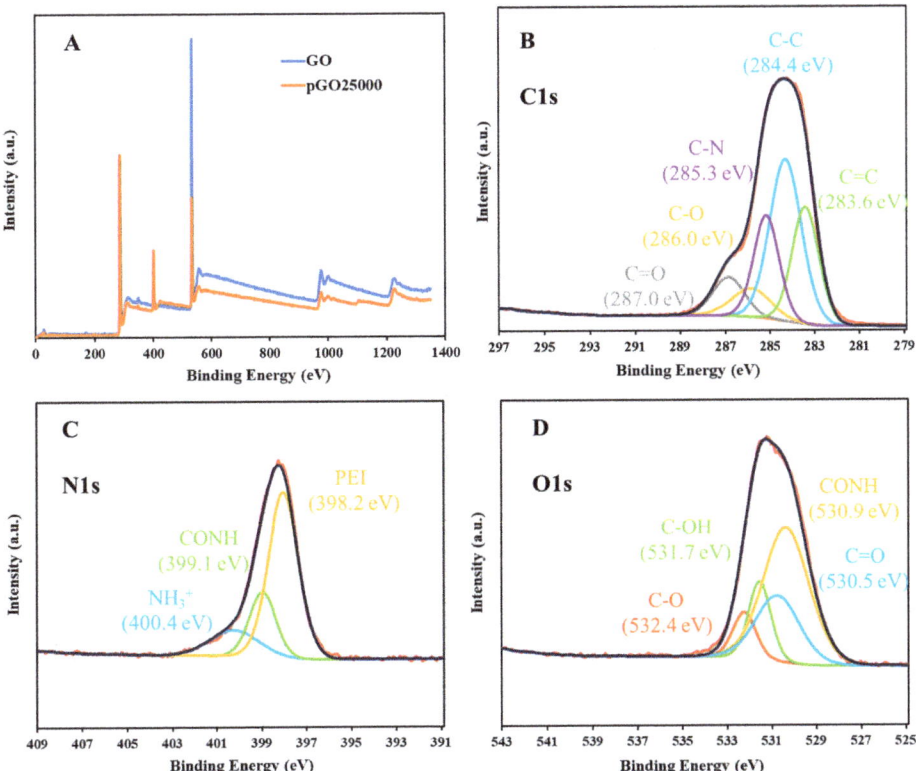

Figure 3. High-resolution XPS spectra of GO and pGO25000: (**A**) wide scan, (**B**) C1s spectrum of pGO25000, (**C**) N1s spectrum of pGO25000, and (**D**) O1s spectrum of pGO25000.

2.3. Fluorescence Quenching of FAM–ssDNA by pGO25000

To better understand the fluorescence quenching efficiency of positively charged GO depending on the pH values, a solution with a pH range of 7.5–9.0 was investigated. The fluorescence intensity of FAM–ssDNA in a 600 nM Hg^{2+} solution significantly increased and then slightly decreased with increasing pH values, as shown in Figure S2A. At pH 8.5, the maximum fluorescence signal was obtained. After adding 0.1 μg/mL pGO25000, there was a sharp reduction in fluorescence intensity due to fluorescence quenching caused by pGO25000, and there was no large difference under various pH values from 7.5 to 9.0. Hence, pH 8.5 was the optimum pH according to the ratio of the fluorescence intensity without pGO25000 and in the presence of pGO25000. It has been mentioned that fluorescence quenching by pGO25000 can be completed immediately; thus, fluorescence detection was instantly performed after adding pGO25000.

To demonstrate that the positively charged GO (pGO25000) is more efficient in fluorescence quenching for DNA probes, GO and pGO25000 were analyzed. The fluorescence of FAM–ssDNA was quenched by various concentrations of GO from 0.01 to 30 μg/mL, as shown in Figure S3B. As the amount of GO increased, the fluorescent signal was reduced, and fluorescence quenching was not efficient even if the concentration was enhanced to 30 μg/mL. However, the fluorescence was almost quenched by pGO25000 at 0.1 μg/mL (Figure 4). Thus, positively charged PEI plays a crucial role in the affinity between pGO25000 and FAM–ssDNA, which can promote the fluorescent dyes to be close to pGO25000, thus increasing the FRET efficiency to quench fluorescence. The fluorescence spectra of FAM–ssDNA quenched by pGO25000 in the range from 0 to 30 μg/mL with excitation at 495 nm are shown in Figure 4 and Figure S3A. The signal intensity of the FAM–ssDNA probes was moderately reduced with increasing pGO25000 concentration, clearly increased with a pGO25000 concentration higher than 0.1 μg/mL, and then decreased again until the concentration of pGO25000 was greater than 6 μg/mL. The fluorescence quenching ability of pGO25000 was induced by electrostatic and π–π stacking interactions, which were dependent on the concentration of pGO25000. This result indicated that pGO25000 had a better binding affinity with the FAM–ssDNA probes at very low concentrations from 0 to 0.1 μg/mL due to electrostatic interactions. As pGO25000 increased from 0.1 μg/mL to 6 μg/mL, fluorescence quenching was not efficient. This was possibly caused by the steric hindrance of PEI in pGO25000, which impeded the interaction of FAM–ssDNA with GO in pGO25000, reducing π–π stacking and hydrogen bonding between nucleobases and GO. However, the fluorescence quenching efficiency was improved when the pGO25000 concentration was more than 6 μg/mL. This phenomenon was observed because the electrostatic interaction between FAM–ssDNA and pGO25000 was increased, and fluorescence quenching was mainly dependent on the electrostatic interaction.

The influence of various metal ions on the fluorescence quenching ability of pGO25000 was also assessed by measuring the fluorescence intensity, and different metal ions were used, including Hg^{2+}, K^+, Sn^{2+}, Al^{3+}, Ni^{2+}, Mn^{2+}, Mg^{2+}, Cu^{2+}, and Co^{2+}. Figure S4 shows that there was no influence on the fluorescence quenching ability. Therefore, pGO25000 was synthesized as an efficient nanoquencher of fluorescence.

2.4. Fluorescence Detection of Hg^{2+}

Fluorescence quenching efficiency is dependent on the interaction between DNA and GO, which is determined by the length of DNA [45,46], GO surface modification [25], size of GO [30], and concentrations of DNA and GO [47]. Compared to GO, the positively charged modified GO (pGO25000) had a perfect fluorescence quenching ability at very low concentrations. However, there was no large change in the fluorescence quenching of FAM–ssDNA by 0.1 μg/mL pGO25000 with or without Hg^{2+}. However, FAM–ssDNA exists in the hairpin structure due to the formation of T–Hg^{2+}–T after adding Hg^{2+}. It was indicated that fluorescence quenching was efficient for the same sequences of DNA with different structures. Based on this result, an Exo I-assisted strategy to detect Hg^{2+} is proposed.

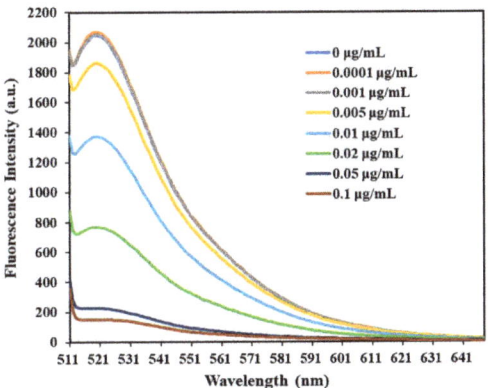

Figure 4. Effect of pGO25000 concentrations from 0 to 0.1 µg/mL on fluorescence quenching of FAM–ssDNA.

To demonstrate the effect of pH values on T–Hg^{2+}–T complex formation and the activity of Exo I, the Exo I/pGO25000-assisted FAM–ssDNA sensor was studied for Hg^{2+} detection, as demonstrated in Figure S2B. The fluorescent signal was significantly improved with increasing pH values in the absence of Hg^{2+}. It is well known that FAM is a pH-dependent dye, and the optimal pH value is >8.5 [48,49]. FAM–ssDNA can be digested by the 3′–5′ Exo I and releases FAM dyes in a range of pH 7.5~9.0; thus, the fluorescence response can be recovered without Hg^{2+}. When Hg^{2+} was present, all samples at various pH values were fluorescence-quenched except for the conditions at pH 7.5 and 9.0. Thus, the FAM–ssDNA probe could be subjected to the conditions at pH 8.0 and 8.5. The fluorescence signal was not restored because of the FAM–hairpin DNA structure formation, which was caused by the T–Hg^{2+}–T construction, which prevented digestion by Exo I. However, at pH 7.5 and 9.0, the fluorescence was slightly recovered, and the reason could be attributed to the overactivity of Exo I under these conditions. As a result, 10 mM Tris–HNO_3 buffer (40 mM $NaNO_3$) with a pH value of 8.5 was used during Hg^{2+} detection.

2.5. Sensitivity of Hg^{2+} Detection

The Hg^{2+} concentration has a large effect on the FAM–ssDNA probe with the pGO25000/Exo I-assisted strategy. The sensitivity of this proposed method was determined using various concentrations of Hg^{2+} solution. Figure 5A shows that the emission signal was gradually reduced by excitation at 495 nm with increasing Hg^{2+} concentrations from 0 to 800 nM, and fluorescence was nearly quenched when the concentration of Hg^{2+} was greater than 600 nM. In a certain range, the higher the concentration of Hg^{2+}, the more efficient the fluorescence quenching. This occurred because Exo I activity was restricted by the DNA hairpin structure, which was formed by the T–Hg^{2+}–T complexes, and the FAM dye could not be released from the DNA strand, which was quenched by pGO25000. Without Hg^{2+} added, the fluorescence signal was perfectly recovered after digestion by Exo I. It was attributed to the destruction of the interaction between the DNA strand and pGO25000 during the hydrolysis of DNA; thus, the FAM dye was released from the DNA strand and kept far away from pGO25000; then, the fluorescence was restored.

The relative fluorescence intensity (F/F_0) decreased proportionally as the Hg^{2+} concentration increased. It showed excellent analytical performance with a linear relationship in the range of 0 to 250 nM Hg^{2+}, following a linear correlation equation described as y = −0.0031x + 0.9804 (y represents F/F_0, x represents the concentration of Hg^{2+} in solution, R^2 = 0.9955). Based on the 3σ slope, the limit of detection (LOD) for the Exo I/pGO25000-assisted FAM–ssDNA sensor was estimated to be 3.93 nM, which was far below the largest permissible dose of Hg^{2+} in potable water (10 nM) by the U.S. Environmental Protection Agency (EPA) [21,50]. The obtained result indicates that this biosensor strategy has potential applications in the quantitative analysis of Hg^{2+} at certain concentrations. Various nanomaterials were used as the nanoquencher of DNA probes, which were designed for Hg(II) detection, and the results are shown in Table 1. The sensitivity and fluorescence quenching efficiency of the nanoquenchers were compared using the previously presented analytical methods. Note that the proposed scheme presented a lower LOD and higher sensitivity compared with that using GO, SWCNTs, or MWCNTs as the nanoquencher, and pGO25000 had a more efficient fluorescence quenching ability compared to C_{60}.

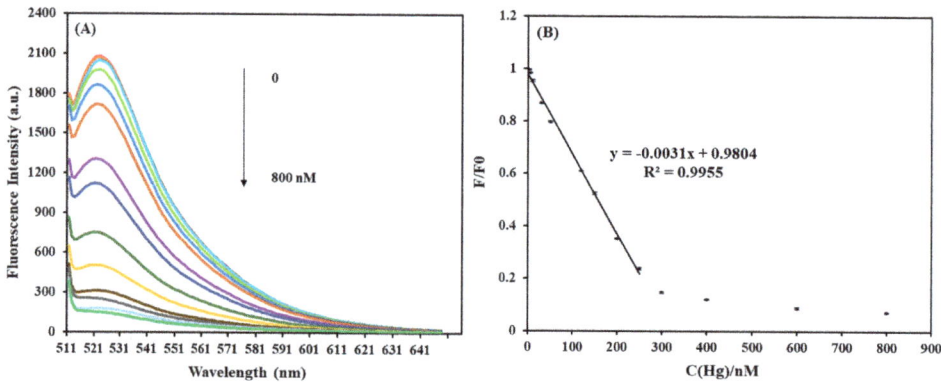

Figure 5. (**A**) Fluorescence emission spectra upon addition of various Hg^{2+} concentrations ranging from 0 to 800 nM in Tris–HNO$_3$ buffer (10 mM, pH 8.5) containing 40 mM NaNO$_3$. The Hg^{2+} concentrations were as follows: 0, 2, 5, 10, 30, 50, 120, 150, 200, 250, 300, 400, 600, and 800 nM. (**B**) Linear response of the relative fluorescence intensity (F/F$_0$, F: fluorescence-detected, with various Hg^{2+} concentrations from 0 to 800 nM, F$_0$: fluorescence-initial, without Hg^{2+}). An amount of 50 nM FAM–ssDNA was used for each reaction, and the reactions were performed in Tris–HNO$_3$ buffer (10 mM, 40 mM NaNO$_3$, pH 8.5). Ex = 495 nm, Em = 520 nm.

Table 1. Comparison of different nanomaterials as nanoquencher for Hg(II) detection.

Nanomaterials	Fluorescence Quenching Ability (µg/mL·nM)	Quenching Efficiency	Liner Range (µM)	LOD (nM)	Reference
GO	0.4 µg/mL·nM	90%	0–1.0	/	[9]
SWCNTs	0.4 µg/mL nM	72%	0.05–8.0	14.5	[24]
MWCNTs	0.2 µg/mL nM	91%	0–0.5	/	[9]
C_{60}	3.0 µg/mL nM	88.4%	0.03–0.15	0.5	[26]
pRGO$_3$	0.075 µg/mL nM	97%	/	/	[25]
P-CDs	0.125 µg/mL nM	79~84%	/	/	[28]
pGO25000	0.002 µg/mL nM	95.5%	0–0.25	3.93	This work

2.6. Selectivity of Hg^{2+}

Confirming the selectivity for Hg^{2+} is the key point to evaluate the performance of the developed DNA biosensor. To estimate the selectivity of Exo-I/pGO25000 by the FAM–ssDNA sensing system, experiments were performed to detect the fluorescence intensities of the Exo I/pGO25000-assisted FAM–ssDNA probes in the presence or absence of Hg^{2+} (600 nM) solution mixed with other metal ions (K^+, Fe^{2+}, Sn^{2+}, Al^{3+}, Ni^{2+}, Mn^{2+}, Mg^{2+}, Cu^{2+}, and Co^{2+}) at a concentration of 6 mM, as shown in Figure 6. Hg^{2+} produced a remarkable decrease in fluorescence intensity, indicating that only Hg^{2+} could bind to two thymine bases, and T–Hg^{2+}–T mismatched base pairs formed, resulting in the formation of a hairpin structure; thus, degradation by Exo I was hindered. As a result, the FAM–

hairpin DNA remained for fluorescence quenching by pGO25000. The obtained results also revealed that the proposed method could still detect Hg^{2+}, which mixed with other probable interference metals. Clearly, the biosensor based on FAM–ssDNA probes and Exo I/pGO25000 has an excellent selectivity towards Hg^{2+}.

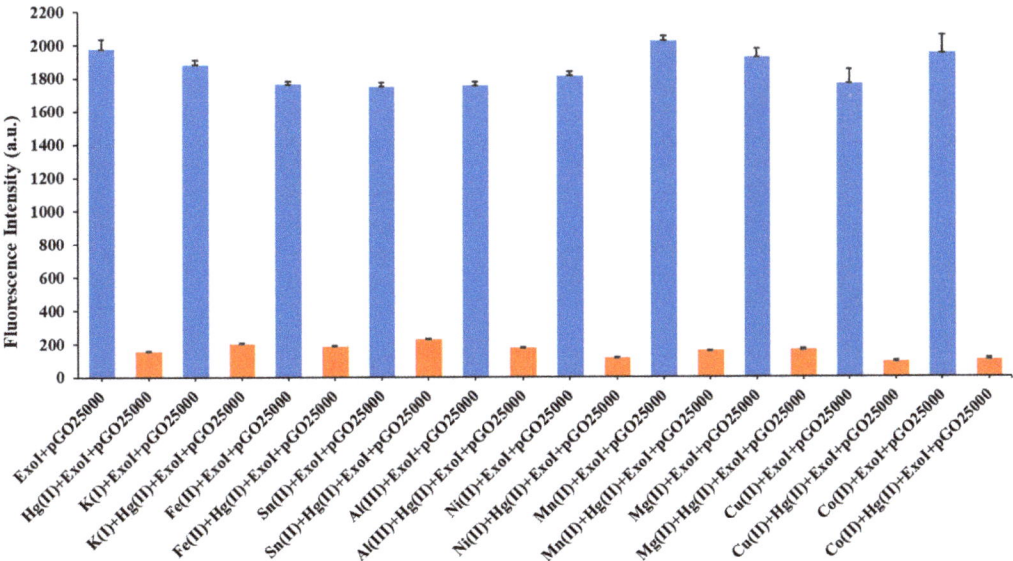

Figure 6. Selectivity of Hg^{2+} detection by the FAM–ssDNA probe. The fluorescence intensities of the Exo I (2 U)/pGO25000 (0.1 μg/mL)-assisted FAM–ssDNA probes were recorded without or with Hg^{2+} (600 nM) and mixed with other possible interfering metal ions (K^+, Fe^{2+}, Sn^{2+}, Al^{3+}, Ni^{2+}, Mn^{2+}, Mg^{2+}, Cu^{2+}, and Co^{2+}, 6 mM). $Ex = 495$ nm, $Em = 520$ nm. Each sample was repeated three times.

2.7. Application in Real Samples

To assess the approach feasibility, the Exo I/pGO25000-assisted FAM–ssDNA biosensor strategy was used for the Hg^{2+} analysis of pond water samples collected from a pond at Zhejiang Sci-Tech University. The impurities of pond water were removed by filtration using NY 0.22 μm, and three concentrations of Hg^{2+} (25 nM, 50 nM, 200 nM) were diluted by the pond water, which were individually detected by the sensor system. The recoveries of Hg^{2+} in pond water were in the range of 99.6%–101.1%, and all relative standard deviations (RSDs) were as low as 5% ($n = 3$) (Table 2). It is revealed that the quantitative detection of Hg^{2+} can be performed by this sensitive biosensor method. Meanwhile, it illustrated that the proposed sensor has good feasibility and accuracy to measure the Hg^{2+} concentration in real samples.

Table 2. Average recoveries of Hg^{2+} in the pond water samples ($n = 3$).

Added/nM	Repeats	F/F_0	Calculated Value/nM	Avg./nM	Recovery
	1	0.900	25.9		
25	2	0.904	24.6	24.9	99.6%
	3	0.905	24.3		
	1	0.819	52.1		
50	2	0.825	50.1	49.9	99.8%
	3	0.833	47.5		
	1	0.340	206.6		
200	2	0.357	201.1	202.2	101.1%
	3	0.364	198.8		

3. Experimental

3.1. Chemicals and Materials

The oligonucleotide (FAM–ssDNA probe: 5′-FAM-TA<u>T</u>CG<u>T</u>GC<u>T</u>CCCC<u>T</u>GC<u>T</u>CG<u>T</u>TA) was purified by HPLC and purchased from Sangon Biotech Co., Ltd. (Shanghai, China). The oligonucleotide stock solution (50 µM) was prepared with double distilled water (ddH$_2$O). Exonuclease I (Exo I, 20 U/µL) was provided by Bio Basic Inc. (Canada). A 10 mM Tris–HNO$_3$ buffer including 40 mM NaNO$_3$ (pH 8.5) was used in the experiments. Hg^{2+} solutions with different concentrations were prepared from a standard mercury ion solution (5.0 mM), which was purchased from Aoke Biology Research Co., Ltd. (Beijing, China). All metal salts and tris(hydroxymethyl)aminomethane (Tris) were obtained from Macklin (Shanghai, China) and Aladdin (Beijing, China). Graphene oxide (GO) was supplied by XFNano Material Tech Co., Ltd. (Nanjing, China). PEI (average MW = 25,000) was purchased from Sigma-Aldrich. All other chemicals and reagents used in this work were of analytical grade and used without further purification. The pond water was prefiltered with an NY 0.22 µm filter.

3.2. Apparatus

Fluorescence spectra were measured by an F-4600 fluorescence spectrometer (Hitachi, Japan) at RT (room temperature, approximately 27 °C). The emission spectra data were acquired from 511 nm to 620 nm after illuminating at the maximum excitation wavelength of 495 nm, and the fluorescence intensity was measured at 520 nm (Em$_{max}$). Both the excitation and emission slit widths were set at 5 nm and 10 nm, respectively. High-resolution X-ray photoelectron spectroscopy (XPS) was performed using a Thermo Scientific K-Alpha XPS spectrometer (Kratos Analytical, Manchester, UK). The charge polarity and density of GO/pGO25000 colloids were obtained by zeta potential measurement (Malvern Zetasizer Nano ZS90, Great Malvern, UK). The Raman spectrum was recorded by an Optosky ATP3007 (Xiamen, China) Raman spectrometer with a 785 nm excitation lase. A Nicolet iS50 spectrometer (Thermo Fisher Scientific, Waltham, USA) using KBr pellets was used for Fourier-transform infrared (FT-IR) characterization.

3.3. Preparation of pGO25000

Positively charged GO (pGO25000) was synthesized by grafting PEI (MW = 25,000) onto GO nanosheets via a condensation reaction between amino groups and carboxyl groups. Briefly, 120 mg of GO, 1.0 g of PEI, and 300 mg of EDC were sequentially dissolved in 40 mL of ddH$_2$O. Then, the pH value of this mixture was adjusted to pH 7.0 by adding a certain amount of diluted HCl (1.0 M) and stored at 4 °C for 24 h. Then, the mixture was dialyzed against ddH$_2$O for two weeks to remove foreign ions. After diluting this mixture with ddH$_2$O and ultrasonication, the pGO solution (1 mg/mL) was successfully prepared.

3.4. Fluorescence Quenching Assay

A 50 nM FAM–ssDNA probe was incubated in 200 µL of Tris–HNO$_3$ buffer (10 mM, 40 mM NaNO$_3$, pH 8.5) with different concentrations of pGO25000/GO (0, 0.0001, 0.001, 0.005, 0.01, 0.02, 0.05, 0.1, 1, 3, 6, 15, and 30 µL/mL). The mixtures were immediately measured by a fluorescence spectrometer, and the fluorescence emission was monitored in the wavelength range from 511 to 620 nm. Each sample solution was repeated and measured at least three times.

3.5. Study of Exo I Activity in Hg^{2+} Detection

A 0.01 nmole FAM–ssDNA probe (50 µM, 0.2 µL) and 1 µL of Hg^{2+} solution (600 nM) were added to a 0.5 mL centrifuge tube, followed by dilution to 10 µL with Tris–HNO$_3$ buffer (10 mM, 40 mM NaNO$_3$, pH 8.5). The FAM–ssDNA probe/Hg^{2+} solution was added to 2 U of Exo I and incubated for 30 min at 27 °C (RT) or 5 min at 40 °C in an oven. Then, pGO25000 (0.1 µL/mL) was mixed with the FAM–ssDNA probe/Hg^{2+}/Exo I mixture and diluted with Tris–HNO$_3$ buffer (10 mM, 40 mM NaNO$_3$, pH 8.5) to a final volume of 200 µL.

Exo I nuclease was denatured by heating at 80 °C for 15 min before the fluorescence test. To evaluate the sensitivity of Hg^{2+} detection, the final concentrations of Hg^{2+} were 0, 2, 5, 10, 30, 50, 120, 150, 200, 250, 300, 400, 600, and 800 nM. To evaluate the selectivity of Hg^{2+} detection, 6 μM K^+, Na^+, Cu^{2+}, Mn^{2+}, Ni^{2+}, Pb^{2+}, and Fe^{3+} and 600 nM Hg^{2+} were used. The fluorescence emission was analyzed from 511 to 620 nm for all samples with an excitation wavelength of 495 nm, and the maximum fluorescence intensity was measured at 520 nm. To assess the application of the DNA sensor in the real samples, a 50 nM FAM–ssDNA probe was incubated in 10 μL of Tris–HNO_3 buffer (10 mM, 40 mM $NaNO_3$, pH 8.5) with three different concentrations of Hg^{2+} solution (25 nM, 50 nM, 200 nM), which were prepared from the pond water. Each sample solution was analyzed three times.

4. Conclusions

In summary, positively charged GO (pGO25000) was synthesized by modification of the GO surface with PEI and used in fluorescence quenching of a DNA probe. Compared to GO, the fluorescence quenching efficiency of pGO25000 was dramatically improved due to the additional electrostatic interaction induced by PEI. Electrostatic attraction plays a vital role in the interaction between pGO25000 and DNA strands, which increases the affinity of pGO25000 to the DNA strands. As a result, when pGO25000 is at a very low concentration (0.1 μg/mL), it possesses a higher and more efficient fluorescence quenching ability compared to GO. In view of the perfect fluorescence quenching efficiency of pGO25000, the pGO25000 and FAM–ssDNA probes were designed for Hg^{2+} detection with the assistance of Exo I, and fluorescence was specifically prevented by adding Hg^{2+}. The FAM–ssDNA probe was formed in a hairpin structure in the presence of Hg^{2+} due to the formation of T–Hg^{2+}–T complexes, which imposed restrictions on the degradation by Exo I. However, the fluorescence was recovered without Hg^{2+} because of the hydrolysis of FAM–ssDNA caused by Exo I. Therefore, the limit of detection of 3.93 nM was obtained. Compared to other metal ions, the Exo I/pGO25000-assisted FAM–ssDNA sensor has a great selectivity for Hg^{2+}. In addition, the designed strategy is applicable for Hg^{2+} detection in real samples with satisfactory recoveries. In consideration of the efficient fluorescence quenching property of pGO25000, as well as the absolute quantification ability by the Exo I-assisted FAM–ssDNA sensor, the biosensor mechanism can be applied in more toxic substances and in gene mutation analysis (e.g., SNP).

Supplementary Materials: The following supporting information can be downloaded at: https://www.mdpi.com/article/10.3390/ijms23116326/s1.

Author Contributions: Conceptualization, J.Q.; methodology, X.J. and Z.W.; software, X.L.; validation, T.S. and X.L.; formal analysis, X.J.; investigation, Z.W.; resources, X.L.; data curation, Z.W.; writing—original draft preparation, T.S.; writing—review and editing, J.Q. and X.C.; visualization, J.Q.; supervision, J.Q.; project administration, J.Q.; funding acquisition, J.Q. All authors have read and agreed to the published version of the manuscript.

Funding: This research was funded by Zhejiang Sci-Tech University grant number 16042017-Y.

Institutional Review Board Statement: Not applicable.

Informed Consent Statement: Not applicable.

Data Availability Statement: Not applicable.

Acknowledgments: We thank all members of the Xiachao Chen Laboratory for many productive discussions.

Conflicts of Interest: The authors declare no conflict of interest.

References

1. Boening, D.W. Ecological effects, transport, and fate of mercury: A general review. *Chemosphere* **2000**, *40*, 1335–1351. [CrossRef]
2. Valko, M.; Morris, H.; Cronin, M.T.D. Metals, toxicity and oxidative stress. *Curr. Med. Chem.* **2005**, *12*, 1161–1208. [CrossRef]
3. Duan, J.L.; Zhan, J.H. Recent developments on nanomaterials-based optical sensors for Hg^{2+} detection. *Sci. China-Mater.* **2015**, *58*, 223–240. [CrossRef]
4. Covaci, E.; Senila, M.; Tanaselia, C.; Angyus, S.B.; Ponta, M.; Darvasi, E.; Frentiu, M.; Frentiu, T. A highly sensitive eco-scale method for mercury determination in water and food using photochemical vapor generation and miniaturized instrumentation for capacitively coupled plasma microtorch optical emission spectrometry. *J. Anal. At. Spectrom.* **2018**, *33*, 799–808. [CrossRef]
5. Baika, L.M.; Dos Santos, E.J.; Herrmann, A.B.; Grassi, M.T. Simultaneous determination of As, Hg, Sb, and Se in mineral fertilizers using ultrasonic extraction and CVG-ICP OES. *Anal. Methods* **2016**, *8*, 8362–8367. [CrossRef]
6. Xing, Y.Q.; Han, J.; Wu, X.; Pierce, D.T.; Zhao, J.X. Aggregation-based determination of mercury(II) using DNA-modified single gold nanoparticle, T-Hg(II)-T interaction, and single-particle ICP-MS. *Microchim. Acta* **2020**, *187*, 56. [CrossRef] [PubMed]
7. Chan, M.H.M.; Chan, I.H.S.; Kong, A.P.S.; Osaki, R.; Cheung, R.C.K.; Ho, C.S.; Wong, G.W.K.; Tong, P.C.Y.; Chan, J.C.N.; Lam, C.W.K. Cold-vapour atomic absorption spectrometry underestimates total mercury in blood and urine compared to inductively-coupled plasma mass spectrometry: An important factor for determining mercury reference intervals. *Pathology* **2009**, *41*, 467–472. [CrossRef] [PubMed]
8. Liu, S.J.; Nie, H.G.; Jiang, J.H.; Shen, G.L.; Yu, R.Q. Electrochemical Sensor for Mercury(II) Based on Conformational Switch Mediated by Interstrand Cooperative Coordination. *Anal. Chem.* **2009**, *81*, 5724–5730. [CrossRef]
9. Liu, W.; Yang, L.J.; Yu, K.L.; Li, Y.Y. Enzyme-Assisted Cyclic Signal Amplification by Using Carbon Nanomaterials for Hg^{2+} Detection. *ChemistrySelect* **2020**, *5*, 4267–4272. [CrossRef]
10. Xie, S.B.; Tang, C.Y.; Liu, H.; Zhang, T.E.; Tang, Y.; Teng, L.M.; Zhang, J. An electroanalytical platform for nereistoxin-related insecticide detection based on DNA conformational switching and exonuclease III assisted target recycling. *Analyst* **2020**, *145*, 946–952. [CrossRef] [PubMed]
11. Wei, Z.W.; Lan, Y.F.; Zhang, C.; Jia, J.; Niu, W.F.; Wei, Y.L.; Fu, S.L.; Yun, K.M. A label-free Exonuclease I-assisted fluorescence aptasensor for highly selective and sensitive detection of silver ions. *Spectroc. Acta Pt. A-Molec. Biomolec. Spectr.* **2021**, *260*, 119927. [CrossRef]
12. Xing, X.J.; Zhou, Y.; Liu, X.G.; Pang, D.W.; Tang, H.W. Graphene Oxide and Metal-Mediated Base Pairs Based "Molecular Beacon" Integrating with Exonuclease I for Fluorescence Turn-on Detection of Biothiols. *Small* **2014**, *10*, 3412–3420. [CrossRef] [PubMed]
13. Wang, S.C. Construction of DNA Biosensors for Mercury (II) Ion Detection Based on Enzyme-Driven Signal Amplification Strategy. *Biomolecules* **2021**, *11*, 399. [CrossRef]
14. Wei, Y.L.; Chen, Y.X.; Li, H.H.; Shuang, S.M.; Dong, C.; Wang, G.F. An exonuclease I-based label-free fluorometric aptasensor for adenosine triphosphate (ATP) detection with a wide concentration range. *Biosens. Bioelectron.* **2015**, *63*, 311–316. [CrossRef]
15. Lan, Y.F.; Qin, G.J.; Wei, Y.L.; Wang, L.; Dong, C. Exonuclease I-assisted fluorescence aptasensor for tetrodotoxin. *Ecotoxicol. Environ. Saf.* **2020**, *194*, 110417. [CrossRef]
16. Zhang, J.X.; Ma, X.; Chen, W.H.; Bai, Y.F.; Xue, P.L.; Chen, K.H.; Chen, W.; Bian, L.J. Bifunctional single-labelled oligonucleotide probe for detection of trace Ag(I) and Pb(II) based on cytosine-Ag(I)-cytosine mismatches and G-quadruplex. *Anal. Chim. Acta* **2021**, *1151*, 338258. [CrossRef]
17. Wang, H.M.; Hou, S.N.; Wang, Q.Q.; Wang, Z.W.; Fan, X.; Zhai, J. Dual-response for Hg^{2+} and Ag^{+} ions based on biomimetic funnel-shaped alumina nanochannels. *J. Mater. Chem. B* **2015**, *3*, 1699–1705. [CrossRef] [PubMed]
18. Zhang, H.; Wang, Q.; Yang, M.; Fu, X. Exonuclease III-assisted Dual-Cycle Isothermal Signal Amplification for Highly Sensitive "Turn-on" Type Detection of Mercury Ion (II). *Chin. J. Anal. Chem.* **2019**, *47*, 899–906.
19. Yuan, M.; Zhu, Y.G.; Lou, X.H.; Chen, C.; Wei, G.; Lan, M.B.; Zhao, J.L. Sensitive label-free oligonucleotide-based microfluidic detection of mercury (II) ion by using exonuclease I. *Biosens. Bioelectron.* **2012**, *31*, 330–336. [CrossRef]
20. Huang, H.L.; Shi, S.; Zheng, X.Y.; Yao, T.M. Sensitive detection for coralyne and mercury ions based on homo-A/T DNA by exonuclease signal amplification. *Biosens. Bioelectron.* **2015**, *71*, 439–444. [CrossRef]
21. Sahin, S.; Caglayan, M.O.; Ustundag, Z. A review on nanostructure-based mercury (II) detection and monitoring focusing on aptamer and oligonucleotide biosensors. *Talanta* **2020**, *220*, 121437. [CrossRef]
22. Qi, Y.Y.; Ma, J.X.; Chen, X.D.; Xiu, F.R.; Chen, Y.T.; Lu, Y.W. Practical aptamer-based assay of heavy metal mercury ion in contaminated environmental samples: Convenience and sensitivity. *Anal. Bioanal. Chem.* **2020**, *412*, 439–448. [CrossRef]
23. Tan, L.; Chen, Z.; Zhang, C.; Wei, X.; Lou, T.; Zhao, Y. Colorimetric Detection of Hg^{2+} Based on the Growth of Aptamer-Coated AuNPs: The Effect of Prolonging Aptamer Strands. *Small* **2017**, *13*, 1603370. [CrossRef]
24. Zhang, L.B.; Tao, L.; Li, B.L.; Jing, L.; Wang, E.K. Carbon nanotube-DNA hybrid fluorescent sensor for sensitive and selective detection of mercury(II) ion. *Chem. Commun.* **2010**, *46*, 1476–1478. [CrossRef]
25. Wang, Y.H.; Deng, H.H.; Liu, Y.H.; Shi, X.Q.; Liu, A.L.; Peng, H.P.; Hong, G.L.; Chen, W. Partially reduced graphene oxide as highly efficient DNA nanoprobe. *Biosens. Bioelectron.* **2016**, *80*, 140–145. [CrossRef]
26. Li, H.L.; Zhai, J.F.; Sun, X.P. Nano-C-60 as a novel, effective fluorescent sensing platform for mercury(II) ion detection at critical sensitivity and selectivity. *Nanoscale* **2011**, *3*, 2155–2157. [CrossRef]
27. Liu, Y.; Kannegulla, A.; Wu, B.; Cheng, L.J. Quantum Dot Fullerene-Based Molecular Beacon Nanosensors for Rapid, Highly Sensitive Nucleic Acid Detection. *ACS Appl. Mater. Interfaces* **2018**, *10*, 18524–18531. [CrossRef]

28. Guo, R.B.; Chen, B.; Li, F.L.; Weng, S.H.; Zheng, Z.F.; Chen, M.; Wu, W.; Lin, X.H.; Yang, C.Y. Positive carbon dots with dual roles of nanoquencher and reference signal for the ratiometric fluorescence sensing of DNA. *Sens. Actuator B-Chem.* **2018**, *264*, 193–201. [CrossRef]
29. Cui, X.; Zhu, L.; Wu, J.; Hou, Y.; Wang, P.Y.; Wang, Z.N.; Yang, M. A fluorescent biosensor based on carbon dots-labeled oligodeoxyribonucleotide and graphene oxide for mercury (II) detection. *Biosens. Bioelectron.* **2015**, *63*, 506–512. [CrossRef]
30. Zhang, H.; Jia, S.; Lv, M.; Shi, J.; Zuo, X.; Su, S.; Wang, L.; Huang, W.; Fan, C.; Huang, Q. Size-Dependent Programming of the Dynamic Range of Graphene Oxide–DNA Interaction-Based Ion Sensors. *Anal. Chem.* **2014**, *86*, 4047–4051. [CrossRef]
31. Bayraktutan, T.; Meral, K. Merocyanine 540 adsorbed on polyethylenimine-functionalized graphene oxide nanocomposites as a turn-on fluorescent sensor for bovine serum albumin. *Phys. Chem. Chem. Phys.* **2016**, *18*, 23400–23406. [CrossRef]
32. Sun, Y.; Wang, Y.A.; Lau, C.; Chen, G.L.; Lu, J.Z. Hybridization-initiated exonuclease resistance strategy for simultaneous detection of multiple microRNAs. *Talanta* **2018**, *190*, 248–254. [CrossRef]
33. Oh, B.; Lee, C.H. Development of Man-rGO for Targeted Eradication of Macrophage Ablation. *Mol. Pharm.* **2015**, *12*, 3226–3236. [CrossRef] [PubMed]
34. Eigler, S.; Dotzer, C.; Hirsch, A. Visualization of defect densities in reduced graphene oxide. *Carbon* **2012**, *50*, 3666–3673. [CrossRef]
35. Jiang, F.W.; Zhao, W.J.; Wu, Y.M.; Wu, Y.H.; Liu, G.; Dong, J.D.; Zhou, K.H. A polyethyleneimine-grafted graphene oxide hybrid nanomaterial: Synthesis and anti-corrosion applications. *Appl. Surf. Sci.* **2019**, *479*, 963–973. [CrossRef]
36. Tong, W.; Zhang, Y.; Zhang, Q.; Luan, X.; Duan, Y.; Pan, S.; Lv, F.; An, Q. Achieving significantly enhanced dielectric performance of reduced graphene oxide/polymer composite by covalent modification of graphene oxide surface. *Carbon* **2015**, *94*, 590–598. [CrossRef]
37. Yuan, Z.; Tai, H.L.; Ye, Z.B.; Liu, C.H.; Xie, G.Z.; Du, X.S.; Jiang, Y.D. Novel highly sensitive QCM humidity sensor with low hysteresis based on graphene oxide (GO)/poly(ethyleneimine) layered film. *Sens. Actuators B-Chem.* **2016**, *234*, 145–154. [CrossRef]
38. Krishnamoorthy, K.; Veerapandian, M.; Yun, K.; Kim, S.J. The chemical and structural analysis of graphene oxide with different degrees of oxidation. *Carbon* **2013**, *53*, 38–49. [CrossRef]
39. Chen, J.H.; Xing, H.T.; Guo, H.X.; Weng, W.; Hu, S.R.; Li, S.X.; Huang, Y.H.; Sun, X.; Su, Z.B. Investigation on the adsorption properties of Cr(vi) ions on a novel graphene oxide (GO) based composite adsorbent. *J. Mater. Chem. A* **2014**, *2*, 12561–12570. [CrossRef]
40. Pan, N.; Li, L.; Ding, J.; Wang, R.B.; Jin, Y.D.; Xia, C.Q. A Schiff base/quaternary ammonium salt bifunctional graphene oxide as an efficient adsorbent for removal of Th(IV)/U(VI). *J. Colloid Interface Sci.* **2017**, *508*, 303–312. [CrossRef] [PubMed]
41. Khatri, P.K.; Choudhary, S.; Singh, R.; Jain, S.L.; Khatri, O.P. Grafting of a rhenium-oxo complex on Schiff base functionalized graphene oxide: An efficient catalyst for the oxidation of amines. *Dalton Trans.* **2014**, *43*, 8054–8061. [CrossRef]
42. Lin, Z.Y.; Yao, Y.G.; Li, Z.; Liu, Y.; Li, Z.; Wong, C.P. Solvent-Assisted Thermal Reduction of Graphite Oxide. *J. Phys. Chem. C* **2010**, *114*, 14819–14825. [CrossRef]
43. Liu, H.; Zhou, Y.; Yang, Y.; Zou, K.; Wu, R.; Xia, K.; Xie, S. Synthesis of polyethylenimine/graphene oxide for the adsorption of U(VI) from aqueous solution. *Appl. Surf. Sci.* **2019**, *471*, 88–95. [CrossRef]
44. Li, F.; Yang, Z.; Weng, H.; Chen, G.; Lin, M.; Zhao, C. High efficient separation of U(VI) and Th(IV) from rare earth elements in strong acidic solution by selective sorption on phenanthroline diamide functionalized graphene oxide. *Chem. Eng. J.* **2018**, *332*, 340–350. [CrossRef]
45. Ding, S.; Cargill, A.A.; Das, S.R.; Medintz, I.L.; Claussen, J.C. Biosensing with Förster Resonance Energy Transfer Coupling between Fluorophores and Nanocarbon Allotropes. *Sensors* **2015**, *15*, 14766–14787. [CrossRef]
46. Nitu, F.R.; Burns, J.S.; Ionita, M. Oligonucleotide Detection and Optical Measurement with Graphene Oxide in the Presence of Bovine Serum Albumin Enabled by Use of Surfactants and Salts. *Coatings* **2020**, *10*, 420. [CrossRef]
47. Reina, G.; Chau, N.D.Q.; Nishina, Y.; Bianco, A. Graphene oxide size and oxidation degree govern its supramolecular interactions with siRNA. *Nanoscale* **2018**, *10*, 5965–5974. [CrossRef]
48. Sjöback, R.; Nygren, J.; Kubista, M. Absorption and fluorescence properties of fluorescein. *Spectrochim. Acta Part A Mol. Biomol. Spectrosc.* **1995**, *51*, L7–L21. [CrossRef]
49. Le Guern, F.; Mussard, V.; Gaucher, A.; Rottman, M.; Prim, D. Fluorescein Derivatives as Fluorescent Probes for pH Monitoring along Recent Biological Applications. *Int. J. Mol. Sci.* **2020**, *21*, 9217. [CrossRef]
50. Nolan, E.M.; Lippard, S.J. Tools and tactics for the optical detection of mercuric ion. *Chem. Rev.* **2008**, *108*, 3443–3480. [CrossRef]

Article

Transformation of Cellulose via Two-Step Carbonization to Conducting Carbonaceous Particles and Their Outstanding Electrorheological Performance

Tomas Plachy, Erika Kutalkova *, David Skoda and Pavlina Holcapkova

Centre of Polymer Systems, University Institute, Tomas Bata University in Zlin, Trida Tomase Bati 5678, 76001 Zlin, Czech Republic; plachy@utb.cz (T.P.); dskoda@utb.cz (D.S.); holcapkova@utb.cz (P.H.)
* Correspondence: ekutalkova@utb.cz; Tel.: +420-576-031-730

Abstract: In this study, cellulose was carbonized in two-steps using hydrothermal and thermal carbonization in sequence, leading to a novel carbonaceous material prepared from a renewable source using a sustainable method without any chemicals and, moreover, giving high yields after a treatment at 600 °C in an inert atmosphere. During this treatment, cellulose was transformed to uniform microspheres with increased specific surface area and, more importantly, conductivity increased by about 7 orders of magnitude. The successful transition of cellulose to conducting carbonaceous microspheres was confirmed through SEM, FTIR, X-ray diffraction and Raman spectroscopy. Prepared samples were further used as a dispersed phase in electrorheological fluids, exhibiting outstanding electrorheological effects with yield stress over 100 Pa at an electric field strength 1.5 kV mm^{-1} and a particle concentration of only 5 wt%, significantly overcoming recent state-of-the-art findings. Impedance spectroscopy analysis showed clear interfacial polarization of this ER fluid with high dielectric relaxation strength and short relaxation time, which corresponded to increased conductivity of the particles when compared to pure cellulose. These novel carbonaceous particles prepared from renewable cellulose have further potential to be utilized in many other applications that demand conducting carbonaceous structures with high specific surface area (adsorption, catalyst, filtration, energy storage).

Keywords: cellulose; renewable; carbonization; electrorheology; suspension; conducting

1. Introduction

Nowadays, sustainability demands and the emphasis on utilizing renewable material sources are increasing in many applications. Cellulose is the most abundant and renewable natural polymer, and is used in many applications as a fiber material, including wood, construction material, cotton, cellulose particles, etc. It is well known that many organic materials can be carbonized in an inert atmosphere to produce novel carbonaceous conducting structures, further expanding their possible application areas. The carbonization of cellulose, however, commonly gives very low yields, even during its treatment in an inert atmosphere [1], and cellulose is, therefore, mainly carbonized hydrothermally at higher temperatures and pressures [2–6]. Such approach gives, however, rise to non-conducting particles due to the disorder in the structure caused by the transformation of cellulose to polyaromatic hydrochar [5,7]. Since cellulose is an abundant and cheap source, which is often a part of waste material that is not further effectively utilized, it is crucial to introduce a way to transform cellulose particles to conducting carbonaceous structures that could be further widely utilized in many applications instead of conducting polymers, as many hazardous chemicals are used during their synthesis. Contrarily, during carbonization the process, only heat is utilized without any other potential risk.

One of the proposed applications for cellulose particles is their utilization as a dispersed phase in electrorheological (ER) fluids [8–10], which are systems whose rheological behavior can be controlled via applied external electric fields [11]. While, in the absence of an electric field, electrically polarizable particles are randomly dispersed in continuous non-conducting medium and the fluid exhibits Newtonian behavior, in its presence, particles align to its direction creating rigid chain-like structures [12]. The fluid then behaves as viscoplastic material with a certain yield stress. This transition from a liquid-like to solid-like state, accompanied with an abrupt change in rheological parameters (viscosity, elastic modulus, yield stress, etc.) is called the ER effect, and can be utilized in many applications to suppress vibration (dampers), to serve as a medium for motion control (haptic devices, brakes, joints), or for controlled lubrication [13].

Since unmodified cellulose-based ER fluids exhibit only slight changes in viscosity, their derivatives, such as cellulose tartrate, carbamate, cellulose phosphate, carboxymethyl-cellulose, etc., are commonly used. These systems exhibit a moderate ER effect confirmed by a high difference between viscosity in the presence of an electric field and its absence [14–18]. For moderate effects of ER fluids based on pure cellulose, some moisture or adsorbed water on cellulose particles must be present [19–21]. Zhang et al. found that the optimal content of water in cellulose particles, regarding values of the yield stress, is around 8.5 wt% [21]. From the application point of view, however, the present moisture is undesirable and can cause the corrosion of the device, or other unpleasant phenomena can occur due to low boiling temperature of water. Therefore, the ER fluids based on cellulose derivatives utilized as an anhydrous dispersed phase are prioritized. Liu et al. combined cellulose and laponite into composite particles leading to values of yield stress of 150.1 Pa at an electric field of strength 2 kV mm^{-1} [22], and modified cellulose with urea for even higher yield stresses of over 270 Pa in the same field and a mass fraction of 15 wt% in anhydrous ER fluids [23]. Cellulose can also be prepared in various forms, such as nanocrystals and nanofibers [13], microfibrillated [24], microcrystalline [25], etc. Recently, an ER fluid based on nanocellulose particles dispersed in castor oil has been used as a biolubricant for the electroactive control of friction behavior [13]. Moreover, as a renewable material, cellulose obtained from spent coffee grounds [26], rice husk-based microcrystalline cellulose [25,27], and kenaf cellulose [17] have been utilized as the dispersed phase in ER fluids.

Besides using derivatives of conventional ER materials to enhance their ER performance, carbonization is another process to develop new carbon-based structures with high specific surface and significantly enhanced electric and dielectric properties. Carbonaceous materials such as carbonaceous nanotubes [28,29], graphene-enriched carbonaceous sheets [30], or carbon nanotubes [31] are commonly used as a dispersed phase in ER fluids due their mild conductivity and excellent dielectric properties giving high ER effects of their silicone–oil fluids. Furthermore, in electrorheology, carbonized particles of polyaniline [32], its oligomers [33], poly (p-phenylendiamine) [34], polyaniline-derived carbonaceous nanotubes [35], and carbonaceous materials prepared by the carbonization of starch/silica precursor [36] have been utilized as a dispersed phase. These particles have exhibited outstanding ER performance and, moreover, can be utilized in many applications implementing conducting or semiconducting materials, or generally as carbon materials with high specific surface area, such as catalysts [37], energy storage material [38], filtration [39], etc.

This study introduces a unique two-step carbonization process that combines hydrothermal carbonization with subsequent carbonization in an inert atmosphere at high temperature to obtain carbonized cellulose. Due to the former treatment, the thermal carbonization gives significantly high yields (>40%) of novel conducting carbonaceous structures from hydrochar-prepared cellulose. Its utilization as a renewable dispersed phase in electrorheology led to outstanding ER effects of their fluids even at low concentration (5 wt%). It should be further mentioned that there are many other applications besides electrorheology demanding conducting and semiconducting particles that could be obtained from green sources by inexpensive and ecological synthesis.

2. Results and Discussion

Carbonization is a process commonly used for the transformation of carbon-containing materials into novel materials with enhanced properties. In the case of cellulose, unfortunately, thermal carbonization even in an inert atmosphere gave very low yields of the carbonaceous traces (Figure 1). Before thermal carbonization, therefore, cellulose was hydrothermally carbonized, giving rise to particles further labeled as HTC-Cellulose, which were subsequently thermally carbonized in an argon atmosphere at 600 °C (HTC-TC600-Cellulose), giving a significant yield of novel carbonaceous materials prepared using a green synthesis without any chemicals. For a closer description of the particle preparation process, please see the Section 3. Materials and Methods.

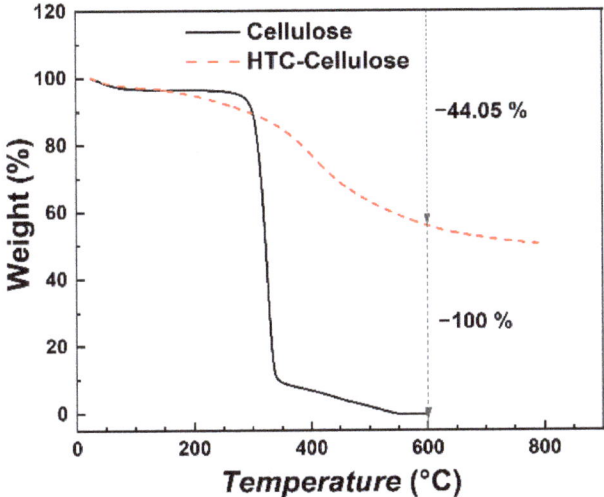

Figure 1. TGA analysis of the cellulose and HTC-Cellulose samples. Measurement ran in a nitrogen atmosphere.

HTC-Cellulose during TGA analysis exhibited a low decrease in weight (~44%), even at 600 °C, thus giving the opportunity to meaningfully use this material as a precursor for the preparation of thermally carbonized materials without the needed addition of any other substances (Figure 1). Since cellulose is one of the most abundant renewable materials, it is desirable to find an approach to transform it to carbonaceous structures with various morphologies, conductivities, etc., that can be further utilized in many applications. It should be mentioned that a yield of the HTC-TC600-Cellulose after the carbonization process in a furnace was slightly lower (~45%) when compared to the TGA analysis, since the sample was held at 600 °C for one hour. Thus, according to mass balance, the hydrothermal carbonization gave a yield of 34.35% of HTC-Cellulose from cellulose. After thermal carbonization, 44.7% of HTC-Cellulose was transformed to HTC-TC600-Cellulose. Overall, approximately 15.35% of the original cellulose was successfully transformed to HTC-TC600-Cellulose using above-mentioned two-step carbonization process. Further optimization of the hydrothermal conditions could lead to significantly higher yields of both HTC-Cellulose and HTC-TC600-Cellulose [6].

While, in the case of pure cellulose, it is hard to obtain yields over 40% after thermal carbonization due to its limited carbon content (commonly 44% at the highest), during hydrothermal carbonization, the carbon amount in the sample increased to approximately 70% depending on the hydrothermal carbonization temperature. This enabled us to obtain high yields after thermal carbonization [4].

XRD analysis was performed to describe the transformation of cellulose during its carbonization. The pristine cellulose exhibited, in its diffractogram, three typical peaks (Figure 2), where the former two represent the crystalline part of the cellulose and the former stands for its longitudinal structure [5]. For the HTC-Cellulose sample, no clear peak was observed since, during hydrothermal carbonization, crystalline structures within the sample were disordered and the amorphous irregular structure was obtained. It has been shown [5] that the critical temperature for hydrothermal carbonization of pristine cellulose is at least 220 °C, and that when the temperature is only slightly lower (210 °C), almost no transformation occurs and the XRD pattern of such material is very close to the XRD pattern of pristine cellulose. The XRD pattern of the HTC-TC600-Cellulose sample exhibited broad diffraction peaks, indicating the amorphous character of the sample. The slightly different shape of the diffractogram may be explained as a consequence of carbonization at the higher temperature that leads to the short-range order of the amorphous structure [40]. Moreover, the additional features of morphology and porosity can be considered to affect the shape of the XRD pattern. The carbonization of the particles further significantly affected the conductivity of the particles. While pristine cellulose and HTC-Cellulose particles were non-conducting with conductivities of 1.09×10^{-13} S cm^{-1} and 2.81×10^{-13} S cm^{-1}, respectively, their carbonized analogue HTC-TC600-Cellulose exhibited a conductivity of 3.48×10^{-6} S cm^{-1}. This increase in conductivity enables the utilization of HTC-TC600-Cellulose in many applications where conducting polymers are commonly used, and the obtain conductivity is, for example, desirable for ER fluids to exhibit high ER effects.

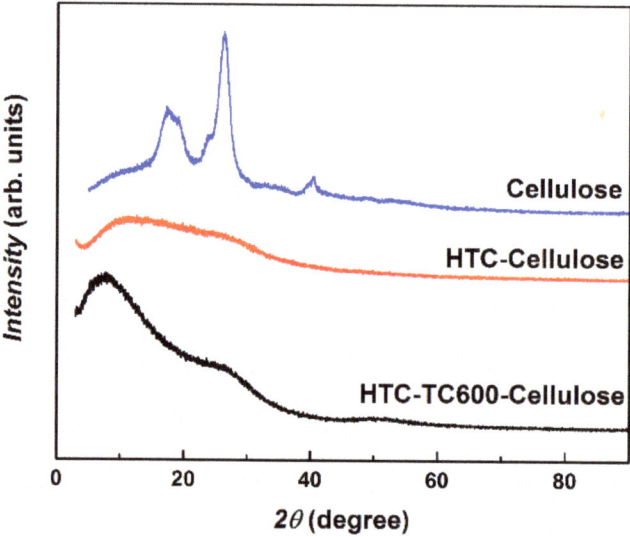

Figure 2. XRD patterns of the prepared particles: cellulose, HTC-Cellulose and HTC-TC600-Cellulose.

Fourier transform infrared spectroscopy and Raman spectroscopy were used to analyze the chemical composition changes in the carbonized cellulose when compared to pure cellulose. For pure cellulose particles, a typical spectrum representing cellulose was observed. Namely, the broad spectral band at 3334 cm^{-1} and the spectral band at 2896 cm^{-1} represent hydroxyl group and C–H stretching, respectively, while a dominant band at 1029 cm^{-1} stands for the etheric C–O group present in the ring (Figure 3a) [3]. After the hydrothermal carbonization process (Figure 3a), the intensity of the cellulose wide stretching band attributing to the vibrations of the hydroxyl group at 3320 cm^{-1} decrease, and during the thermal carbonization the peak disappears completely. Furthermore, during hydrothermal carbonization, benzene units are possibly created which is reflected by the bands at 1606 cm^{-1} representing C–C vibration at the aromatic ring [1], a low disordered peak around 1520 cm^{-1}, and a peak at 796 cm^{-1} standing for the C–H bond at the aromatic ring [4–6]. The presence of the carbonyl group C=O at 1701 cm^{-1} further corresponds well with the wide band at ca 3700–3000 cm^{-1}, representing the stretching of the O–H group in the hydroxyl or carboxyl groups, which are possibly formed [4,5] (Figure 3a). The spectrum for HTC-TC600-Cellulose is very disordered, confirming the significant change in the structure during thermal carbonization. A Raman spectrum of cellulose exhibits typical spectral bands (C–O, C–H stretching and bending); however, after carbonization, the Raman spectra of both samples of HTC-Cellulose and HTC-TC600-Cellulose contain bands that can be assigned to carbon-like materials with two characteristic bands representing graphitic (G) and disordered (D) states of carbon (Figure 3b). For HTC-Cellulose, the D peak shoulder can only be found, which can be connected with the incomplete transformation into carbonaceous structures.

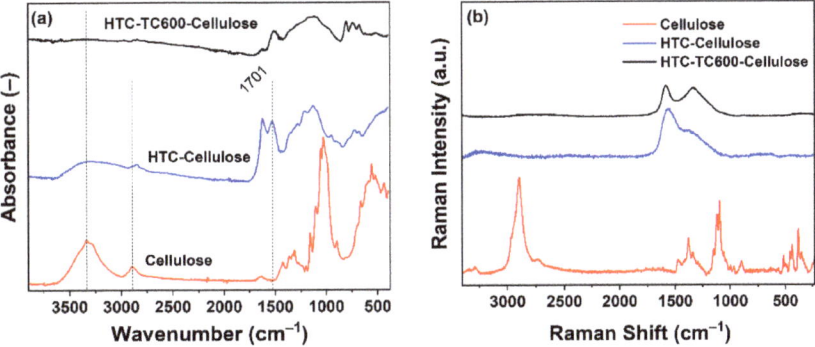

Figure 3. FTIR (**a**) and Raman (**b**) spectra of the cellulose, HTC-Cellulose and HTC-TC600-Cellulose particles.

The carbonization process significantly affected particle size and morphology (Figure 4). While particles of pure cellulose were of irregular shape up to 100 µm (Figure 4a), after the hydrothermal process the HTC-Cellulose particles were uniform microspheres sized from hundreds of nanometers up to micrometers (Figure 4b). After thermal carbonization, the particles preserved their spherical shape and size and became more uniform (Figure 4c); nevertheless, it has to be said that the transition does not seem to have been completed, since some irregular complexes were found within the sample (Figure 4d).

Nitrogen adsorption/desorption isotherms are displayed in Figure 5 and they exhibit type II shape [41] characteristics for microporous carbon-based materials. Based on the results from the BET method, it is shown that the carbonization process increased the surface area, a_s, of the samples by about two orders of magnitude. While pristine cellulose possessed a_s only 1.22 m^2 g^{-1}, its carbonized analogue HTC-TC600-Cellulose had a_s 103.7 m^2 g^{-1}, which can significantly increase the ER effect of its ER fluids. Since the adsorption of nitrogen takes place at low pressures and the hysteresis loop was open, it can be assumed that HTC-TC600-Cellulose sample exhibited microporous character [41] (Figure 5). The surface of the HTC-Cellulose sample was 26.6 m^2 g^{-1}, which is only slightly higher than the reference sample.

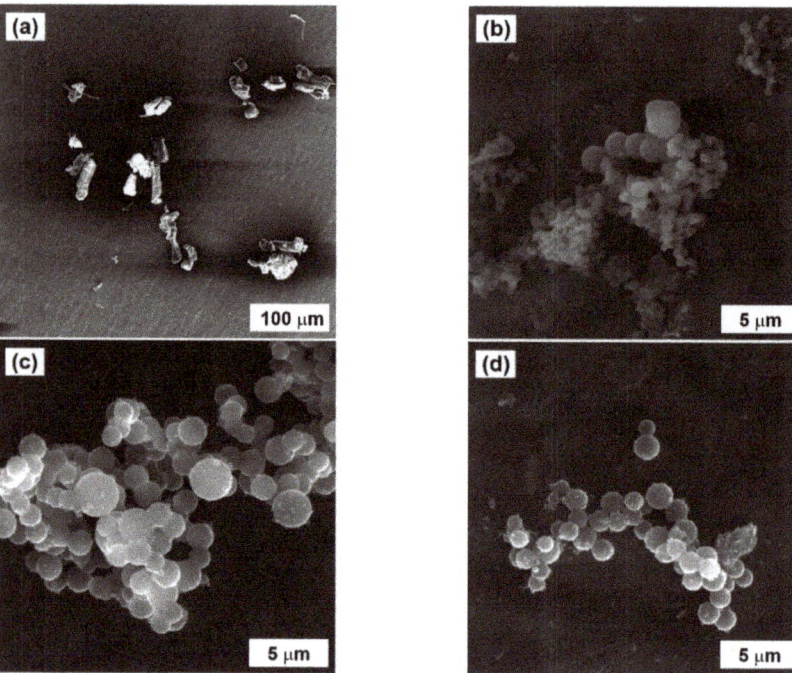

Figure 4. SEM images of (**a**) cellulose, (**b**) HTC-Cellulose, and (**c**,**d**) HTC-TC600-Cellulose particles.

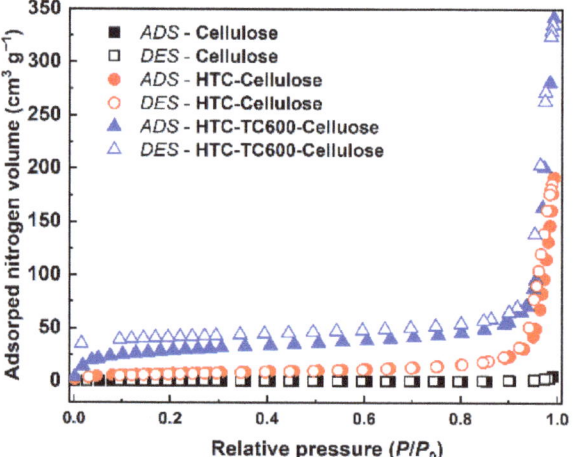

Figure 5. Nitrogen adsorption–desorption isotherms of the tested samples: cellulose, HTC-Cellulose, HTC-TC600-Cellulose. Abbreviations: ADS and DES stands for adsorption and desorption cycle, respectively.

The rheological behavior of the prepared ER fluids at the concentration of 5 wt%, under various external electric field strengths in the range of 0.5–1.5 kV mm^{-1}, as well as without it, is presented in Figure 6. In the off-state, the ER fluids displayed almost Newtonian behavior where shear stress increased nearly linearly with the applied shear rate. In the on-state, particles created the internal chain-like structure within the silicone oil leading to an increase in the rheological parameters of the prepared ER fluids, which was most prominent in the ER fluids based on HTC-TC600-Cellulose particles. For the ER fluids based on pristine cellulose or HTC-Cellulose, a slight increase in shear stress at low shear rates reflecting the low ER effect due to poor electrostatic forces among the particles, and thus the low reorganization of the structures, was observed (Figure 6a,b). The ER fluid based on the HTC-TC600-Cellulose particles exhibited a wide plateau of shear stress independent of the shear rate, reflecting a high ER effect due to the increased conductivity of the carbonized particles and the strong electrostatic forces among the particles, which were stronger than the hydrodynamic forces in a broad range of shear rate values.

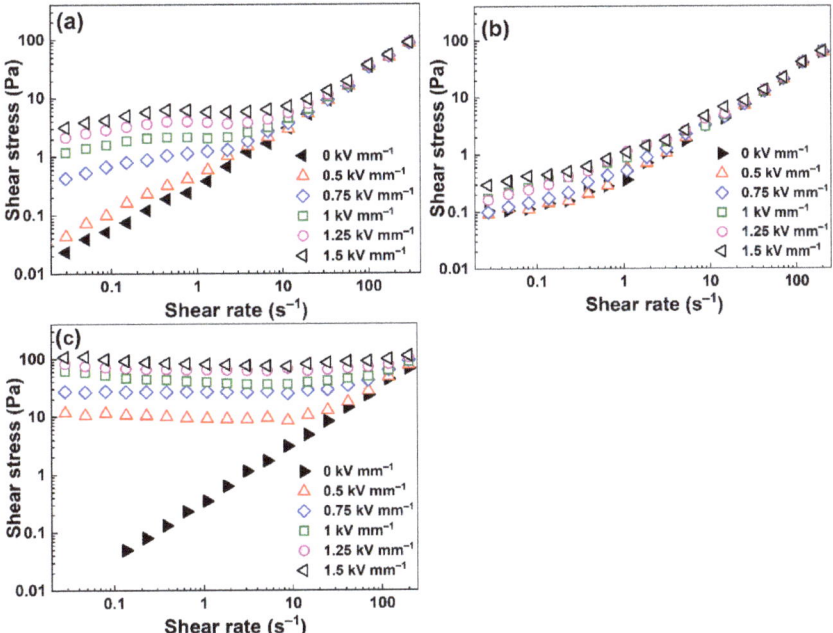

Figure 6. Log–log dependence of shear stress on the shear rate for prepared ER fluids based on (**a**) cellulose, (**b**) HTC-Cellulose and (**c**) HTC-TC600-Cellulose. The concentration of the particles was 5 wt%.

The fluid containing HTC-Cellulose (Figure 6b) showed a very low ER response due to the low conductivity and disorder of the structure character of cellulose by the hydrothermal carbonization. On the other hand, due to the enhancement of the conductivity after the carbonization process, the ER fluid based on HTC-TC600-Cellulose reached ER effects two orders of magnitude higher than the fluids based on pure cellulose. The value of yield stress, τ_y, expressing the stress which the liquid had to overcome to begin to flow, was more than 100 Pa for the ER fluid based on HTC-TC600-Cellulose at the electric field strength of 1.5 kV mm^{-1} (Figure 6c). The found ER performance is outstanding in comparison to the other published cellulose-based works of research [26]—where such results have not been achieved at such concentrations—and even exceeds most of the so-far-presented results for ER fluids based on other carbonized systems [32,34,42] or state-of-the-art materials [12,43,44]. The increased conductivity of HTC-TC600-Cellulose further influences the leaking current through its ER fluid, determining interactions between the particles. While in the case of ER fluids based on cellulose and HTC-Cellulose, the measured leaking current densities were ~0 μA cm^{-2}, in the case of ER fluids based on HTC-TC600-Cellulose, the high leaking current occurred due to the conductivity of the particles, which led to a strong interaction between the particles and the further high ER effects (Table 1). With a higher leaking current, stronger interactions between particles occur and the structure can then withstand higher stress before its distortion.

Table 1. Leaking current density of the prepared ER fluids.

Electric Field Strength (kV mm^{-1})	Leaking Current Density (μA cm^{-2})		
	Cellulose	HTC-Cellulose	HTC-TC600-CelluLose
0.50	0	0	63.03
0.75	0	0	176.03
1.00	0	0	319.90
1.25	0	0	× *
1.50	0	0	× *

* Leaking currents were over the limit of the measuring device.

The yield stress of ER fluids is proportional to the applied electric field and obeys a power law dependence (Equation (1)) where symbols τ_y, q, E, and a represent yield stress value, internal rigidity of the system, electric field strength, and slope of the curve fitting the data, respectively. In our work, due to simplicity and the low ER effect of ER fluids based on pristine cellulose or HTC-Cellulose, shear stress values at a low shear rate (~0.03 s^{-1}) were taken as the values representing yield stress. For the ER fluid based on HTC-TC600-Cellulose, the slope had a value 2.07, which is close to the theoretical polarization model, indicating that the polarization between the particles was the most relevant force directing its ER effect. For the ER fluids based on pristine cellulose or HTC-Cellulose, on the other hand, the values of the slope were 2.88 and 1.36, respectively, which were not possible to connect with either the conduction (slope ~1.5) or polarization models (slope ~2) (Figure 7) [45]. This can be explained by a very weak or nearly absent ER effect; thus, the values of shear stress at low shear rates correspond to chain-like structures that are not fully developed, and which later undergo their reformation and distort the measured values.

$$\tau_y \sim q \times E^a \tag{1}$$

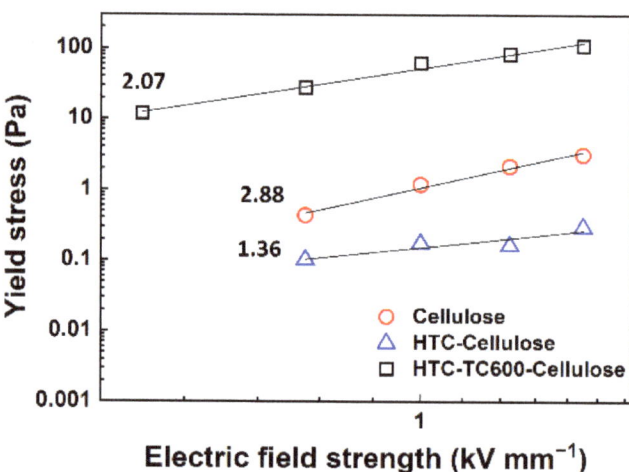

Figure 7. A log–log dependence of yield stress values taken as shear stress at low shear rates on electric field strength for the prepared ER fluids.

Dielectric spectroscopy is a suitable tool for the evaluation of ER fluids, as it is closely related to ER activity. After the application of the external electric field, the interfacial polarization occurs in the ER fluids at the particle interface and carrier medium. The polarization rate and the forces generated between the polarized particles are described by a complex permittivity. It has already been proposed [46,47] that the most important quantities connected with ER effects are dielectric relaxation strength, $\Delta\varepsilon'$, and relaxation time, t_{rel}, which should be as high and as short as possible, respectively. The Havriliak–Negami model was used to analyze the dielectric behavior of ER fluids according to the following equation (Equation (2)) [48–50]:

$$\varepsilon^*_{(\omega)} = \varepsilon'_\infty + \frac{(\varepsilon'_0 - \varepsilon'_\infty)}{\left(1 + (i\omega \cdot t_{rel})^\alpha\right)^\beta} \quad (2)$$

where $\varepsilon^*_{(\omega)}$ is a complex permittivity, ε'_0 is a relative permittivity at zero frequency, ε'_∞ is a relative permittivity at infinite frequency, ω is angular frequency ($2\pi f$), and α and β are shape parameters both in the range of 0–1, describing the asymmetry of the dielectric function. In the case of the α value differing significantly from zero and the β value from one, the relaxation is more asymmetrical.

Frequency dependencies of relative permittivity and the dielectric loss factor of the prepared ER fluids at the concentration of 5 wt% are shown in Figure 8. Among the prepared ER fluids, only the one based on HTC-TC600-Cellulose particles displayed a clear relaxation process in the region of interfacial polarization at higher frequencies, which is desirable for a high ER effect. Its relaxation time was estimated to be 2.52×10^{-7} s using the Havriliak–Negami model, with a very high value of dielectric loss factor (~0.4). Hao [51] has proposed that, for a high ER effect, this value should be >0.1 (Figure 8b). The find value ~0.4 is further comparable, and even exceeds values found for most carbonized materials used in electrorheology, showing high ER effects [33,34,36]. For cellulose or HTC-Cellulose-based ER fluids, no dielectric relaxation appears at all in the measured frequency range, which likewise affirms a weak or even non-existent ER effect under the electric field applied. The two-step carbonization of cellulose thus confirmed a proper approach for the preparation of electrically polarizable particles used in ER fluids.

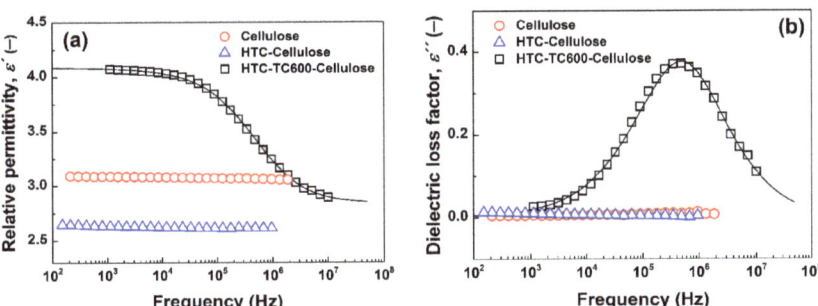

Figure 8. The frequency dependence of relative permittivity, ε' (**a**) and dielectric loss factor, ε'' (**b**), for 5 wt% silicone oil ER suspensions based on prepared particles. Solid lines represent the Havriliak–Negami model fit.

3. Materials and Methods

3.1. Cellulose Carbonization

In this study, cellulose particles (fibers, (medium size), moisture content < 10%, Sigma Aldrich; St. Louis, MI, USA) were carbonized in two steps, giving high yields, in order to obtain novel carbonaceous particles from a renewable source (Figure 9). The commercial cellulose was characterized by a content of 67.54% α-cellulose and a degree of polymerization (DP = 451) determined using the methods described in the Section 3.2. Particle Characterization. Firstly, hydrothermal carbonization was performed using a teflon autoclave with a total volume of 100 mL. Then, 2 g of cellulose was dispersed in 60 mL of demineralized water and poured into an autoclave. After its fixation in a steel autoclave, the autoclave was put into the oven and heated to 220 °C, which has been demonstrated as the lowest temperature for the transformation of cellulose into microspheres [5]. The hydrothermal carbonization ran for 14 h; afterwards, the autoclave was left to cool down to room temperature in the oven and the yield was thoroughly filtered with demineralized water and ethanol, and subsequently dried at 70 °C in a vacuum. Such particles were labelled as HTC-Cellulose.

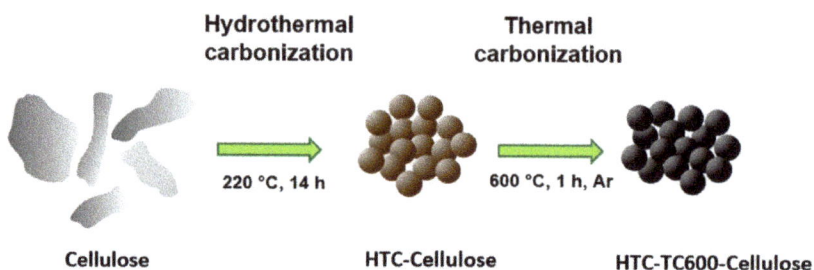

Figure 9. Illustration of the two-step carbonization process of cellulose particles utilizing consecutive hydrothermal and thermal carbonization.

Subsequently, the thermal carbonization of HTC-cellulose particles was performed using a tube furnace (Compact 1600c Tube Furnace; MTI Corporation, Richmond, CA, USA). The oven was heated to 600 °C in an argon atmosphere at a heating rate 3.3 K/min and the temperature was kept for 1 h; then, the oven was switched off. The obtained particles were labelled as HTC-TC600-Cellulose. The carbonization temperature of 600 °C was sufficient for the transformation of HTC-Cellulose particles to particles with conductivity high enough for an outstanding ER performance, and mild enough not to short-circuit the measuring instruments [31,34]. A brief scheme of the preparation is captured in Figure 9.

3.2. Particle Characterization

The powder XRD diffractograms were recorded on a Rigaku MiniFlex 600 diffractometer (Tokyo, Japan) equipped with a CoKα (λ = 1.7903 Å) X-ray tube (40 kV, 15 mA). Data processing was performed using Rigaku PDXL2 software. The surface area value of the samples was obtained by the physical adsorption of nitrogen gas using a surface area analyzer (Belsorp-mini II, BEL Japan, Inc., Osaka, Japan) using the Brunauer, Emmett and Teller (BET) multipoint method. Samples were outgassed for 18 h at 90 °C prior to the measurements. The prepared particles were further characterized using Fourier transform infrared spectroscopy (FTIR; Nicolet 6700, Thermo Scientific, Waltham, MA, USA) and dispersive Raman microscopy (Nicolet DXR, Thermo Scientific, Waltham, MA, USA). The morphology and size of the particles were observed with a scanning electron microscope (SEM; Vega II LMU, Tescan, Brno, Czech Republic). Thermogravimetric (TGA) analysis was performed in a temperature range of 25–800 °C at a heating rate of 10 K/min under a nitrogen atmosphere using a thermogravimeter (TGA Q500; TA Instruments, New Castle, DE, USA). Conductivity of the compressed powders was measured via a two-point method using an electrometer Keithley (Keithley 6517B, Cleveland, OH, USA).

In the case of pure cellulose particles, their content of α-cellulose was determined using the simplified ISO Test Method 692 Pulps-determination of alkali solubility. Briefly, pure cellulose was treated with 17.5% NaOH solution (w/w) for 1 h at 25 °C. A weight of delignified cellulose ($m_{cellulose}$) can be calculated as $m_{cellulose} = w_\alpha + w_\beta + w_\gamma$, where w_α, w_β and w_γ stand for weight of α-cellulose, β-cellulose and γ-cellulose, respectively [52]. Since the latter two are soluble in the alkali mixture, the α-cellulose was obtained as a solid fracture using centrifugation and decantation methods after the treatment. The sample was washed several times with 17.5% NaOH solution (w/w) to remove all the traces, and further washed several times with water and dried overnight at 60 °C. It was found that the used cellulose consisted of 67.54% α-cellulose.

The DP of cellulose was determined using Mark–Houwink equation (Equation (3)), where $[\eta]$ stands for intrinsic viscosity and K and α are parameters describing certain polymer–solvent systems.

$$[\eta] = K \times DP^\alpha \qquad (3)$$

In this work, a solution consisting of dimethyl sulfoxide/tetrabutylammonium hydroxide/H$_2$O in a ratio of 8:1:1 (w/w) was used as a suitable solvent for cellulose. The solvent dimethyl sulfoxide with purity ≥99.5% was supplied by Sigma Aldrich; St. Louis, MI, USA, as well as tetrabutylammonium hydroxide (~40% in H$_2$O). Bu et al. [53] found that the parameters K and α possess values of 0.24 and 1.21 for the solvent used in Equation (3). The intrinsic viscosity was found to be 391.15 mL/g and, subsequently, the DP of cellulose was calculated to be 451.

3.3. Preparation of Electrorheological Fluids and Their Characterization

Electrorheological fluids at the concentration of 5 wt% were prepared by dispersing the cellulose, HTC-Cellulose or HTC-TC600-Cellulose particles within the silicone oil (Lukosiol M200, Chemical Work Kolín, Czech Republic; dynamic viscosity = 194 mPa s at 25 °C). Before the measurements, the prepared fluids were mechanically stirred and then treated using an ultrasonicator (UP400S, Hielscher, Teltow, Germany) at the 30% vibration amplitude after 0.4 cycles for 1 min, which ensured a thorough homogenization. A rotational rheometer Bohlin Gemini (Malvern Instruments, Malvern, UK) with plate–plate geometry (20 mm in diameter with a gap of 0.5 mm) was used to investigate the ER behavior of the prepared ER fluids. The steady shear testing was performed in a controlled shear rate mode in the absence of the external electric field, as well as in the presence of the electric field at the strengths of 0.5–1.5 kV mm^{-1}. Before each subsequent measurement, the sample was sheared for 1 min at the shear rate of 20 s^{-1} to destroy any residual structures, and after that the electric field was applied for 1 min to form chain-like structures within a fluid.

Novocontrol Concept 50 (Novocontrol, Montabaur, Germany), an impedance analyzer, was used to investigate the dielectric properties of the prepared ER fluids in the frequency range of 0.003 Hz to 10 MHz. The Havriliak–Negami model was used to analyze the obtained dielectric spectra to obtain the relevant dielectric parameters, as discussed.

4. Conclusions

In this study, novel conducting carbonaceous material from the renewable and abundant source of cellulose was prepared by implementing a two-step carbonization process. Firstly, hydrothermal carbonization at 220 °C for 14 h was used, followed by thermal carbonization at 600 °C for 1 h in an argon atmosphere. During thermal carbonization, cellulose is commonly fully combusted even in an inert atmosphere; in our study, the yield of the conducting cellulose-based carbonaceous structure after thermal carbonization was around 45%. The transformation of cellulose into the amorphous carbonaceous microspheres during hydrothermal carbonization was confirmed via XRD analysis, FTIR and Raman spectroscopy. The particles preserved their size and shape after the thermal treatment; however, their surface area increased about two orders of magnitude when compared to pure cellulose and, most importantly, increased from 1.09×10^{-13} S cm^{-1} to 3.48×10^{-6} S cm^{-1}. The particles were further used as a dispersed phase in electrorheological fluids at the concentration of 5 wt%. The suspensions based on two-step carbonized cellulose exhibited a considerably enhanced ER effect when compared to pure cellulose and the values of shear stress were more than 100 Pa at the electric field strength of 1.5 kV mm^{-1}, which surpasses the so-far-published results for cellulose particles. The observed electrorheological effects corresponded well to the dielectric properties, which further confirmed the transformation from non-conducting particles to semiconducting carbonized microspheres. The two-step carbonization process of cellulose thus seems to be a novel approach to prepare new carbonaceous materials from renewable sources, with the potential to be used in a variety of industrial applications besides electrorheology, such as adsorbents, catalysts, the stationary phase of liquid chromatography, or electrode materials. The main benefits of the carbonaceous material come from its enhanced conductivity and large surface area, which are desirable in the mentioned applications.

Author Contributions: T.P., conceptualization, formal analysis, methodology, visualization, validation, writing—original draft, writing—review and editing; E.K., conceptualization, formal analysis, methodology, visualization, validation, writing—original draft; D.S., formal analysis, methodology; P.H., formal analysis. All authors have read and agreed to the published version of the manuscript.

Funding: This work was supported by the Ministry of Education, Youth and Sports of the Czech Republic—DKRVO (RP/CPS/2022/003).

Institutional Review Board Statement: Not applicable.

Informed Consent Statement: Not applicable.

Data Availability Statement: The data presented in this study are available on request from the corresponding author.

Conflicts of Interest: The authors declare no conflict of interest.

References

1. Bober, P.; Kovarova, J.; Pfleger, J.; Stejskal, J.; Trchova, M.; Novak, I.; Berek, D. Twin carbons: The carbonization of cellulose or carbonized cellulose coated with a conducting polymer, polyaniline. *Carbon* **2016**, *109*, 836–842. [CrossRef]
2. Gericke, M.; Trygg, J.; Fardim, P. Functional Cellulose Beads: Preparation, Characterization, and Applications. *Chem. Rev.* **2013**, *113*, 4812–4836. [CrossRef] [PubMed]
3. Feng, Z.X.; Odelius, K.; Rajarao, G.K.; Hakkarainen, M. Microwave carbonized cellulose for trace pharmaceutical adsorption. *Chem. Eng. J.* **2018**, *346*, 557–566. [CrossRef]
4. Sheng, K.C.; Zhang, S.; Liu, J.L.; Shuang, E.; Jin, C.D.; Xu, Z.H.; Zhang, X.M. Hydrothermal carbonization of cellulose and xylan into hydrochars and application on glucose isomerization. *J. Clean. Prod.* **2019**, *237*, 117831. [CrossRef]
5. Sevilla, M.; Fuertes, A.B. The production of carbon materials by hydrothermal carbonization of cellulose. *Carbon* **2009**, *47*, 2281–2289. [CrossRef]
6. Kang, S.M.; Li, X.H.; Fan, J.; Chang, J. Characterization of hydrochars produced by hydrothermal carbonization of lignin, cellulose, d-xylose, and wood meal. *Ind. Eng. Chem. Res.* **2012**, *51*, 9023–9031. [CrossRef]
7. Fakkaew, K.; Koottatep, T.; Polprasert, C. Effects of hydrolysis and carbonization reactions on hydrochar production. *Bioresour. Technol.* **2015**, *192*, 328–334. [CrossRef]
8. Rejon, L.; Ramirez, A.; Paz, F.; Goycoolea, F.M.; Valdez, M.A. Response time and electrorheology of semidiluted gellan, xanthan and cellulose suspensions. *Carbohydr. Polym.* **2002**, *48*, 413–421. [CrossRef]
9. Ramos-Tejada, M.M.; Arroyo, F.J.; Delgado, A.V. Negative electrorheological behavior in suspensions of inorganic particles. *Langmuir* **2010**, *26*, 16833–16840. [CrossRef]
10. Tilki, T.; Yavuz, M.; Karabacak, C.; Cabuk, M.; Uluturk, M. Investigation of electrorheological properties of biodegradable modified cellulose/corn oil suspensions. *Carbohydr. Res.* **2010**, *345*, 672–679. [CrossRef]
11. Kuznetsov, N.M.; Zagoskin, Y.D.; Bakirov, A.V.; Vdovichenko, A.Y.; Malakhov, S.N.; Istomina, A.P.; Chvalun, S.N. Is chitosan the promising candidate for filler in nature-friendly electrorheological fluids? *Acs Sustain. Chem. Eng.* **2021**, *9*, 3795–3803. [CrossRef]
12. Gao, C.Y.; Kim, M.H.; Jin, H.J.; Choi, H.J. Synthesis and Electrorheological Response of Graphene Oxide/Polydiphenylamine Microsheet Composite Particles. *Polymers* **2020**, *12*, 1984. [CrossRef] [PubMed]
13. Delgado-Canto, M.A.; Fernandez-Silva, S.D.; Roman, C.; Garcia-Morales, M. On the electro-active control of nanocellulose-based functional biolubricants. *ACS Appl. Mater. Interfaces* **2020**, *12*, 46490–46500. [CrossRef] [PubMed]
14. Kordonsky, V.I.; Korobko, E.V.; Lazareva, T.G. Electrorheological polymer-based suspensions. *J. Rheol.* **1991**, *35*, 1427–1439. [CrossRef]
15. Ahn, B.G.; Choi, U.S.; Kwon, O.K. Electro-rheological properties of anhydrous ER suspensions based on phosphoric ester cellulose particles. *Polym. Int.* **2000**, *49*, 567–573. [CrossRef]
16. Kim, S.G.; Kim, J.W.; Jang, W.H.; Choi, H.J.; Jhon, M.S. Electrorheological characteristics of phosphate cellulose-based suspensions. *Polymer* **2001**, *42*, 5005–5012. [CrossRef]
17. Gan, S.; Piao, S.H.; Choi, H.J.; Zakaria, S.; Chia, C.H. Synthesis of kenaf cellulose carbamate and its smart electric stimuli-response. *Carbohydr. Polym.* **2016**, *137*, 693–700. [CrossRef]
18. Choi, K.; Gao, C.Y.; Nam, J.D.; Choi, H.J. Cellulose-based smart fluids under applied electric fields. *Materials* **2017**, *10*, 1060. [CrossRef]
19. Kawai, A.; Uchida, K.; Kamiya, K.; Gotoh, A.; Yoda, S.; Urabe, K.; Ikazaki, F. Effect of dielectric property of hydrous dispersoid on electrorheology. *Int. J. Mod. Phys. B* **1996**, *10*, 2849–2855. [CrossRef]

20. Ikazaki, F.; Kawai, A.; Uchida, K.; Kawakami, T.; Edamura, K.; Sakurai, K.; Anzai, H.; Asako, Y. Mechanisms of electrorheology: The effect of the dielectric property. *J. Phys. D-Appl. Phys.* **1998**, *31*, 336–347. [CrossRef]
21. Zhang, S.; Winter, W.T.; Stipanovic, A.J. Water-activated cellulose-based electrorheological fluids. *Cellulose* **2005**, *12*, 135–144. [CrossRef]
22. Liu, Z.; Zhao, Z.J.; Jin, X.; Wang, L.M.; Liu, Y.D. Preparation of cellulose/laponite composite particles and their enhanced electrorheological responses. *Molecules* **2021**, *26*, 1482. [CrossRef] [PubMed]
23. Liu, Z.; Chen, P.P.; Jin, X.; Wang, L.M.; Liu, Y.D.; Choi, H.J. Enhanced electrorheological response of cellulose: A double effect of modification by urea-terminated silane. *Polymers* **2018**, *10*, 867. [CrossRef] [PubMed]
24. Choi, K.; Nam, J.D.; Kwon, S.H.; Choi, H.J.; Islam, M.S.; Kao, N. Microfibrillated Cellulose Suspension and Its Electrorheology. *Polymers* **2019**, *11*, 2119. [CrossRef]
25. Sim, B.; Bae, D.H.; Choi, H.J.; Choi, K.; Islam, M.S.; Kao, N. Fabrication and stimuli response of rice husk-based microcrystalline cellulose particle suspension under electric fields. *Cellulose* **2016**, *23*, 185–197. [CrossRef]
26. Chun, Y.; Ko, Y.G.; Do, T.; Jung, Y.; Kim, S.W.; Choi, U.S. Spent coffee grounds: Massively supplied carbohydrate polymer applicable to electrorheology. *Colloid Surf. A-Physicochem. Eng. Asp.* **2019**, *562*, 392–401. [CrossRef]
27. Bae, D.H.; Choi, H.J.; Choi, K.; Nam, D.J.; Islam, M.S.; Kao, N. Fabrication of phosphate microcrystalline rice husk based cellulose particles and their electrorheological response. *Carbohydr. Polym.* **2017**, *165*, 247–254. [CrossRef]
28. Yin, J.B.; Wang, X.X.; Zhao, X.P. Silicone-grafted carbonaceous nanotubes with enhanced dispersion stability and electrorheological efficiency. *Nanotechnology* **2015**, *26*, 065704. [CrossRef]
29. Lin, C.; Shan, J.W. Electrically tunable viscosity of dilute suspensions of carbon nanotubes. *Phys. Fluids* **2007**, *19*, 121702. [CrossRef]
30. Yin, J.B.; Shui, Y.J.; Chang, R.T.; Zhao, X.P. Graphene-supported carbonaceous dielectric sheets and their electrorheology. *Carbon* **2012**, *50*, 5247–5255. [CrossRef]
31. Yin, J.B.; Xia, X.A.; Xiang, L.Q.; Zhao, X.P. Conductivity and polarization of carbonaceous nanotubes derived from polyaniline nanotubes and their electrorheology when dispersed in silicone oil. *Carbon* **2010**, *48*, 2958–2967. [CrossRef]
32. Sedlacik, M.; Pavlinek, V.; Mrlik, M.; Moravkova, Z.; Hajna, M.; Trchova, M.; Stejskal, J. Electrorheology of polyaniline, carbonized polyaniline, and their core-shell composites. *Mater. Lett.* **2013**, *101*, 90–92. [CrossRef]
33. Plachy, T.; Sedlacik, M.; Pavlinek, V.; Trchová, M.; Morávková, Z.; Stejskal, J. Carbonization of aniline oligomers to electrically polarizable particles and their use in electrorheology. *Chem. Eng. J.* **2014**, *256*, 398–406. [CrossRef]
34. Plachy, T.; Sedlacik, M.; Pavlinek, V.; Moravkova, Z.; Hajna, M.; Stejskal, J. An effect of carbonization on the electrorheology of poly(p-phenylenediamine). *Carbon* **2013**, *63*, 187–195. [CrossRef]
35. Yin, J.B.; Xia, X.A.; Xiang, L.Q.; Zhao, X.P. Temperature effect of electrorheological fluids based on polyaniline derived carbonaceous nanotubes. *Smart Mater. Struct.* **2011**, *20*, 015002. [CrossRef]
36. Qiao, Y.P.; Zhao, X. Electrorheological effect of carbonaceous materials with hierarchical porous structures. *Colloid Surf. A-Physicochem. Eng. Asp.* **2009**, *340*, 33–39. [CrossRef]
37. Prekob, A.; Hajdu, V.; Muranszky, G.; Fiser, B.; Sycheva, A.; Ferenczi, T.; Viskolcz, B.; Vanyorek, L. Application of carbonized cellulose-based catalyst in nitrobenzene hydrogenation. *Mater. Today Chem.* **2020**, *17*, 100337. [CrossRef]
38. Ruan, C.Q.; Wang, Z.H.; Lindh, J.; Stromme, M. Carbonized cellulose beads for efficient capacitive energy storage. *Cellulose* **2018**, *25*, 3545–3556. [CrossRef]
39. Kotia, A.; Yadav, A.; Raj, T.R.; Keischgens, M.G.; Rathore, H.; Sarris, I.E. Carbon nanoparticles as sources for a cost-effective water purification method: A comprehensive review. *Fluids* **2020**, *5*, 230. [CrossRef]
40. Wang, W.; Zhou, J.H.; Wang, Z.P.; Zhao, L.Y.; Li, P.H.; Yang, Y.; Yang, C.; Huang, H.X.; Guo, S.J. Short-range order in mesoporous carbon boosts potassium-ion battery performance. *Adv. Energy Mater.* **2018**, *8*, 1701648. [CrossRef]
41. Lee, J.; Choi, Y.C. Pore Structure Characteristics of Foam Composite with Active Carbon. *Materials* **2020**, *13*, 4038. [CrossRef] [PubMed]
42. Plachy, T.; Sedlacik, M.; Pavlinek, V.; Stejskal, J.; Graca, M.P.; Costa, L.C. Temperature-dependent electrorheological effect and its description with respect to dielectric spectra. *J. Intell. Mater. Syst. Struct.* **2016**, *27*, 880–886. [CrossRef]
43. Saabome, S.M.; Park, Y.S.; Ko, Y.G. Designing particle size of aminated polyacrylonitrile spheres to enhance electrorheological performances of their suspensions. *Powder Technol.* **2021**, *394*, 986–995. [CrossRef]
44. Li, L.Z.; Gao, S.J. Polyaniline (PANI) and $BaTiO_3$ composite nanotube with high suspension performance in electrorheological fluid. *Mater. Today Commun.* **2020**, *24*, 100993. [CrossRef]
45. Davis, L.C. Polarization forces and conductivity effects in electrorheological fluids. *J. Appl. Phys.* **1992**, *72*, 1334–1340. [CrossRef]
46. Mrlik, M.; Sedlacik, M.; Pavlinek, V.; Bober, P.; Trchova, M.; Stejskal, J.; Saha, P. Electrorheology of aniline oligomers. *Colloid Polym. Sci.* **2013**, *291*, 2079–2086. [CrossRef]
47. Zhao, J.; Lei, Q.; He, F.; Zheng, C.; Liu, Y.; Zhao, X.P.; Yin, J.B. Interfacial polarization and electroresponsive electrorheological effect of anionic and cationic poly(ionic liquids). *Acs Appl. Polym. Mater.* **2019**, *1*, 2862–2874. [CrossRef]
48. Stenicka, M.; Pavlinek, V.; Saha, P.; Blinova, N.; Stejskal, J.; Quadrat, O. The electrorheological efficiency of polyaniline particles with various conductivities suspended in silicone oil. *Colloid Polym. Sci.* **2009**, *287*, 403–412. [CrossRef]
49. Egorysheva, A.V.; Kraev, A.S.; Gajtko, O.M.; Kusova, T.V.; Baranchikov, A.E.; Agafonov, A.V.; Ivriov, V.K. High electrorheological effect in Bi1.8Fe1.2SbO7 suspensions. *Powder Technol.* **2020**, *360*, 96–103. [CrossRef]

50. Havriliak, S.; Negami, S. A complex plane analysis of alpha-dispersions in some polymer systems. *J. Polym. Sci. Part C Polym. Symp.* **1966**, *14*, 99–117. [CrossRef]
51. Hao, T. Electrorheological suspensions. *Adv. Colloid Interface Sci.* **2002**, *97*, 1–35. [CrossRef]
52. He, L.; Xin, L.P.; Chai, X.S.; Li, J. A novel method for rapid determination of alpha-cellulose content in dissolving pulps by visible spectroscopy. *Cellulose* **2015**, *22*, 2149–2156. [CrossRef]
53. Bu, D.Q.; Hu, X.Z.; Yang, Z.J.; Yang, X.; Wei, W.; Jiang, M.; Zhou, Z.W.; Zaman, A. Elucidation of the Relationship between Intrinsic Viscosity and Molecular Weight of Cellulose Dissolved in Tetra-N-Butyl Ammonium Hydroxide/Dimethyl Sulfoxide. *Polymers* **2019**, *11*, 1605. [CrossRef] [PubMed]

Article

Functionalization of Tailored Porous Carbon Monolith for Decontamination of Radioactive Substances

Joonwon Bae [1,†], Gyo Eun Gu [2,†], Yeon Ju Kwon [2], Jea Uk Lee [3,*] and Jin-Yong Hong [2,*]

1. Department of Applied Chemistry, Dongduk Women's University, Seoul 02748, Korea; redsox7@dongduk.ac.kr
2. Center for C1 Gas & Carbon Convergent Research, Korea Research Institute of Chemical Technology (KRICT), 141 Gajeong-ro, Yuseong-gu, Daejeon 34114, Korea; goo9@aekyung.com (G.E.G.); kyj0905@krict.re.kr (Y.J.K.)
3. Department of Advanced Materials Engineering for Information and Electronics, Integrated Education Institute for Frontier Science & Technology (BK21 Four), Kyung Hee University, 1732 Deogyeong-daero, Giheung-gu, Yongin-si 17104, Korea
* Correspondence: leeju@khu.ac.kr (J.U.L.); jyhong@krict.re.kr (J.-Y.H.); Tel.: +82-31-201-3655 (J.U.L.); +82-42-860-7591 (J.-Y.H.)
† These authors contributed equally to this work.

Abstract: As the control over radioactive species becomes critical for the contemporary human life, the development of functional materials for decontamination of radioactive substances has also become important. In this work, a three-dimensional (3D) porous carbon monolith functionalized with Prussian blue particles was prepared through removal of colloidal silica particles from exfoliated graphene/silica composite precursors. The colloidal silica particles with a narrow size distribution were used to act a role of hard template and provide a sufficient surface area that could accommodate potentially hazardous radioactive substances by adsorption. The unique surface and pore structure of the functionalized porous carbon monolith was examined using electron microscopy and energy-dispersive X-ray analysis (EDS). The effective incorporation of PB nanoparticles was confirmed using diverse instrumentations such as X-ray diffraction (XRD), Fourier-transform infrared (FT-IR), and X-ray photoelectron spectroscopy (XPS). A nitrogen adsorption/desorption study showed that surface area and pore volume increased significantly compared with the starting precursor. Adsorption tests were performed with ^{133}Cs ions to examine adsorption isotherms using both Langmuir and Freundlich isotherms. In addition, adsorption kinetics were also investigated and parameters were calculated. The functionalized porous carbon monolith showed a relatively higher adsorption capacity than that of pristine porous carbon monolith and the bulk PB to most radioactive ions such as ^{133}Cs, ^{85}Rb, ^{138}Ba, ^{88}Sr, ^{140}Ce, and ^{205}Tl. This material can be used for decontamination in expanded application fields.

Keywords: exfoliated graphene; porous carbon; functionalization; radionuclide; decontamination

Citation: Bae, J.; Gu, G.E.; Kwon, Y.J.; Lee, J.U.; Hong, J.-Y. Functionalization of Tailored Porous Carbon Monolith for Decontamination of Radioactive Substances. *Int. J. Mol. Sci.* **2022**, *23*, 5116. https://doi.org/10.3390/ijms23095116

Academic Editor: Ana María Díez-Pascual

Received: 29 March 2022
Accepted: 2 May 2022
Published: 4 May 2022

Publisher's Note: MDPI stays neutral with regard to jurisdictional claims in published maps and institutional affiliations.

Copyright: © 2022 by the authors. Licensee MDPI, Basel, Switzerland. This article is an open access article distributed under the terms and conditions of the Creative Commons Attribution (CC BY) license (https://creativecommons.org/licenses/by/4.0/).

1. Introduction

Recently, the need for intensive energy sources has increased dramatically, because human life and activities are becoming complex and energy-consuming. At the same time, a significant interest has been drawn to nuclear energy and technology dealing with dangerous radioactive substances to increase energy production capacity/efficiency, leading to a rapid increase in nuclear waste production. Therefore, it becomes even more important for us to control the potentially hazardous nuclear wastes effectively. Even if several methods have been developed to remove the nuclear wastes, it is still a challenging task to remove the radioactive substances or produce functional media materials for contamination control [1–6].

Since the Fukushima accidents, a desire for the development of environmentally benign and functional materials for decontamination of radioactive substances is ever increasing. Among various strategies to remove the radioactive species, adsorption with porous materials can be promising, because of easy preparation and simple use [7–12]. It is known that the porous materials can be manufactured easily from various precursor materials and accommodate a large amount of substances due to high surface area and pore volume [13–18]. To date, numerous porous materials have been produced using diverse methods, such as the template approach, porogen addition, and self-assembly [19,20]. The selection of precursor and methodology has become wider and the resulting porous materials can be customized depending on their usage and purpose, for example, storage and release of chemical species such as energy, gas molecules, and drugs [21–27].

In this study, a functional porous carbon monolith (PCM) decorated with the Prussian blue (PB) nanoparticles (PB@PCM) was produced for removal of the radioactive substances by selective removal of colloidal silica microparticles working as the hard template and porogen from exfoliated graphene/silica composite precursors, which was easily obtained by direct mixing of graphene and colloidal silica particles. This approach was simple because the preparation and removal of the hard template with a complicated and tailored structure could be avoided. In addition, it was expected that an unprecedented pore structure could be generated because graphene/silica composite precursors possessing an irregular pore structure were introduced. An interconnected 3D pore structure with an increased surface area was generated after removal of the colloidal silica particles, as a controlled pore structure with a similar pore size was not mandatory for this study. However, it was important to analyze the pore structure and surface properties to see if the obtained material could be used for the decontamination of radioactive species. As the removal of radioactive materials was strongly dependent on the adsorption mechanism, a high surface area of the PCM was one of the most critical prerequisites. Moreover, the surface properties and pore structure were analyzed extensively using various instrumentations.

On the other hand, the adsorption of radioactive species was promoted by the addition of PB nanoparticles to the internal structure of the PCM material, because it was reported that the PB, known as ferric(III) hexacyanoferrate(II), has been the most effective adsorptive material for cesium (Cs^+) ions. [28–32]. In particular, the cage size of PB crystal is similar to the hydration radius of a cesium ion, resulting in excellent binding ability and selectivity of the Cs^+ ions [33–39]. Due to the unique internal pore structure and increased surface area of the PCM material, the incorporation of the PB nanoparticles into the PCM materials could be promoted. Subsequently, adsorption tests were conducted with a typical radioactive ion, ^{133}Cs, to observe adsorption isotherms using both Langmuir and Freundlich isotherms and calculate kinetic parameters associated with the adsorption behavior. The PB-decorated PCM showed a higher adsorption capacity to typical radioactive ions ^{133}Cs, ^{85}Rb, ^{138}Ba, ^{88}Sr, ^{140}Ce, and ^{205}Tl than the precursors and the bulk PB. The newly developed material and method might offer an opportunity for future research and investigation of nuclear waste control. Moreover, it can be transplanted to diverse technologies for waste removal and treatment.

2. Results and Discussion
2.1. Preparation of the Porous Carbon Monolith with Tailored Pore Structure

The overall production procedure to obtain the PB@PCM is presented in Figure 1. In general, it is important to disperse the silica particles into the exfoliated graphene (EG) network. This was difficult to achieve because the EG-silica interactions were unfavorable and uniform mixing was almost impossible due to mild self-aggregation of the silica particles. To improve the EG-silica interactions, the surface of the silica particles was modified with amino silane coupling agent. The zeta potential value of silica particles increased from −4.03 to +6.17 mV after surface treatment (Supplementary Information, S1). Hence, the aggregation of silica particles could be suppressed and dispersion was promoted. The SEM image in Figure 1a shows no apparent trace of silica particle aggregation.

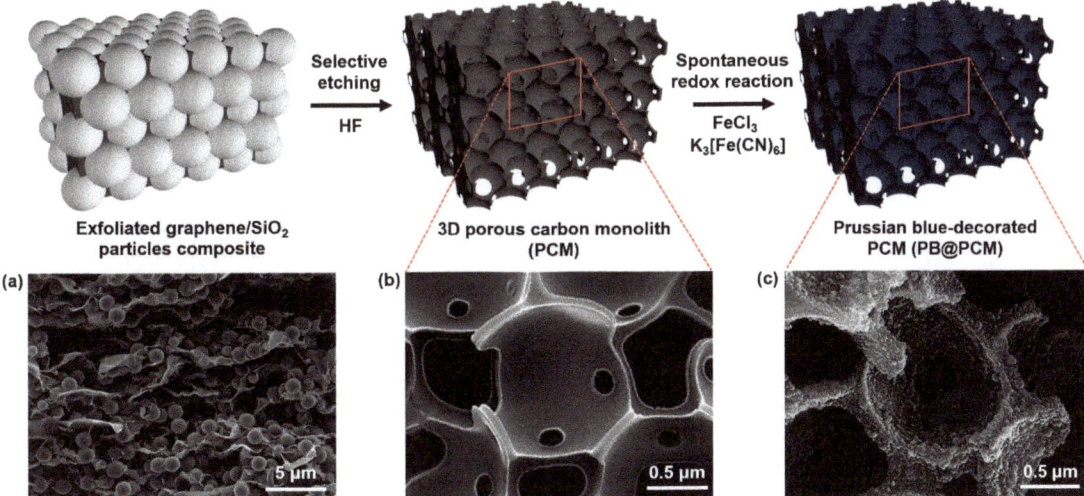

Figure 1. A schematic illustration for fabricating PCM through etching of the silica particle templates in EG/SiO$_2$ composite and obtaining PB@PCM by a subsequent decoration of PB nanoparticles into the PCM. FE-SEM images of (**a**) EG/SiO$_2$ particles composite, (**b**) PCM, and (**c**) PB@PCM.

The particle locations were random and any regular texture was not observed. The external appearance of the EG/silica composite precursor was monolithic, because stacking of the EG sheets occurred spontaneously. In addition, this shape could be retained in the final product, PB@PCM.

Subsequently, the silica particles were removed by etching with HF and the PCM material was produced as shown in Figure 1b. A silica trace was clearly observed in the center of the FE-SEM image. Note that the PCM possessed an interconnected open pore structure and relatively small window pores. The formation of the window pores might be associated with local dewetting of the thin EG sheets, because the EG-silica particle interface was still less friendly after silane treatment. However, it could be inferred that the transport of radioactive ion species could be facilitated due to the presence of the window pores. The irregular pores might have originated from the inherent internal voids. It could be expected that the surface area increased after the silica etching.

The PB decoration into the PCM materials led to the formation of the PB@PCM monolithic material, which would be used for removal of radioactive species. After the decorating with the PB nanoparticles, the rough surface of PCM material becomes smooth compared with that before the decoration, but there is no obvious difference in the structure or pore shape (Figure 1c). It was known that the PB decoration process was spontaneous and the average size of the PB nanoparticles was indiscernible [40–43]. The detailed internal structure of the PB@PCM is observed further in Figures 2 and 3.

Figure 2 shows the macro and microscopic morphology of the PCM, which was uniform, and variation in appearance was negligible. This was because the components, EG and silica, in the precursor were homogeneous. It was difficult to observe any regular pore pattern and the pores were open and interconnected. The features shown in Figure 1b were observed analogously at the center and outer regions of the PCM monolith. It was concluded that the internal and external structure of the PCM was identical. This seemed desirable, because adsorption performance would be constant and independent.

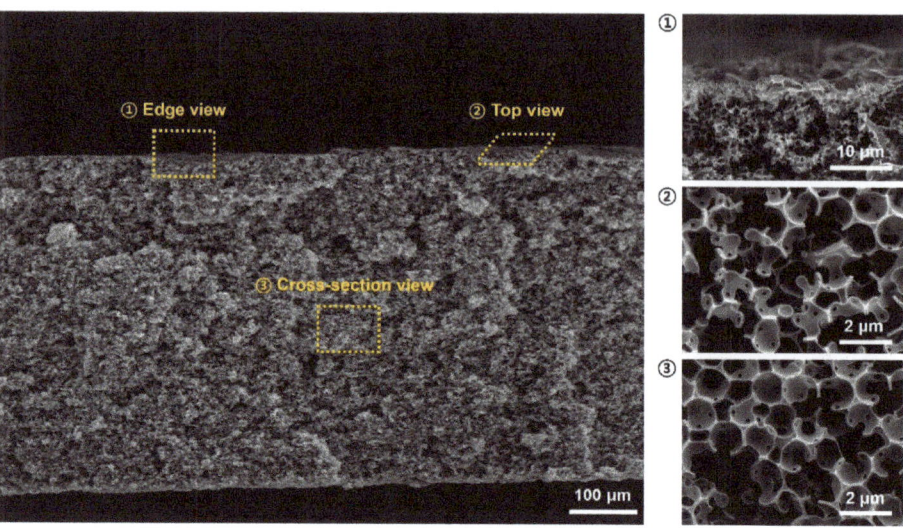

Figure 2. FE-SEM images of the 3D porous carbon monolith at different sections (edge, top, cross-section).

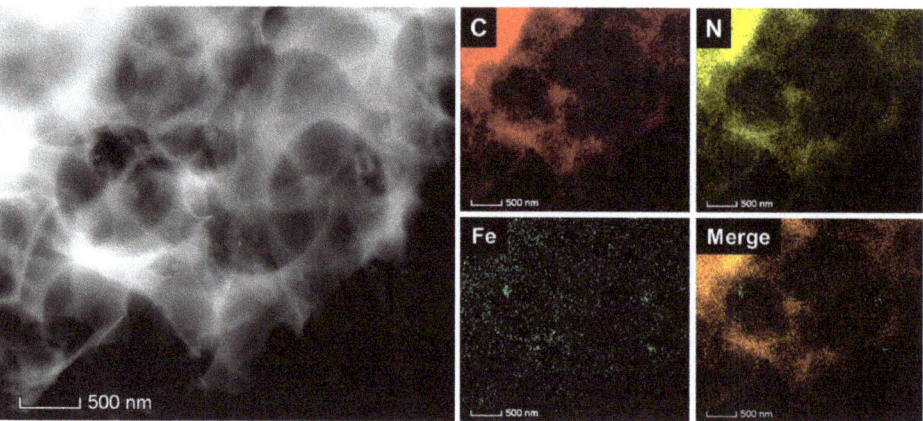

Figure 3. High-angle annular dark-field scanning TEM (HAADF-STEM) image and EDS elemental mapping images of PB@PCM for carbon (C), nitrogen (N), and iron (Fe). PB particles uniformly distributed over the whole PCM surface.

2.2. Characterization of the Functionalized Porous Carbon Monolith

In this work, effective introduction of the PB nanoparticles and retention of porous structure were critical for PB@PCM production. A high-angle annular dark-field scanning TEM (HAADF-STEM) image of the PB@PCM indicates that the porous structure remained intact after PB addition (Figure 3). In addition, successful introduction of the PB nanoparticles was examined with EDS mapping analysis. It was verified that a coating layer of PB nanoparticles was introduced along the surface of the PCM. Several major elements such as C, N, O, and Fe were observed and traces of K and Cl also appeared (Supplementary Information, S2). The EDS mapping also provided the elemental compositions of the PB and PB@PCM. The Fe composition was high for PB nanoparticles, whereas C ratio increased for PB@PCM due to the presence of EG. The N and O compositions were almost constant regardless of the PB nanoparticle addition.

The PB@PCM was further characterized using diverse instrumentations. The FT-IR spectrum of the EG showed weak peaks for C-O and C-C bending because there was a small amount of carbonyl groups on the EG (Figure 4a). A sharp peak for C≡N stretching appeared in the PB spectrum. All those features were seen in the PB@PCM spectrum, indicating that PB nanoparticles were successfully introduced to the surface of PCM. A similar trend was observed from the XRD profiles in Figure 4b. Peaks for highly crystalline PB nanoparticles were observed at 17(200), 25(220), 36(400), and 40(420) degrees, whereas that for crystalline EG was seen at 27(002) degrees [44].

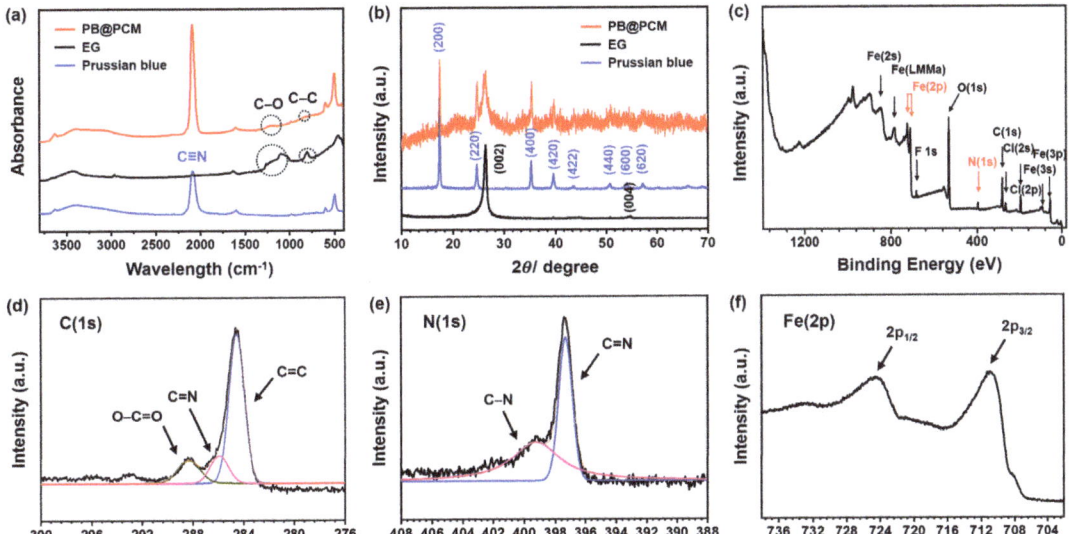

Figure 4. Characterization of Prussian blue, exfoliated graphene and PB@PCM with (**a**) FT-IR, (**b**) XRD. XPS survey spectra of PB@PCM (**c**) wide scan, (**d**) C(1s), (**e**) N(1s) and (**f**) Fe(2p).

Those representative diffraction peaks of PB and EG could also be assigned in the PB@PCM profile. The formation of a PB nanoparticle layer on the PCM was further confirmed by XPS. The survey spectrum of the PB@PCM in Figure 4c exhibited C(1s) (284.6 eV), O(1s) (531.2 eV), N(1s) (397.4 eV), and Fe(2p) (710.5 eV) peaks, indicating the substantial introduction of the PB nanoparticles into the PCM. The Gaussian fitting of the C(1s) peak in Figure 4d led to the occurrence of the three peaks: non-oxygenated C=C (288.3 eV, 69.41 at%), C=N (285.3 eV, 21.8 at%), and O-C=O (284.5 eV, 69.4 at%). The second one was highly associated with the presence of cyanide (C≡N) groups in the PB nanoparticles. The N(1s) peak was also fitted using the Gaussian method (Figure 4e) and showed the presence of two peaks: C-N (399.31 eV, 57.1 at%) and C=N (391.2 eV, 42.9 at%). In the fitted curve of the Fe(2p) in Figure 4f, two characteristic peaks of Fe (2p1/2) and Fe (2p3/2) were observed at 724.6 and 710.9 eV, respectively, both indicating the presence of α-FeOOH, whereas the peak at 708.2 eV was attributed to [Fe(CN)6] [45,46].

As an increase in surface area was important to achieve a high removal capacity, porous structure was introduced to the PB@PCM. N_2 adsorption/desorption isotherms (BET) provided critical information regarding surface characteristics of the PB@PCM for practical application (Supplementary Information, S3). Note that the PB@PCM showed a stable adsorption behavior, indicating that it could adsorb a significantly higher amount of volume at relatively low pressure than the PCM. The structural parameters closely related to the surface properties were measured and are summarized in Table 1.

Table 1. Comparison of the parameters obtained from BET and Mercury Porosity Analysis.

Sample [a]	Bulk Density (g/mL)	Total Pore Area (m²/g)	Specific Surface Area (m²/g)	Porosity (%)	Avg. Pore Diameter (μm)
PCM	0.41	21.76	43.49	78.04	0.35
PB@PCM	0.46	20.23	128.76	61.82	0.27

[a] These values are estimated by using the BET and mercury porosimeter.

The specific area of the PB@PCM (128.76 m²/g) was approximately three times larger than that of PCM (43.49 m²/g), owing to the PB nanoparticle addition. Average pore diameter decreased from 0.35 to 0.27 μm after the PB addition. Even if the pore structure was irregular and wide open, the decrease in the average pore size was meaningful for adsorption capacity improvement. A significant change also occurred to other parameters except the total pore area after the PB introduction, suggesting that the change was attributed to the PB nanoparticle addition.

2.3. Adsorption/Decontamination Performances of the Functionalized Porous Carbon Monolith

To investigate the radioactive ion removal capacity of the PB@PCM, extensive adsorption tests were performed with the stable ^{133}Cs isotope instead of the radioactive ^{137}Cs. The ^{133}Cs ion uptake behavior by the PB@PCM was monitored at pH 7.0 and 20 °C with varying ^{133}Cs equilibrium concentration, as shown in Figure 5a.

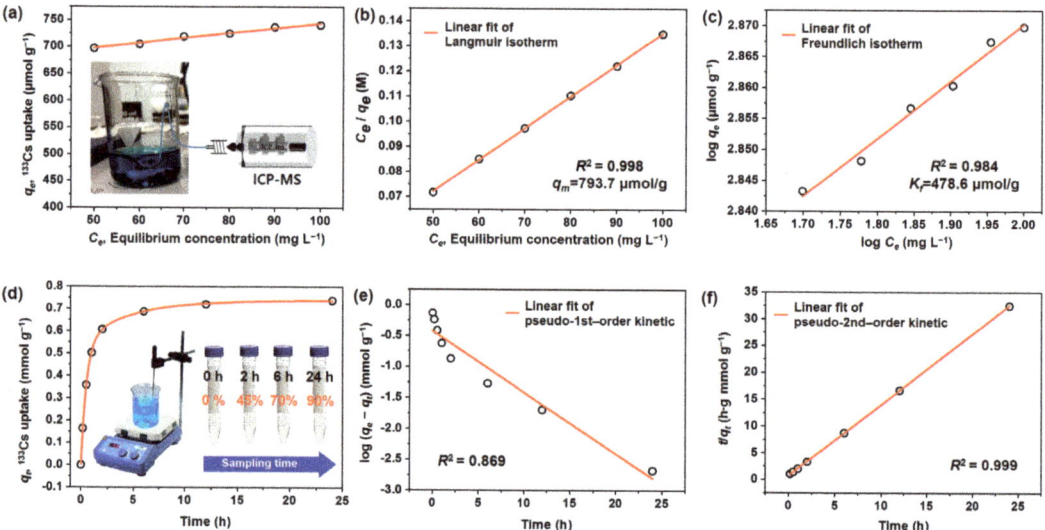

Figure 5. (a) ^{133}Cs ion uptake behavior as a function of equilibrium ion concentration and adsorption isotherms for ^{133}Cs ion uptake by the PB@PCM, linearly fitted to the (b) Langmuir and (c) Freundlich isotherm equation. (d) ^{133}Cs ion uptake behavior as a function of contact time and adsorption isotherms for ^{133}Cs ion uptake by the PB@PCM, linearly fitted to (e) pseudo-first-order and (f) pseudo-second-order kinetics.

In these experiments, the adsorption equilibrium was gradually established for 24 h after the addition of a designated amount of the PB@PCM to the test solutions. Accordingly, the adsorption capacity of the PB@PCM for ^{133}Cs ion was measured after 24 h contact time. The ^{133}Cs uptake by adsorption was calculated by the following equation:

$$q_e = \frac{(C_o - C_e)V}{AW} \quad (1)$$

where q_e is the equilibrium adsorption capacity (mmol/g) of adsorbent, C_o and C_e the initial and equilibrium concentration (mg/L) of the ^{133}Cs ion, V the total volume (L) of solution, A the atomic weight (g/mol) of ^{133}Cs, and W is the weight (g) of the PB@PCM.

To analyze the adsorption behavior quantitatively, the equilibrium adsorption data was fitted linearly according to the Langmuir and Freundlich isotherms. Note that the Langmuir isotherm is dependent on an assumption that all the surface adsorption sites have identical affinity toward the adsorbate; hence, adsorption at one site does not affect the adsorption at an adjacent site [47]. In addition, each adsorbate molecule tends to occupy a single site; hence, monolayer formation is promoted on the adsorbent surface. The linear and nonlinear form of the Langmuir equations are as follows:

$$\frac{C_e}{q_e} = \left(\frac{1}{q_m b}\right) + \left(\frac{1}{q_m}\right) C_e \text{ (linear form) or } q_e = \frac{q_m b C_e}{1 + b C_e} \text{ (nonlinear form)} \quad (2)$$

where q_m is the maximum ^{133}Cs uptake (mmol/g) and b the constant (L/mg) that refers to the bonding energy of adsorption related to free energy and net enthalpy. On the other hand, the Freundlich model includes the reversible adsorption at a heterogeneous surface; therefore, it is not restricted to monolayer formation [48]. The linear and nonlinear forms of the Freundlich adsorption isotherms are as follows:

$$\log q_e = \log K_f + \frac{1}{n} \log C_e \text{ (linear form) or } q_e = K_f C_e^{1/n} \text{ (linear form)} \quad (3)$$

where K_f is the constant (mmol/g) related to the adsorption capacity of the adsorbent, and $1/n$ the intensity of the adsorption constant. Compared with the Langmuir isotherm, K_f stands for analogous physical meaning to q_m, even if those parameters are fundamentally different. Linear fitting was conducted according to the linear equation forms; consequently, the fitting results are shown in Figure 5b,c. Additional regression analysis was carried out by computer software (OriginPro 8.0, Originlab corporation, Northampton, MA, USA) to obtain the accurate parameters, and values are summarized in Table 2. It is clear that the fitting to the Langmuir isotherm showed a higher consistency with the adsorption data from the experiments with a regression coefficient of $R^2 = 0.999$ and reasonable errors for parameters.

Table 2. The adsorption parameters obtained according to the Langmuir and Freundlich isotherms at room temperature for the adsorption of ^{133}Cs to the PB@PCM.

	Langmuir [a]			Freundlich [a]		
	q_m (μmol/g)	b (L/mg)	R^2	K_f (μmol/g)	n	R^2
^{133}Cs [b]	793.651	0.136	0.999	478.631	10.782	0.984

[a] R^2 = regression coefficient. The q_m, b, K_f, n, values and the nonlinear regression correlations for Langmuir and Freundlich isotherms were measured by nonlinear regression analysis using OriginPro 8.0. [b] Three sets of valid data were collected, and the C_e and q_e values for ^{133}Cs were calculated to be in the error range of ±5%.

A supplementary parameter was suggested to further characterize the ^{133}Cs adsorption behavior by the PB@PCM, which was separation factor (R_L), a dimensionless constant [49]. R_L value was calculated by the following equation:

$$R_L = \frac{1}{1 + bC_0} \quad (4)$$

where b and C_o were the Langmuir constant and initial ^{133}Cs concentration. The type of adsorption isotherm can be classified as unfavorable ($R_L > 1$), linear ($R_L = 1$), and favorable ($R_L = 0$), depending on its value. The parameter obtained in this study was approximately 0.068, supporting the favorable adsorption of ^{133}Cs on the PB@PCM. Therefore, it was concluded that the PB@PCM possessed desirable structural and surface features for ^{133}Cs ion adsorption.

The adsorption behavior was also investigated as a function of contact time between ^{133}Cs ions and PB@PCM. The curve in Figure 5d shows that the adsorption capacity increased rapidly during the initial 4 h and reached a maximum value (saturation) after 24 h. For quantitative analysis of the adsorption mechanism, the experimental data were fitted to the pseudo-first-order [50] and pseudo-second-order [51] kinetics and the results are exhibited in Figure 5e,f. The kinetics equations are as follows:

$$\log(q_e - q_t) = \log q_e - \frac{K_1}{2.303} t \text{ (pseudo-first-order equation)} \quad (5)$$

$$\frac{t}{q_t} = \frac{1}{K_2 q_e^2} + \frac{t}{q_e} \text{ (pseudo-second-order equation)} \quad (6)$$

where q_e and q_t are the ^{133}Cs uptake at equilibrium and time t, respectively, K_1 the constant (1/min) for first-order adsorption, and K_2 the rate constant (g/mmol·min) for second-order adsorption. The parameter values are summarized in Table 3 and it shows that the pseudo-second-order model described the adsorption mechanism better than the pseudo-first-order model with regression coefficients $R^2 = 0.999$. The pseudo-second-order model is dependent on the assumption that the adsorption occurs on the adsorbent without interactions between adsorbates, and the desorption rate is negligible compared to the adsorption rate. It was inferred that the interactions between PB nanoparticles in PB@PCM and incoming radioactive ions could be facilitated due to the permanent/induced electrostatic forces between partially negative cyano groups (–C≡N) of PB nanoparticles and positive ^{133}Cs ions. On the other hand, the interactions between the ions are supposed to be inconspicuous due to electrostatic repulsion.

Table 3. The kinetic parameters obtained using pseudo 1st-order and 2nd-order models at room temperature for the adsorption of ^{133}Cs to the PB@PCM.

	Pseudo-First-Order [a]			Pseudo-Second-Order [a]		
	K_1 (1/min)	q_{e1} (mmol/g)	R^2	K_2 (g/mmol·min)	q_{e2} (mmol/g)	R^2
^{133}Cs [b]	0.228	0.424	0.936	2.395	0.755	0.999

[a] R^2 = regression coefficient. The K_1, K_2, q_{e1}, q_{e2} values and the nonlinear regression correlations for pseudo-first-order and pseudo-second-order models were measured by nonlinear regression analysis using OriginPro 8.0. [b] Three sets of valid data were collected, and the q_e and q_t values for ^{133}Cs were calculated to be in the error range of ±5%.

To reveal the effectiveness of synergistic interaction between PB nanoparticles and PCM material, the ^{133}Cs ion uptake test was performed using the PCM, bulk PB and PB@PCM (Supplementary Information, S4). It was important that the PB@PCM provided an improved uptake by adsorption compared with the PCM and bulk PB (A compilation of investigations carried out by using various PB-based adsorbents is also presented in Supplementary Information, S5 [52–57]). The maximum uptake value of PCM, bulk PB, and PB@PCM were ca. 0.0018, 0.0069, and 0.7225 mmol/g, respectively. It is noteworthy that the maximum ^{133}Cs adsorption capacity of the PB@PCM is approximately 400 and 100 times higher than that of pristine PCM and bulk PB, respectively. This striking difference could be interpreted by the enhanced porosity and permeability of PCM and high surface area and large amount of PB nanoparticles in the synthesized PB@PCM, providing abundant active sites for ^{133}Cs adsorption. Compared to the bulk PB, the PB@PCM containing both interconnected macroporous structure and nanometer-sized PB particles have enhanced adsorption properties due to increased mass transport through the material and maintenance of a specific surface area.

It was also expected that the PB@PCM can adsorb various radioactive species; hence, uptake tests were performed at pH 7.0 and 20 °C with other radioactive ion species such as ^{133}Cs, ^{85}Rb, ^{138}Ba, ^{88}Sr, ^{140}Ce, and ^{205}Tl (Figure 6a). The adsorption capacity value varied from 0.72 to 0.13 mmol/g, with a decreasing order of ^{133}Cs (0.72), ^{85}Rb (0.63), ^{138}Ba (0.32), ^{88}Sr (0.26), ^{140}Ce (0.18), and ^{205}Tl (0.13). This trend might have originated from the degree of interaction between PB nanoparticles in PB@PCM and individual radioactive ion species. More positively charged ions could interact more favorably with the PB nanoparticles. It can be inferred that ^{133}Cs might have a highly positive nature, because its electronegativity and ionization energy was lowest compared with other ion species (Supplementary Information, S6). However, the underlying correlation needs to be examined systematically in a separate research project.

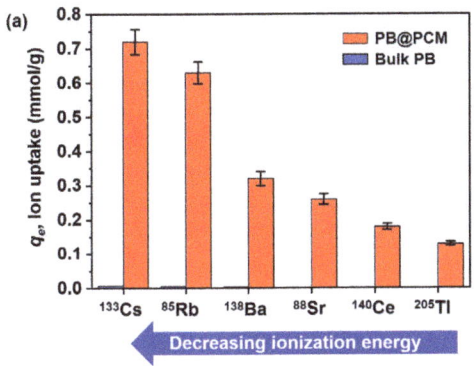
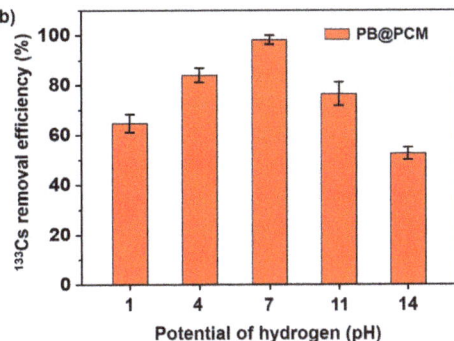

Figure 6. (a) Measurement of radioactive ion species uptake by the bulk PB and the PB@PCM to 6 radioactive ion species (^{133}Cs, ^{87}Rb, ^{88}Sr, ^{137}Ba, ^{140}Ce, and ^{205}Tl) at pH 7.0 and 20.0°C for 24 h. (b) The pH effect on the ^{133}Cs removal efficiency of the PB@PCM at 20.0 °C for 24 h. Initial ^{133}Cs ion concentration is 100 ppm.

In addition, the effect of the pH on the ^{133}Cs adsorption capacity was also investigated (Figure 6b). Interestingly, the removal efficiency of the PB@PCM increased in acidic conditions and decreased in basic media, and the PB@PCM exhibited the highest uptake capacity at pH 7.0. This phenomenon can be ascribed to the competitive interaction between hydrogen (H+) and hydroxyl (OH−) ions during the adsorption process. In acidic condition (<pH 7.0), the H+ ion competes with the ^{133}Cs ions towards the active sites of PB@PCM, which results in a relatively lower adsorption capacity. As the pH increased (≈pH 7.0), the decrease of the hydrogen ion concentration made it possible to increase the concentration of ^{133}Cs ions to bind electrons, leading to a high adsorption uptake. However, when the basification was in progress (>pH 7.0), the stabilization attributed to the reaction between ^{133}Cs ions and increased OH− ions occurred, and then the adsorption capacity decreased.

3. Materials and Methods

3.1. Materials

Tetraethyl orthosilicate (TEOS, 98%), 3-aminopropyltrimethoxysilane (APS), iron(III) chloride (FeCl$_3$, 97%), potassium hexacyanoferrate(III) (K$_3$Fe(CN)$_6$, ≥99%), cesium, barium, strontium, and rubidium standard solution (1000 ppm) for inductively coupled plasma-mass spectrometer (ICP-MS) analyses were purchased from Sigma Aldrich (Milwaukee, WI, USA). Sodium chloride (NaCl), Ammonia solutions (NH$_4$OH, 28.0~30.0%), hydrochloric acid (HCl), Ethanol (C$_2$H$_5$OH, ≥99%), and N-Methyl-2-pyrrolidone (NMP) were supplied from Samchun Chemical (Seoul, Korea). Hydrofluoric acid (HF, 40%) was purchased from

J. T. Baker (Phillipsburg, NJ, USA). All chemicals were used as received without further purification. Electrochemically exfoliated graphene (EG) was produced according to the method reported previously [58].

3.2. Synthesis of Monodispersed Silica Microparticles

A certain amount of TEOS (36 g) was vigorously mixed with ethanol (200 mL) (Solution 1). An amount of ammonia solution (54 mL) and NaCl (0.1 g) was dissolved in a mixture of water (40 mL) and ethanol (190 mL) (Solution 2). Subsequently, solution 1 was slowly added to solution 2 at a speed of 1.2 µL/min using a micro syringe pump at 30 °C. Then the mixture solution was stirred for 1 h at 800 rpm. The produced microparticles were purified by centrifugation and washed with ethanol and distilled water 3 times. Finally, the obtained homogeneous silica particles were stored at room temperature.

3.3. Surface Modification of the Silica Particles

The obtained silica particles (1 g) were dispersed in ethanol (10 mL) by ultra-sonication for 10 min, resulting in a milky suspension. A small amount of glacial acetic acid (60 µL) was added to a mixture of ethanol (95 mL) and water (5 mL) to make the solution acidic (pH 5.0) and then APS (4.2 g, 19 mM) was added dropwise to the acidic solution. The resulting clear colorless solution was stirred (400 rpm) at RT for 15 min in order to form reactive silanol groups by hydrolysis.

3.4. Preparation of Porous Exfoliated Graphene Monolith

A certain amount of exfoliated graphene (EG, 240 mg) was dispersed in 20 mL of NMP and sonicated for 90 min to achieve a homogeneous mixing, which was critical for this experiment. Then, the silica particles were added to the EG solution, and the EG to silica weight ratio was 10:90. The obtained solution was filtered with a nylon membrane filter (47 mm in diameter, 0.2 µm pore size, Whatman) under vacuum suction to obtain an EG/silica composite precursor. The film type EG/silica composite was retrieved by peeling it off from the nylon membrane filter. A film type porous EG was obtained by removing the silica particles with HF solution (10%) for a few minutes. The PCM was dried in a vacuum oven at 120 °C overnight.

3.5. Fabrication of Prussian Blue Decorated Porous Carbon Monolith

The PB@PCM was manufactured by a redox reaction in an aqueous mixture solution of $FeCl_3$, $K_3[Fe(CN)_6]$, and PCM. Typically, PCM was soaked in 10 M $FeCl_3$ solution for 15 min and the solution in sample was completely eliminated with a filter paper. Then, the PCM decorated with Fe^{3+} ions was dried in a convection oven at 80 °C for at least 3 h. Then, the preprocessed PCM was immersed in a $K_3[Fe(CN)_6]$ (0.64 g, 100 mL) solution and stirred at 55 °C for 12 h. The final PB@PCM product was obtained and washed with distilled water 3 times and dried under vacuum at 80°C overnight to complete evaporation of the solvent. The mass of the finally obtained PB@PCM sample was 2.4 g, which can be scaled up by increasing the amount of the EG/silica composite precursor. The bulk PB was fabricated by direct mixing of equimolar $FeCl_3$ and $K_4[Fe(CN)_6]$ solutions at room temperature for 6 h. The final products were washed with ethanol several times to remove residual reagent. Subsequently, the resulting bulk PB was dried under a vacuum oven and pressed into pellets.

3.6. Radioactive Material Extraction Test

The adsorption behavior of the PB@PCM for ^{133}Cs ion was monitored as follows: the cesium standard solutions with varying concentration from 0.01 to 100 mg/L (ppm) were produced in 100 mL of distilled water. To control the pH, ammonia or hydrochloric acid solution was injected into the aqueous solutions. Then, 0.1 g of adsorbents (PB@PCM) was introduced into the solution and stirred at 300 rpm, which would be suitable for ^{133}Cs ion adsorption. Adsorption experiments were carried out at 20 °C and pH 7.0, and the adsorption capacity for ^{133}Cs ion was monitored from 0 min to 24 h. After adsorption, 3 mL of the solution was filtered through a syringe filter (PTFE, pore diameter 0.2 μm), and then an ICP-MS test was performed to determine the ^{133}Cs ion concentration as a function of adsorption time. To obtain accurate results, the ^{133}Cs ion concentration was measured three times and the average values were collected. The adsorption capacity to the other ions—rubidium, barium, strontium, cerium, and titanium—was also measured identically.

3.7. Characterization

The morphology was confirmed using a field emission scanning electron microscopy at 10 kV (Tescan Mira-3 FEG, Brno, Czech Republic) and transmission electron microscopy (FEI, Tecnai G2-20, Hillsboro, OR, USA). A particle size analyzer (Malvern, Zetasizer Nano-ZS, Worcestershire, United Kingdom) was used to determine the silica particle size and zeta potential. The FT-IR spectra were recorded on Alpha-P (Bruker, Ettlingen, Germany). The XRD patterns were collected using an Ultima IV with Cu Kα radiation (Rigaku, Tokyo, Japan). The BET analyses were performed with an ASAP 2020 (Micromeritics, Gwinnett County, GA, USA) using the N2 adsorption–desorption isotherms. The pore size and distribution in the porous carbon was analyzed with an AutoPore IV 9500 mercury porosimeter at room temperature (Micromeritics, Gwinnett County, GA, USA). The XPS data were obtained using an Axis Nova (KRATOS, Tokyo, Japan) with monochromatic Al-Kα X-ray source under 10-8 Torr vacuum analysis chamber. The ICP-MS was used to analyze and confirm the cesium concentration in solution (iCAP RQ, Thermo Fisher Scientific, Waltham, MA, USA).

4. Conclusions

In this study, convenient manufacture of monolithic porous carbon material embedded with Prussian blue particles was demonstrated and its radioactive substance removal performance was also examined using various radioactive ion species. The functional porous carbon was prepared by etching of silica template and subsequent Prussian blue nanoparticle decoration. The obtained porous carbon possessed an increased surface area compared with the precursor and the intermediate. In addition, its pore structure was open and interconnected with small window pores, which was advantageous for ion transport and movement. The radioactive substance removal was achieved by the incorporation of Prussian blue nanoparticles, which can adsorb radioactive ion species effectively. The ^{133}Cs adsorption behavior could be described by both Langmuir and Freundlich isotherms, and fitting to the Langmuir isotherm provided a closer correlation with the experimental data. The porous carbon showed a better adsorption capacity than the precursor and an intermediate capacity for radioactive ion species such as ^{133}Cs, ^{87}Rb, ^{88}Sr, ^{137}Ba, ^{140}Ce, and ^{205}Tl. The maximum and minimum adsorption capacity values were obtained for ^{133}Cs and ^{205}Tl, respectively. As a porous carbon material was obtained by a relatively simple, cost-effective, and environmentally benign method, an interest in analogous functional porous materials for nuclear waste control is expected to increase significantly. Therefore, this research can offer important information for future relevant research and, moreover, development of advanced materials and devices.

Supplementary Materials: The following supporting information can be downloaded at: https://www.mdpi.com/article/10.3390/ijms23095116/s1.

Author Contributions: Conceptualization, J.U.L. and J.-Y.H.; methodology, G.E.G. and Y.J.K.; validation, J.B and J.-Y.H.; formal analysis, G.E.G. and J.B.; investigation, G.E.G. and Y.J.K.; data curation, J.B. and J.-Y.H.; writing—original draft preparation, J.B.; writing—review and editing, J.U.L. and J.-Y.H. All authors have read and agreed to the published version of the manuscript.

Funding: This work was supported by the Technology Innovation Program (20006777, Development of SiOx/artificial graphite composite for anode material by using binder and coating pitch), funded by the Ministry of Trade, Industry & Energy (MOTIE, Korea) and by the Industrial Strategic Technology Development Program (20012763, development of petroleum residue-based porous adsorbent for industrial wastewater treatment), funded by the Ministry of Trade, Industry & Energy (MOTIE, Korea), by the National Research Foundation of Korea, grant numbers NRF-2021R1F1A1061939, and by the National Research Foundation of Korea grant, funded by the Korean government (MSIT) (No. 2021R1F1A1061183).

Institutional Review Board Statement: Not applicable.

Informed Consent Statement: Not applicable.

Data Availability Statement: Data sharing is not applicable to this article.

Acknowledgments: Not applicable.

Conflicts of Interest: The authors declare no conflict of interest.

References

1. A Lyczko, M.; Wiaderek, B.; Bilewicz, A. Separation of radionuclides from spent decontamination fluids via adsorption onto titanium dioxide nanotubes after photocatalytic degradation. *Nanomaterials* **2020**, *10*, 1553. [CrossRef] [PubMed]
2. Obaid, S.S.; Gaikwad, D.K.; Sayyed, M.I.; Al-Rashdi, K.; Pawar, P.P. Heavy metal ions removal from waste water by the nat-ural zeolites. *Mater. Today* **2018**, *5*, 17930–17934.
3. El-Deen, S.E.A.S.; El-Deen, G.E.S.; Jamil, T.S. Sorption behavior of co-radionuclides from radioactive waste solution on gra-phene enhanced by immobilized sugarcane and carboxy methyl cellulose. *Radiochimica Acta* **2019**, *107*, 397–413. [CrossRef]
4. Janusz, W.; Skwarek, E. Study of sorption processes of strontium on the synthetic hydroxyapatite. *Adsoption* **2016**, *22*, 697–706. [CrossRef]
5. Olatunji, M.A.; Khandaker, M.U.; Mahmud, H.E.; Amin, Y.M. Influence of adsorption parameters on caesium uptake from aqueous solutions-a brief review. *RSC Advances* **2015**, *5*, 71658–71683. [CrossRef]
6. Li, J.; Zan, Y.; Zhang, Z.; Dou, M.; Wang, F. Prussian blue nanocubes decorated on nitrogen-doped hierarchically porouscarbon network for efficient sorption of radioactive cesium. *J. Hazard. Mater.* **2020**, *385*, 121568. [CrossRef]
7. Talan, D.; Huang, Q. Separation of radionuclides from a rare earth-containing solution by zeolite Adsorption. *Minerals* **2021**, *11*, 20. [CrossRef]
8. Boulanger, N.; Kuzenkova, A.S.; Iakunkov, A.; Romanchuk, A.Y.; Trigub, A.V.; Egorov, A.V.; Bauters, S.; Amidani, L.; Re-tegan, M.; Kvashnina, K.O.; et al. Enhanced sorption of radionuclides by defect-rich graphene oxide. *ACS Appl. Mater. Interfaces* **2020**, *12*, 45122–45135. [CrossRef]
9. Skwarek, E.; Janusz, W. Adsorption of Ba2+ ions at the hydroxyapatite/NaCl solution interface. *Adsoption* **2019**, *25*, 279–288. [CrossRef]
10. Majidnia, Z.; Idris, A. Evaluation of caesium removal from radioactive waste water using maghemite PVA–alginate beads. *Chem. Eng. J.* **2015**, *262*, 372–382. [CrossRef]
11. Ofomaja, A.E.; Pholosi, A.; Naidoo, E.B. Kinetics and competitive modelling of caesium biosortion onto chemically modified pine cone powder. *J. Taiwan Inst. Chem. Eng.* **2013**, *44*, 943–951. [CrossRef]
12. Wang, J.; Zhuang, S. Caesium separation from radioactive waste by extraction and adsorption based on crown ethers and calixarenes. *Nucl. Eng. Technol.* **2020**, *52*, 328–336. [CrossRef]
13. Hong, J.Y.; Bak, B.M.; Wie, J.J.; Kong, J.; Park, H.S. Reversibly compressible, highly elastic, and durable graphene aerogels for energy storage devices under limiting condition. *Adv. Funct. Mater.* **2015**, *25*, 1053–1062. [CrossRef]
14. Hong, J.Y.; Yu, X.; Bak, B.M.; Pang, C.; Park, H.S. Bio-inspired functionalization and redox charge transfer of graphene oxide sponges for pseudocapacitive electrodes. *Carbon* **2015**, *83*, 71–78. [CrossRef]
15. Hong, J.Y.; Sohn, E.H.; Park, S.; Park, H.S. Highly-efficient and recyclable oil absorbing performance of functionalized graphene aerogel. *Chem. Eng. J.* **2015**, *269*, 229–235. [CrossRef]
16. Hong, J.Y.; Yun, S.; Wie, J.J.; Zhang, X.; Dresselhaus, M.S.; Kong, J.; Park, H.S. Cartilage-inspired superelastic ultradurable graphene aerogels prepared by the selective gluing of intersheet joints. *Nanoscale* **2016**, *8*, 12900–12909. [CrossRef]
17. Tian, W.; Zhang, H.; Duan, X.; Sun, H.; Shao, G.; Wang, S. Porous Carbons: Structure-Oriented Design and Versatile Applications. *Adv. Funct. Mater.* **2020**, *30*, 1909265. [CrossRef]
18. Oh, Y.J.; Shin, M.C.; Kim, J.H.; Yang, S.J. Facile preparation of ZnO quantum dots@porous carbon composites through direct carbonization of metal–organic complex for high-performance lithium ion batteries. *Carbon Lett.* **2021**, *31*, 323–329. [CrossRef]
19. Das, S.; Heasman, P.; Ben, T.; Qiu, S. Porous Organic Materials: Strategic Design and Structure–Function Correlation. *Chem. Rev.* **2017**, *117*, 1515–1563. [CrossRef]
20. Yang, X.Y.; Chen, L.H.; Li, Y.; Rooke, J.C.; Sanchez, C.; Su, B.L. Hierarchically porous materials: Synthesis strategies and structure design. *Chem. Soc. Rev.* **2017**, *46*, 481–558. [CrossRef]
21. Zhang, Z.; He, J.; Tang, X.; Wang, Y.; Yang, B.; Wang, K.; Zhang, D. Supercapacitors based on a nitrogen doped hierarchical porous carbon fabricated by self activation of biomass: Excellent rate capability and cycle stability. *Carbon Lett.* **2019**, *29*, 585–594. [CrossRef]

22. Othman, F.E.C.; Ismail, M.S.; Yusof, N.; Samitsu, S.; Yusop, M.Z.; Arifn, N.F.T.; Alias, N.H.; Jaafar, J.; Aziz, F.; Salleh, W.N.W.; et al. Methane adsorption by porous graphene derived from rice husk ashes under various stabilization temperatures. *Carbon Lett.* **2020**, *30*, 535–543. [CrossRef]
23. Yan, R.; Wang, K.; Tian, X.; Li, X.; Yang, T.; Xu, X.; He, Y.; Lei, S.; Song, Y. Heteroatoms in situ doped hierarchical porous hollow activated carbons for high performance supercapacitor. *Carbon Lett.* **2020**, *30*, 331–344. [CrossRef]
24. Xu, X.; Feng, X.; Liu, Z.; Xue, S.; Zhang, L. 3D flower-liked Fe_3O_4/C for highly sensitive magnetic dispersive solid-phase extraction of four trace non-steroidal anti-inflammatory drugs. *Microchim. Acta* **2021**, *188*, 52. [CrossRef] [PubMed]
25. Wang, Q.; Lv, Y.; Zhu, T. Silane coupling agent assisting dopamine-functionalized biomass porous carbons for enhanced adsorption of organic acids: Effects of acid–alkali activation on microstructure. *Carbon Lett.* **2021**, *31*, 29–37. [CrossRef]
26. Cheng, J.; Lu, Z.; Zhao, X.; Chen, X.; Zhu, Y.; Chu, H. Electrochemical performance of porous carbons derived from needle coke with different textures for supercapacitor electrode materials. *Carbon Lett.* **2021**, *31*, 57–65. [CrossRef]
27. Lee, B.M.; Choi, B.S.; Lee, J.Y.; Hong, S.K.; Lee, J.S.; Choi, J.H. Fabrication of porous carbon beads from polyacrylonitrile as electrode materials for electric double-layer capacitors. *Carbon Lett.* **2021**, *31*, 67–74. [CrossRef]
28. Kim, Y.; Kim, I.; Lee, T.S.; Lee, E.; Lee, K.J. Porous hydrogel contaning Prussian blue nanoparticles for effective cesium ion adsorption in aqueous media. *J. Ind. Eng. Chem.* **2018**, *60*, 146–155. [CrossRef]
29. Vaucher, S.; Li, M.; Mann, S. Synthesis of prussian blue nanoparticles and nanocrystal superlattices in reverse microemulsions. *Angew. Chem. Int. Ed.* **2000**, *39*, 1793–1796. [CrossRef]
30. Yang, H.M.; Jang, S.C.; Hong, S.B.; Lee, K.W.; Roh, C.H.; Huh, Y.S.; Seo, B.K. Prussian blue-functionalized magnetic nanoclusters for the removal of radioactive caesium from water. *J. Alloys Compd.* **2016**, *657*, 387–393. [CrossRef]
31. Samain, L.; Grandjean, F.; Long, G.J.; Martinetto, P.; Bordet, P.; Strivay, D. Relationship between the synthesis of prussian blue pigments, their color, physical properties, and their behavior in paint layers. *J. Phys. Chem. C* **2013**, *117*, 9693–9712. [CrossRef]
32. Oh, D.M.; Kim, B.S.; Kang, S.W.; Kim, Y.S.; Yoo, S.J.; Kim, S.; Chung, Y.S.; Choung, S.W.; Han, J.H.; Jung, S.H.; et al. Enhanced immobilization of prussian blue through hydrogel formation by polymerization of acrylic acid for radioactive caesium adsorption. *Sci. Rep.* **2019**, *9*, 16334. [CrossRef] [PubMed]
33. Takahashi, A.; Tanaka, H.; Minami, K.; Noda, K.; Ishizaki, M.; Kurihara, M.; Ogawac, H.; Kawamoto, T. Unveiling Cs-adsorption mechanism of prussian blue analogs: Cs+-percolation via vacancies to complete dehydrated state. *RSC Adv.* **2018**, *8*, 34808–34816. [CrossRef]
34. Hong, J.Y.; Oh, W.K.; Shin, K.Y.; Kwon, O.S.; Son, S.; Jang, J. Spatially controlled carbon sponge for targeting internalized radioactive materials in human body. *Biomaterials* **2012**, *33*, 5056–5066. [CrossRef] [PubMed]
35. Nilchi, A.; Saberi, R.; Moradi, M.; Azizpour, H.; Zarghami, R. Adsorption of caesium on copper hexacyanoferrate-PAN composite ion exchanger from aqueous solution. *Chem. Eng. J.* **2011**, *172*, 572–580. [CrossRef]
36. Sheha, R.R. Synthesis and characterization of magnetic hexacyanoferrate (II) polymeric nanocomposite for separation of caesium from radioactive waste solutions. *J. Colloid Interface Sci.* **2012**, *388*, 21–30. [CrossRef] [PubMed]
37. Vincent, C.; Barré, Y.; Vincent, T.; Taulemesse, J.M.; Robitzer, M.; Guibal, E. Chitin-prussian blue sponges for Cs(I) recovery: From synthesis to application in the treatment of accidental dumping of metal-bearing solutions. *J. Hazard. Mater.* **2015**, *28*, 171–179. [CrossRef]
38. Chen, G.-R.; Chang, Y.-R.; Liu, X.; Kawamoto, T.; Tanaka, H.; Parajuli, D.; Kawasaki, T.; Kawatsu, Y.; Kobayashi, T.; Chen, M.-L.; et al. Cesium removal from drinking water using Prussian blue adsorption followed by anion exchange process. *Sep. Purif. Technol.* **2017**, *172*, 147–151. [CrossRef]
39. Ishizaki, M.; Akiba, S.; Ohtani, A.; Hoshi, Y.; Ono, K.; Matsuba, M.; Togashi, T.; Kananizuka, K.; Sakamoto, M.; Takahashi, A.; et al. Proton-exchange mechanism of specific Cs+ adsorption via lattice defect sites of Prussian blue filled with coordination and crystallization water molecules. *Dalton Trans.* **2013**, *42*, 16049–16055. [CrossRef]
40. Vipin, A.K.; Fugetsu, B.; Sakata, I.; Isogai, A.; Endo, M.; Li, M.; Dresselhaus, M.S. Cellulose nanofiber backboned Prussian blue nanoparticles as powerful adsorbents for the selective elimination of radioactive cesium. *Sci Rep.* **2016**, *6*, 37009. [CrossRef]
41. Kobayashi, N.; Yamamoto, Y.; Akashi, M. Prussian blue as an agent for decontamination of ^{137}Cs in radiation accidents. *Jpn. J. Health Phys.* **1998**, *33*, 323–330. [CrossRef]
42. Jang, S.C.; Hong, S.B.; Yang, H.M.; Lee, K.W.; Moon, J.K.; Seo, B.K.; Huh, Y.S.; Roh, C. Removal of radioactive cesium using prussian blue magnetic nanoparticles. *Nanomaterials* **2014**, *4*, 894–901. [CrossRef] [PubMed]
43. Gu, G.E.; Bae, J.; Park, H.S.; Hong, J.Y. Development of the functionalized nanocomposite materials for adsorption/decontamination of radioactive pollutants. *Materials* **2021**, *14*, 2896. [CrossRef] [PubMed]
44. Vipin, A.K.; Hu, B.; Fugetsu, B. Prussian blue caged in alginate/calcium beads as adsorbents for removal of caesium ions from contaminated water. *J. Hazard. Mater.* **2013**, *258*, 93–101. [CrossRef] [PubMed]
45. Grandjean, F.; Samain, L.; Long, G.J. Characterization and utilization of prussian blue and its pigments. *Dalton Trans.* **2016**, *45*, 18018–18044. [CrossRef] [PubMed]
46. Kjeldgaard, S.; Dugulan, I.; Mamakhel, A.; Wagemaker, M.; Iversen, B.B.; Bentien, A. Strategies for synthesis of prussian blue analogues. *R. Soc. Open Sci.* **2021**, *8*, 201779. [CrossRef] [PubMed]
47. Langmuir, The adsorption of gases on plane surfaces of glass, mica and platinum. *J. Am. Chem. Soc.* **1918**, *40*, 1361–1403. [CrossRef]
48. Freundlich, H.; Heller, W. The adsorption of cis-and trans-azobenzene. *J. Am. Chem. Soc.* **1939**, *61*, 2228–2230. [CrossRef]

49. Rafatullah, M.; Sulaiman, O.; Hashim, R.; Ahmad, A. Adsorption of copper (II), chromium (III), nickel (II) and lead (II) ions from aqueous solutions by meranti sawdust. *J. Hazard. Mater.* **2009**, *170*, 969–977. [CrossRef]
50. Ho, Y.S.; Wase, D.J.; Forster, C.F. Kinetic studies of competitive heavy metal adsorption by sphagnum moss peat. *Environ. Technol.* **1996**, *17*, 71–77. [CrossRef]
51. Mckay, G. Bino, M.J.; Altamemi, A.R. The adsorption of various pollutants from aqueous solutions on to activated carbon. *Water Res.* **1985**, *19*, 491–495. [CrossRef]
52. Kim, B.; Oh, D.; Kang, S.; Kim, Y.; Kim, S.; Chung, Y.; Seo, Y.; Hwang, Y. Reformation of the surface of powdered activated carbon (PAC) using covalent organic polymers (COPs) and synthesis of a Prussian blue impregnated adsorbent for the decontamination of radioactive cesium. *J. Alloys Compd.* **2019**, *785*, 46–52. [CrossRef]
53. Yang, H.; Sun, L.; Zhai, J.; Li, H.; Zhao, Y.; Yu, H. In situ controllable synthesis of magnetic Prussian blue/graphene oxide nanocomposites for removal of radioactive cesium in water. *J. Mater. Chem. A* **2014**, *2*, 326–332. [CrossRef]
54. Thammawong, C.; Opaprakasit, P.; Tangboriboonrat, P.; Sreearunothai, P. Prussian blue-coated magnetic nanoparticles for removal of cesium from contaminated environment. *J. Nanopart. Res.* **2013**, *15*, 1689–1699. [CrossRef]
55. Jang, S.C.; Haldorai, Y.; Lee, G.W.; Hwang, S.K.; Han, Y.K.; Roh, C.; Huh, Y.S. Porous three-dimensional graphene foam/Prussian blue composite for efficient removal of radioactive (137)Cs. *Sci. Rep.* **2015**, *5*, 17510. [CrossRef] [PubMed]
56. Feng, S.; Li, X.; Ma, F.; Liu, R.; Fu, G.; Xing, S.; Yue, X. Prussian blue functionalized microcapsules for effective removal of cesium in a water environment. *RSC Adv.* **2016**, *6*, 34399–34410. [CrossRef]
57. Jia, Z.; Cheng, X.; Guo, Y.; Tu, L. In-situ preparation of iron(III) hexacyanoferrate nano-layer on polyacrylonitrile membranes for cesium adsorption from aqueous solutions. *Chem. Eng. J.* **2017**, *325*, 513–520. [CrossRef]
58. Kwon, Y.J.; Kwon, Y.; Park, H.S.; Lee, J.U. Mass-produced electrochemically exfoliated graphene for ultrahigh thermally conductive paper using a multimetal electrode system. *Adv. Mater. Interfaces* **2019**, *6*, 1900095. [CrossRef]

Article

Ab Initio Study of the Interaction of a Graphene Surface Decorated with a Metal-Doped C_{30} with Carbon Monoxide, Carbon Dioxide, Methane, and Ozone

Mónica Canales [1], Juan Manuel Ramírez-de-Arellano [2,*], Juan Salvador Arellano [1] and Luis Fernando Magaña [3,*]

[1] Universidad Autónoma Metropolitana Unidad Azcapotzalco, Av. San Pablo Xalpa No. 180, Colonia Reynosa Tamaulipas, Delegación Azcapotzalco, Ciudad de México 02200, Mexico; monic.canales@gmail.com (M.C.); jsap@azc.uam.mx (J.S.A.)
[2] Escuela de Ingeniería y Ciencias, Tecnologico de Monterrey, Av. Eugenio Garza Sada 2501, Monterrey 64849, Mexico
[3] Instituto de Física, Universidad Nacional Autónoma de México, Apartado Postal 20-364, Ciudad de México 01000, Mexico
* Correspondence: jramirezdearellano@tec.mx (J.M.R.-d.-A.); fernando@fisica.unam.mx (L.F.M.)

Abstract: Using DFT simulations, we studied the interaction of a semifullerene C_{30} and a defected graphene layer. We obtained the C_{30} chemisorbs on the surface. We also found the adsorbed C_{30} chemisorbs, Li, Ti, or Pt, on its concave part. Thus, the resulting system (C_{30}-graphene) is a graphene layer decorated with a metal-doped C_{30}. The adsorption of the molecules depends on the shape of the base of the semifullerene and the dopant metal. The CO molecule adsorbed without dissociation in all cases. When the bottom is a pentagon, the adsorption occurs only with Ti as the dopant. It also adsorbs for a hexagon as the bottom with Pt as the dopant. The carbon dioxide molecule adsorbs in the two cases of base shape but only when lithium is the dopant. The adsorption occurs without dissociation. The ozone molecule adsorbs on both surfaces. When Ti or Pt are dopants, we found that the O_3 molecule always dissociates into an oxygen molecule and an oxygen atom. When Li is the dopant, the O_3 molecule adsorbs without dissociation. Methane did not adsorb in any case. Calculating the recovery time at 300 K, we found that the system may be a sensor in several instances.

Keywords: carbon nanostructures; semifullerene; pollutant molecules; adsorption; graphene; carbon monoxide; carbon dioxide; methane; ozone

1. Introduction

Molecules, such as CO, CO_2, CH_4, and O_3, are air and water pollutants that threaten the environment and life, prompting the scientific community to develop technological solutions to such challenges [1–3]. In this study, we are interested in exploring the use of fullerenes for such aims.

Surfaces based on fullerenes and their variations have been widely studied since the prediction and further synthesis of the C_{60} structure [4–6], a highly stable group of molecules consisting of 60 carbon atoms, also named buckminsterfullerene, buckyball, or simply fullerene. Although fullerenes, such as C_{60}, C_{70}, or larger, are the most commonly studied [7,8], smaller fullerenes can also be experimentally produced and are of particular interest due to their curvature [9–11].

Fullerene fragments such as a C_{30} hydrocarbon—i.e., half of the buckminsterfullerene C_{60}—can show some of the properties of their complete counterparts [9] while also offering new possibilities due to their open basket-like shape. Similar nonplanar-related structures are corannulene ($C_{20}H_{10}$) and coronene, known since the 1960s [12–14]. The latter is a bowl carbon structure with 20 atoms or C_{20}, the smallest possible fullerene, which has been experimentally produced [11]. And the discovery of bidimensional, planar structures, such as graphene [15,16] and borophene [17], has also attracted attention because of their attractive properties and potential applications.

Previous investigations from other authors considered fullerenes on a graphene surface, focusing on studying weak interactions at a molecular level [18]. Graphene can accept electrons from a C_{60} fullerene relatively quickly, which, combined with the high transport capability of the former, turns this hybrid material into a good candidate for solar cell technology [19]. The development of hybrid surfaces has also focused on fabricating graphene-C_{60} films on silicon surfaces by a multistep self-assembly process [20]. The potential applications of these systems are promising, especially as lubricating films in electromechanics microsystems. Graphene-C_{60} vertical heterostructures composed of C_{60} thin films have also focused on their structural and electrical properties [21]. The absorption of pollutants, such as $COCl_2$ (phosgene), H_2S, CO, or CO_2, among others, by these hybrid structures has also raised attention. Decorating such arrangements with transition metals usually catalyzes absorption [22–25].

This work studies a mixed surface formed by a semifullerene C_{30} adsorbed on a defective 5 × 5 graphene layer without a hexagonal ring, i.e., six carbon vacancies. The roughness of the surface at several sites and the change in curvature make this an attractive system to dope with different atoms. We considered Li, Ti, and Pt-decorations and then studied the ability of the compound system to capture the pollutant molecules mentioned above. We found that all the molecules reacted with the surface except methane.

2. Results

2.1. Optimization of the Semifullerene C_{30}

We took two different parts when splitting a fullerene C_{60} into two halves ("buckyballs") [10] to obtain a semifullerene C_{30}. One has a pentagon in the base (section P), and the other has a hexagonal base (section H). Figure 1 shows the optimization for each case. Figure 1a,b show the C_{30} with a pentagon at the bottom, and Figure 1c,d show the C_{30} with a hexagonal base. After optimization, we discovered that in the C_{60} molecule, the separation between the carbon atoms is 1.425 Å. For section P, the distance is 1.444 Å at the bottom, and for the rest of the particles, the average separation is 1.375 Å. For section H, the space is 1.485 Å at the base, and the average spacing is 1.436 Å for the other particles. The results from other authors [10] agree with our results.

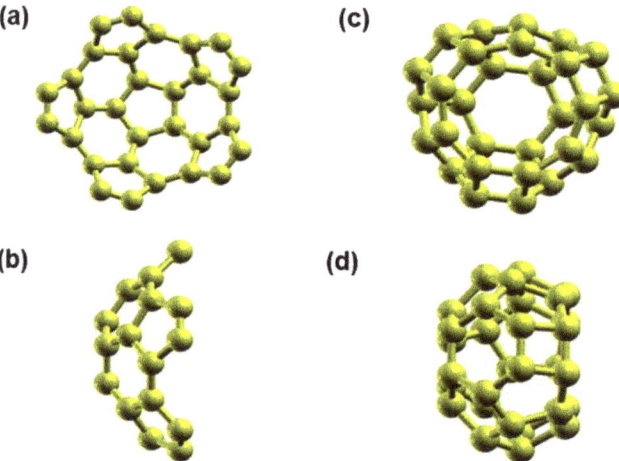

Figure 1. Molecules after optimization of C_{30}. In (**a**,**b**), we have C30 with a pentagonal base, in a front and a side view, respectively. In (**c**,**d**), we show a front and side view for C_{30} with a hexagonal base.

2.2. Optimization of Graphene with a Six-Vacancy Cluster

The vacancies in the graphene layer are necessary for the adsorption of the C30 molecule. We considered a graphene unit cell with 50 atoms and made a six vacancy cluster. Then, we optimized the system. Figure 2 shows the final configuration. We note that there is some distortion in the graphene lattice. The carbon atoms around the vacancies have different separations concerning pristine graphene. The bond lengths marked with A are 1.403 Å, and those marked with X are 1.452 Å. The other bonds are 1.420 Å, which is the same size as pristine graphene.

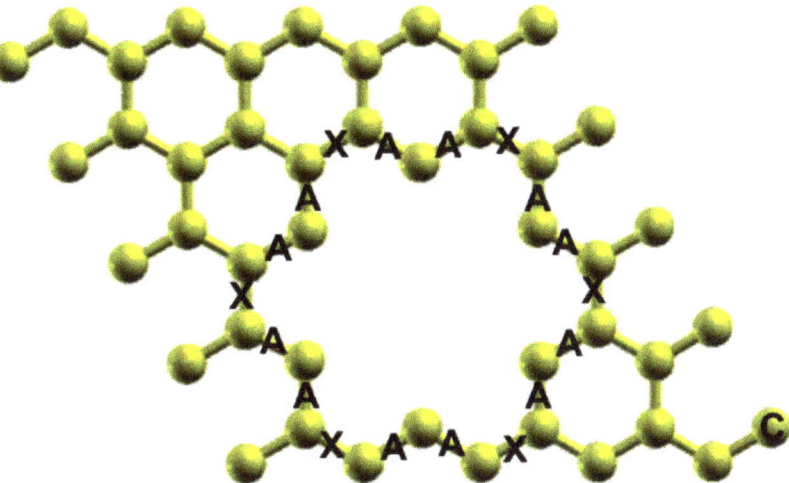

Figure 2. Unit cell after optimization of graphene with a six-vacancy cluster. The bond lengths marked with A are 1.403 Å, and those marked with X are 1.452 Å. The other bonds are the same size as in pristine graphene, 1.420 Å.

2.3. Adsorption of the C_{30} Molecule with a Pentagonal Base

The left column (P) in Figure 3 shows the adsorption of the C_{30} molecule with a pentagonal base in row 1. The initial location of the C_{30} molecule is above the cluster vacancies. Besides, the molecule is, with the closest carbon atom to the surface, at a distance of 3 Å. In the same column, row 2, we can see the system's final configuration. The adsorption energy is −15.29 eV, indicating a powerful graphene reaction. We perceive a view from above, the graphene surface in row 3 of the same column after adsorption using four-unit cells.

2.4. Adsorption of the C_{30} Molecule with a Hexagonal Base

Column H in row 1 shows the initial location of the C_{30} molecule with a hexagonal bottom concerning the graphene layer with the closest carbon atom to the surface at a distance of 3 Å. In the same column, row 2, we can see the system's final configuration. The adsorption energy is −16.410 eV, which is a stronger adsorption than in the pentagonal case. We perceive the graphene surface in row 3 of the same column after adsorption using four-unit cells.

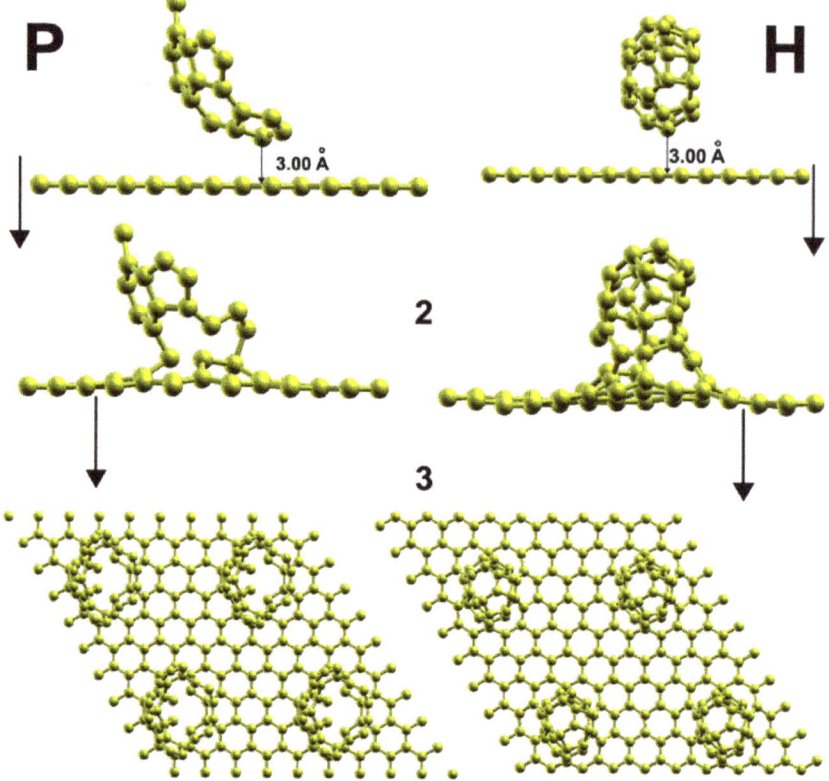

Figure 3. Adsorption of C_{30} on graphene with a six-vacancy cluster. In column P, we show the adsorption of the C_{30} molecule with a pentagonal base. In the same column P, in row 1, we have the initial location of the semifullerene. We have the final configuration after adsorption in the second row of the same column. We view the surface with four-unit cells from above in the last row of this column, P. The corresponding sequence for a C_{30} with a hexagonal base is in column H.

2.5. Adsorption of Metals on the Graphene-C_{30} (P) Surface

2.5.1. Doping with Li

Figure 4a presents the initial and final configuration for the adsorption of a lithium atom on the surface. The initial distance between the metal atom and the plane defined by the opening of C_{30} was 3.27 Å and 5.27 Å from the graphene layer. The lithium atom ends up bound to a carbon atom of the C_{30}. The adsorption energy of Li is −3.686 eV, which indicates a strong reaction with the surface. The Li atom yields 0.0561 electrons.

Figure 5 shows the interaction's projected density of states (PDOS). Note the hybridization of orbitals s and p from carbon with the orbital p from lithium around the Fermi energy at around 4 eV above the Fermi energy and about 2 eV below the Fermi energy.

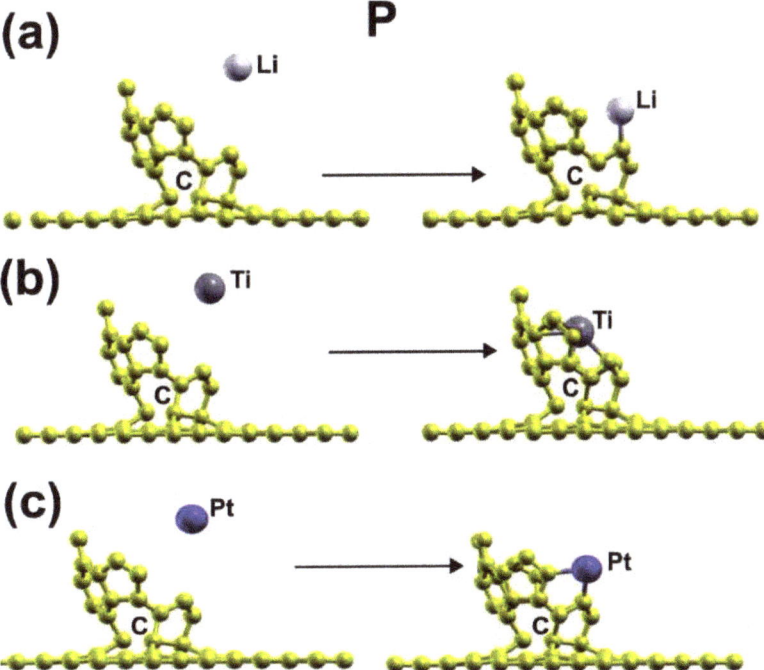

Figure 4. Adsorption of Li, Pt, and Ti on the graphene-C_{30} system for the pentagonal base. The three metals adsorbed with a strong reaction on the surface. (**a**) presents the initial and final configuration for the adsorption of a lithium atom on the surface. (**b**) shows the initial and final configuration for the adsorption of a titanium atom on the surface. (**c**) shows the initial and final configuration for the adsorption of a platinum atom on the surface.

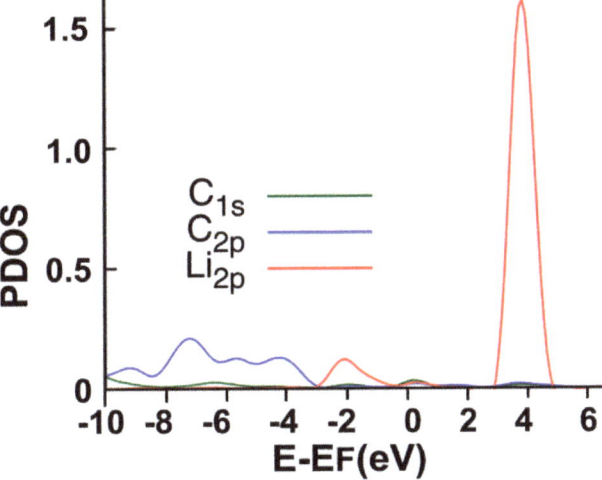

Figure 5. The PDOS for the adsorption of Li on the graphene-C_{30} system for the pentagonal base.

2.5.2. Doping with Ti

Figure 4b shows the initial and final configuration for the adsorption of a titanium atom on the surface. The initial distance between the metal atom and the plane defined by the opening of C_{30} was 3.34 Å and 5.25 Å from the graphene layer. The titanium atom ends up bound to four carbon atoms of the C_{30}. The adsorption energy is = -8.082 eV, implying an intense reaction. The Ti atom yields 0.6129 electrons to the surface.

We can see in Figure 6 the interaction's PDOS. We note the hybridization of orbitals s and d from titanium with the orbitals p from the neighboring carbon atoms between -4 eV, a bit below the Fermi energy, and between 1 eV and 5 eV.

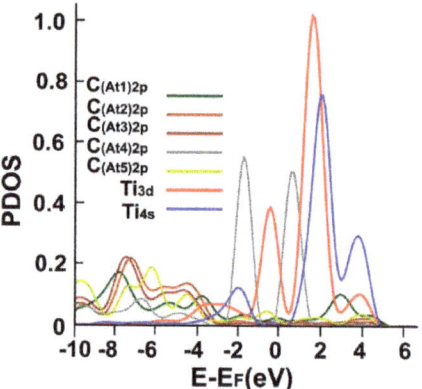

Figure 6. The PDOS for the adsorption of Ti on the graphene-C_{30} system for the pentagonal base.

2.5.3. Doping with Pt

Figure 4c shows the initial and final configuration for the adsorption of a platinum atom on the surface. The initial distance between the metal atom and the plane defined by the opening of C_{30} was 3.36 Å and 5.24 Å from the graphene layer. The platinum atom ends up bound to two carbon atoms of the C_{30}, with an adsorption energy of -5.982 eV, showing a strong reaction with the surface again. The Pt atom yields 0.3910 electrons to the surface.

Figure 7 shows the corresponding PDOS. We note the hybridization of orbital p from carbon with the orbitals s and p from platinum, around the Fermi energy, at around 2 eV above the Fermi energy, at about 2 eV below the Fermi energy, and below -4 eV.

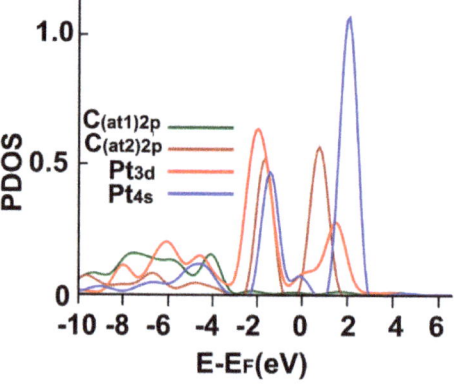

Figure 7. The PDOS for the adsorption of Pt on the graphene-C_{30} system for the pentagonal base.

2.6. Adsorption of Metals on the Graphene-C$_{30}$ (H) Surface

2.6.1. Doping with Li

Figure 8a shows the initial and final configuration for the adsorption of a lithium atom on the surface. The initial distance between the metal atom and the plane defined by the opening of C$_{30}$ was 3.37 Å and 4.57 Å from the graphene layer. The lithium atom ends up bound to a carbon atom of C$_{30}$ with an adsorption energy of -1.551 eV. It is a strong reaction but not as intense as the pentagonal case. The Li atom transfers 0.0364 electrons to the surface.

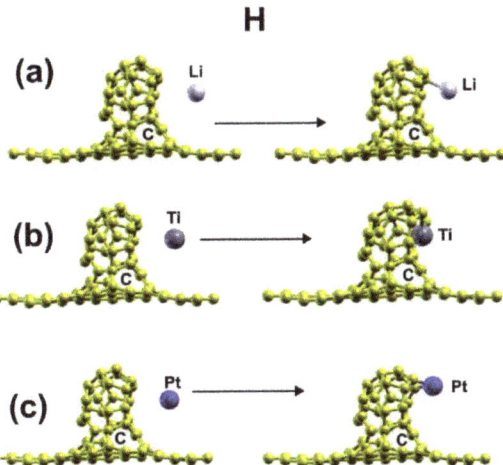

Figure 8. Adsorption of Li, Pt, and Ti on the graphene-C30 system for the hexagonal base. The three metals adsorbed strongly. (**a**) shows the initial and final configuration for the adsorption of a lithium atom on the surface. (**b**) shows the initial and final configuration for the adsorption of a titanium atom on the surface. (**c**) presents the initial and final configuration for the adsorption of a platinum atom on the surface.

Figure 9 shows the interaction's PDOS. Note the hybridization of orbitals s and p from carbon with the orbital p from lithium, between 1eV and 3 eV, around 4 eV, and a weaker hybridization between -2 eV and -1 eV.

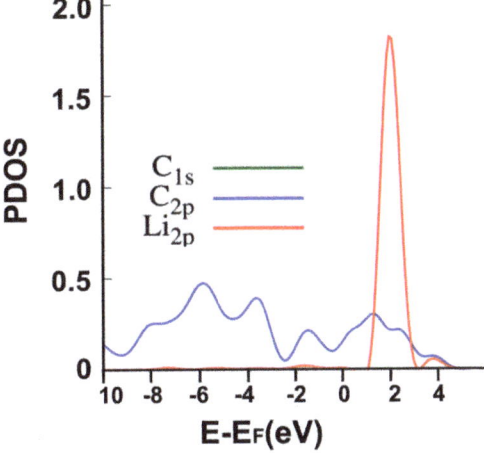

Figure 9. The PDOS for the adsorption of Li on the graphene-C$_{30}$ system for the hexagonal base.

2.6.2. Doping with Ti

Figure 8b shows the initial and final configuration for the adsorption of a titanium atom on the surface. The initial distance between the metal atom and the plane defined by the opening of C_{30} was 3.34 Å and 4.57 Å from the graphene layer. The titanium atom ends up bound to two carbon atoms of C_{30}. The adsorption energy of the titanium atom is −5.435 eV. The Ti atom transfers 0.6179 electrons to the system. The interaction is intense but not as much as in the pentagonal case.

We can see in Figure 10 the interaction's PDOS. We note the hybridization of orbitals s and d from titanium with the orbitals p from the neighboring carbon atoms around −2 eV and between the Fermi energy and 5 eV.

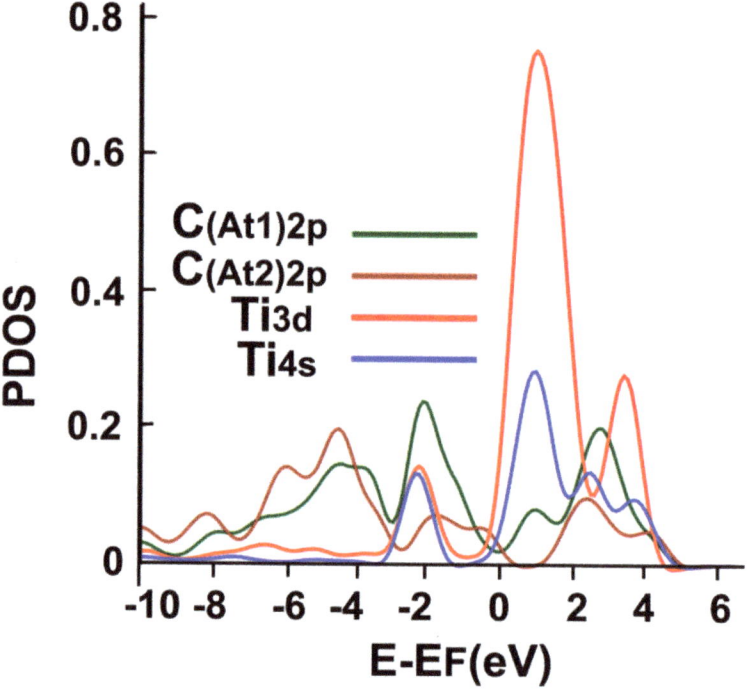

Figure 10. The PDOS for the adsorption of Ti on the graphene-C_{30} system for the hexagonal base.

2.6.3. Doping with Pt

Figure 8c presents the initial and final configuration for the adsorption of a platinum atom on the surface. The initial distance between the metal atom and the plane defined by the opening of C_{30} was 3.34 Å and 4.57 Å from the graphene layer. The platinum atom ends up bound to two carbon atoms of C_{30}. The adsorption energy of the platinum atom is −4.706 eV, which is a strong interaction with the surface but not as intense as in the pentagonal case.

The Pt atom transfers 0.5141 electrons to the surface. Figure 11 shows the corresponding PDOS. We Note the hybridization of orbital p from carbon atoms with the orbitals s and d from platinum, at around −2 eV, about 1.5 eV, and below −4 eV.

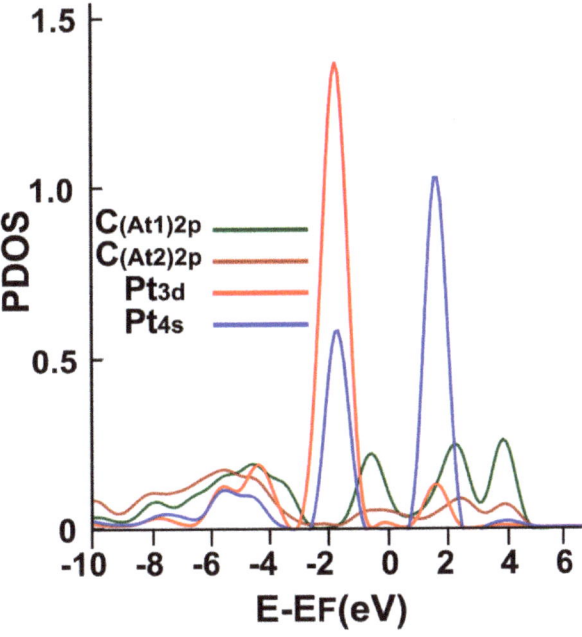

Figure 11. The PDOS for the adsorption of Pt on the graphene-C_{30} system for the hexagonal base.

2.7. *Adsorption of Pollutant Molecules on the Li-doped Graphene-C_{30} (P) Surface*

2.7.1. Adsorption of CO

There is no adsorption in this case.

2.7.2. Adsorption of CO_2

Figure 12a shows the initial and final configuration of the system for the adsorption of a carbon dioxide molecule. The molecule ends up bound to the lithium atom via the oxygen atom with an adsorption energy of -0.373 eV. The molecule transfers 0.04688 electrons to the surface.

Figure 12. (**a**) The adsorption of CO_2 on the Li-doped graphene-C_{30} system for the pentagonal base. The initial distance between the carbon atom of the CO_2 molecule and the Li atom was 3.17 Å, and the distance from the graphene layer was 7.12 Å. The molecule was parallel to the graphene layer. The adsorption is without dissociation. (**b**) The PDOS for the adsorption of CO_2 on the Li-doped graphene-C_{30} system for the pentagonal base.

Figure 12b shows the corresponding PDOS. We note the hybridization of orbital p from the oxygen atom with the orbital s from lithium at around 3 eV.

2.7.3. Adsorption of O_3

Figure 13a shows the initial and final configuration of the system for the adsorption of an ozone molecule. The molecule ends up bound to the lithium atom without dissociation.

The adsorption energy is −1.777 eV, and using MD at 300 K, we found that the particle Li-O_3 remains close to the surface at that temperature.

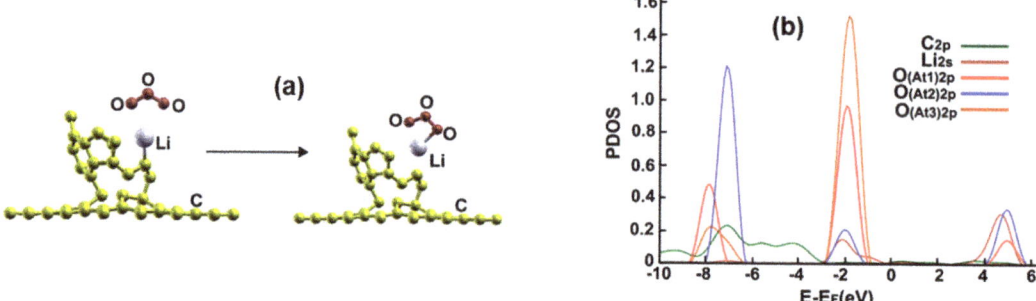

Figure 13. (**a**) The adsorption of O_3 on the Li-doped graphene-C_{30} system for the pentagonal base. The initial distance between the central oxygen atom of the O_3 molecule and the Li atom was 3.015 Å, and the distance from the graphene layer was 7.15 Å. The molecule was perpendicular to the graphene layer. The adsorption is without dissociation. (**b**) The PDOS for the adsorption of O_3 on the Li-doped graphene-C_{30} system for the pentagonal base.

Figure 13b shows the corresponding PDOS. Notice the weak hybridization of orbitals p from the oxygen and carbon atoms with the orbitals s from the lithium between 0 and 2 eV.

2.8. Adsorption of Pollutant Molecules on the Ti-Doped Graphene-C_{30} (P) Surface

2.8.1. Adsorption of CO

Figure 14a shows the initial and final configuration of the system for the adsorption of a carbon monoxide molecule. The molecule ends up bound to the titanium atom without dissociation via the carbon atom. The adsorption energy is −1.21 eV, and the molecule gains 0.0322 electrons from the surface.

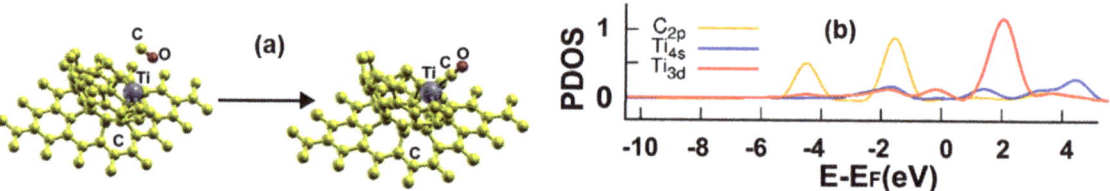

Figure 14. (**a**) Adsorption of CO on the Ti-doped graphene-C_{30} system for the pentagonal base. The initial distance between the carbon atom of the CO molecule and the Ti atom was 4.18 Å, and the distance from the graphene layer was 7.34 Å. The molecule was parallel to the graphene layer. The adsorption is without dissociation. (**b**) The PDOS for the adsorption of CO on the Ti-doped graphene-C_{30} system for the pentagonal base.

Figure 14b shows the corresponding PDOS. Notice the hybridization of orbital p from the carbon atom with the orbitals s and d from the titanium atom at around −2 eV and between 1 eV and 4 eV.

2.8.2. Adsorption of CO_2

There is no adsorption in this case.

2.8.3. Adsorption of CH_4

There is no adsorption in this case.

2.8.4. Adsorption of O_3

Figure 15a shows the initial and final configuration of the system for the adsorption of an ozone molecule. The molecule dissociates into an oxygen atom and an oxygen molecule. The oxygen atom is bound to the titanium, and the oxygen molecule is attached to the titanium atom. The adsorption energy of the ozone molecule is -6.3953 eV. The oxygen atom loses 0.2702 electrons. Besides, the oxygen molecule gains 0.4085 electrons. Using MD at 300 K, we obtained that the particle Ti-O_3 remains close to the surface at that temperature.

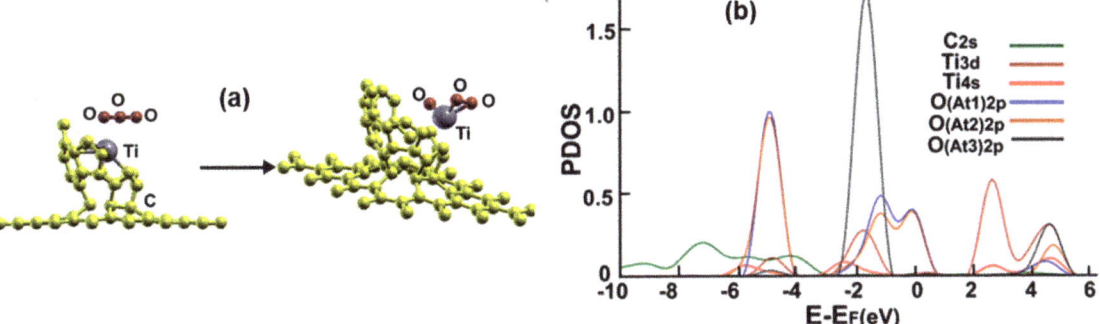

Figure 15. (a) Adsorption of O_3 on the Ti-doped graphene-C_{30} system for the pentagonal base. The initial distance between the central oxygen atom and the Ti atom was 3.0 Å, and the distance from the graphene layer was 7.14 Å. The plane of the ozone molecule was parallel to the graphene layer. The adsorption is with dissociation. (b) The PDOS for the adsorption of O_3 on the Ti-doped graphene-C_{30} system for the pentagonal base.

Figure 15b shows the corresponding PDOS. Notice a weak hybridization of orbitals s from the carbon atom with the orbitals s and d from the titanium atom and p orbitals from the oxygen atoms at around 4 eV and between -6 eV and -4 eV with p orbitals from oxygen atoms and orbitals s from the titanium atom.

2.9. Adsorption of Pollutant Molecules on the Pt-Doped Graphene-C_{30} (P) Surface

2.9.1. Adsorption of CO

There is no adsorption in this case.

2.9.2. Adsorption of CO_2

There is no adsorption in this case.

2.9.3. Adsorption of CH_4

There is no adsorption in this case.

2.9.4. Adsorption of O_3

Figure 16a shows the initial and final configuration of the system for the adsorption of an ozone molecule. The adsorption energy is -0.8521 eV, and the molecule dissociates into an oxygen atom and an oxygen molecule. The oxygen atom ends up bound to a carbon atom. Besides, the oxygen molecule ends up bound to the platinum atom. The oxygen atom, which ends bound to a carbon atom, transfers 0.1207 electrons. The remaining part of the ozone molecule, the oxygen molecule bound to the Pt atom, gains 0.5665 electrons.

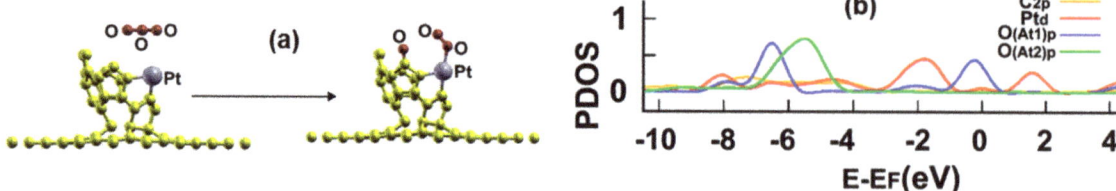

Figure 16. (a) Adsorption of O_3 on the Pt-doped graphene-C_{30} system for the pentagonal base. The initial distance between the central oxygen atom and the Pt atom was 3.60 Å, and the distance from the graphene layer was 7.25 Å. The plane of the ozone molecule was parallel to the graphene layer. The adsorption is with dissociation. (b) The PDOS for the adsorption of O_3 on the Pt-doped graphene-C_{30} system for the pentagonal base.

Figure 16b shows the corresponding PDOS. Notice a weak hybridization of orbitals p from the carbon atom with the orbitals p from the platinum and oxygen atoms at around 4.2 eV. The same hybridization is stronger below −4 eV.

2.10. Adsorption of Pollutant Molecules on the Li-Doped Graphene-C_{30} (H) Surface

2.10.1. Adsorption of CO

There is no adsorption in this case.

2.10.2. Adsorption of CO_2

Figure 17a shows the initial and final configuration of the system for the adsorption of a carbon dioxide molecule. The molecule adsorbs without dissociation, and one oxygen atom ends up bound to the lithium atom. The adsorption energy is −0.6491 eV, and the molecule transfers to the system 0.0803 electrons. The calculated recovery time at 300 K is 0.13 s, a good value for a sensor.

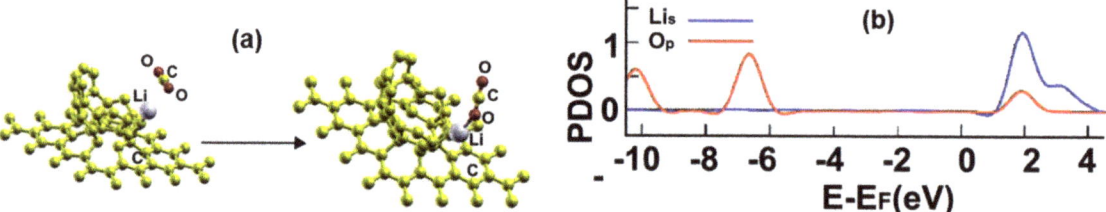

Figure 17. (a) Adsorption of CO_2 on the Li-doped graphene-C_{30} system for the hexagonal base. The initial distance between the carbon atom of the CO_2 molecule and the Li atom was 3.11 Å, and the distance from the graphene layer was 7.15 Å. The molecule was parallel to the graphene layer. The adsorption is without dissociation. (b) The PDOS for the adsorption of CO_2 on the Li-doped graphene-C_{30} system for the hexagonal base.

Figure 17b shows the corresponding PDOS. Notice the hybridization of orbitals p from the oxygen atom with the orbitals s from the lithium atom at around 2 eV.

2.10.3. Adsorption of CH_4

There is no adsorption in this case.

2.10.4. Adsorption of O_3

Figure 18a shows the initial and final configuration of the system for the adsorption of an ozone molecule. The molecule ends up bound to the lithium atom without dissociation. The adsorption energy of the ozone molecule is −2.119 eV, and the surface transfers

0.2883 electrons to the ozone molecule. Using MD at 300 K, we found that the particle Li-O_3 remains close to the surface at that temperature; it does not go away from the surface.

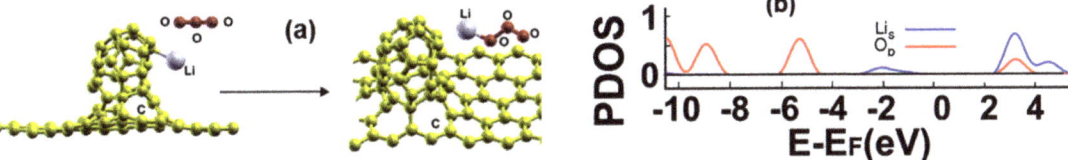

Figure 18. (**a**) Adsorption of O_3 on the Li-doped graphene-C_{30} system for the hexagonal base. The initial distance between the central oxygen atom and the Li atom was 3.26 Å, and the distance from the graphene layer was 7.35 Å. The plane of the ozone molecule was parallel to the graphene layer. The adsorption is without dissociation. (**b**) The PDOS for the adsorption of O_3 on the Li-doped graphene-C_{30} system for the hexagonal base.

Figure 18b shows the corresponding PDOS. Notice the hybridization of orbitals p from the oxygen with the orbitals s from the lithium atom between 3 eV and 4 eV. There is a weaker hybridization below the Fermi energy.

2.11. Adsorption of Pollutant Molecules on the Ti-Doped Graphene-C_{30} (H) Surface

2.11.1. Adsorption of CO

There is no adsorption in this case.

2.11.2. Adsorption of CO_2

There is no adsorption in this case.

2.11.3. Adsorption of CH_4

There is no adsorption in this case.

2.11.4. Adsorption of O_3

Figure 19a shows the initial and final configuration of the system for the adsorption of an ozone molecule. The adsorption energy is -0.8214 eV, and the molecule dissociates into two fractions during adsorption, an oxygen atom and an oxygen molecule. Besides, the first fraction is bound to a carbon atom, and the second remains close to the surface. Using MD at 300 K, we found that the molecule O_2 remains close to the surface at that temperature; it does not go away from the surface.

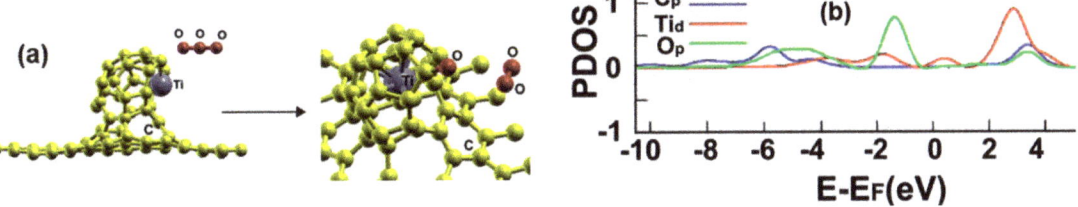

Figure 19. (**a**) Adsorption of O_3 on the Ti-doped graphene-C_{30} system for the hexagonal base. The initial distance between the central oxygen atom and the Ti atom was 3.97 Å, and the distance from the graphene layer was 6.90 Å. The plane of the ozone molecule was parallel to the graphene layer. The adsorption is with dissociation into an oxygen atom and an oxygen molecule. (**b**) The PDOS for the adsorption of O_3 on the Ti-doped graphene-C_{30} system for the hexagonal base.

Figure 19b shows the corresponding PDOS. Notice the hybridization of orbitals p from the carbon and oxygen atoms and the orbitals d from the titanium atom between 2 and 4 eV and below the Fermi energy.

2.12. Adsorption of Pollutant Molecules on the Pt-Doped Graphene-C_{30} (H) Surface

2.12.1. Adsorption of CO

Figure 20a shows the initial and final configuration of the system for the adsorption of a carbon monoxide molecule. The adsorption energy is -1.756 eV without dissociation. The carbon atom ends up bound to the platinum atom.

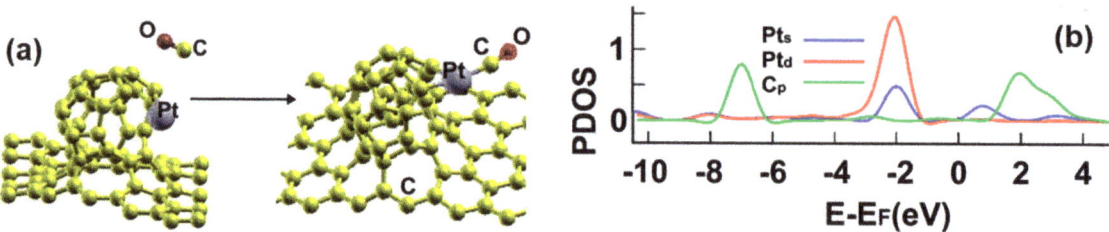

Figure 20. (a) Adsorption of CO on the Pt-doped graphene-C_{30} system for the hexagonal base. The initial distance between the center of the CO molecule and the Pt atom was 3.0 Å, and the distance from the graphene layer was 7.32 Å. The molecule was parallel to the graphene layer. The adsorption is without dissociation; (b) The PDOS for the adsorption of CO on the Pt-doped graphene-C_{30} system for the hexagonal base.

The surface transfers 0.0322 electrons to the carbon monoxide molecule. Figure 20b shows the corresponding PDOS. We can see the hybridization of orbitals p from the carbon atom and the orbitals s from the platinum atom at around 3 eV and about 1.2 eV, respectively. We can also notice a weak hybridization of orbitals p from the carbon atom with orbitals d and s from the platinum atom below -1 eV.

2.12.2. Adsorption of CO_2

There is no adsorption in this case.

2.12.3. Adsorption of CH_4

There is no adsorption in this case.

2.12.4. Adsorption of O_3

Figure 21a shows the initial and final configuration of the system for the adsorption of an ozone molecule that occurs with dissociation and with an adsorption energy of -1.43 eV. The molecule splits into two parts, an oxygen atom and an oxygen molecule. Using MD at 300 K, we found that the particle O_2 remains close to the surface at that temperature; it does not go away from the surface.

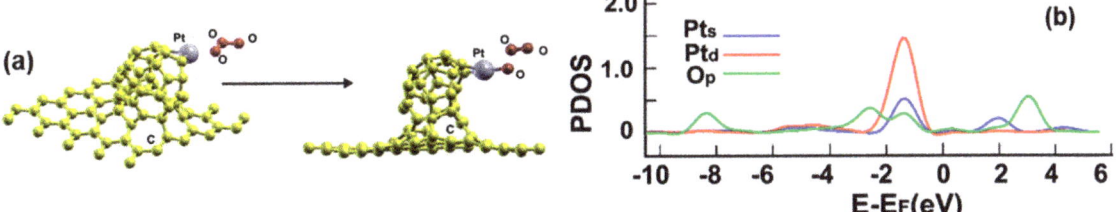

Figure 21. (**a**) Adsorption of O_3 on the Pt-doped graphene-C_{30} system for the hexagonal base. The initial distance between the central oxygen atom and the Pt atom was 3.18 Å, and the distance from the graphene layer was 7.21 Å. The plane of the ozone molecule was parallel to the graphene layer. The adsorption is with dissociation. (**b**) The PDOS for the adsorption of O_3 on the Pt-doped graphene-C_{30} system for the hexagonal base.

The oxygen atom, which ends bound to the platinum atom, transfers 0.1207 electrons. The surface transfers 0.5665 electrons to the remaining fraction of the ozone molecule and the oxygen molecule, which remains close to the surface.

Figure 21b shows the corresponding PDOS. We can see the hybridization of orbitals p from the oxygen atom and the orbitals s and d from the platinum atom at around 2 eV and about −1.75 eV, respectively. We can also notice a weak hybridization of orbitals p from the oxygen atom with orbitals d and s from the platinum atom below −2 eV.

3. Materials and Methods

We used the GGA approximation for the exchange and correlation energies in the Perdew–Burke–Ernzerhohof (PBE) expression [26], using a Martins–Troullier norm-conserving pseudopotential [27]. We performed structural relaxations using the Quantum ESPRESSO code package [28], which uses periodical boundary conditions. We took threshold energy of 1.0×10^{-6} eV for convergence, a cut-off energy point of 1100 eV, and a threshold force of 1.0×10^{-5} eV/Å. We considered 40 k points within the Monkhorst–Pack particular k point scheme for Brillouin-zone integrations [29] with a separation of 0.083 Å^{-1}.

To check the pseudopotentials, we minimized the energy of the different systems. Thus, we obtained the Li lattice parameter 3.495 Å (the experimental value is 3.510 Å) [30]; for titanium, we obtained: a = 2.863 Å, and c = 4.544 Å (the observed values are 2.950 and 4.683 Å, respectively [30]; in the case of Pt, we calculated a lattice parameter of 2.898 Å (the experimental value is 2.924 Å). We obtained the bond lengths and angles of the different pollutant molecules we are considering with the same approach. Figure 22 shows our results, which agree with the experimental values.

In our simulations, the adsorption energy is:

$$E_{ads} = E(Surf + Mol) - [E(Surf) + E(Mol)], \quad (1)$$

where $E(Surf + Mol)$ is the energy corresponding to the final system; $[E(Surf) + E(Mol)]$ corresponds to the initial configuration, which is the energy of the surface, without interaction with the molecule plus the isolated molecule's energy.

We calculated the recovery time (τ) from the Eyring transition state theory using the expression [31,32]:

$$\tau = [h/(k_B T)] e^{-E_{ads}/(k_B T)} \quad (2)$$

In Equation (2), h is the Plank's constant, k_B is the Boltzmann's constant, E_{ads} is the adsorption energy, and T is the absolute temperature.

The desirable set of values for the recovery time is between 10^{-2} and ten seconds, implying at 300 K, adsorption energies in the range (−0.6428, −0.8215) eV.

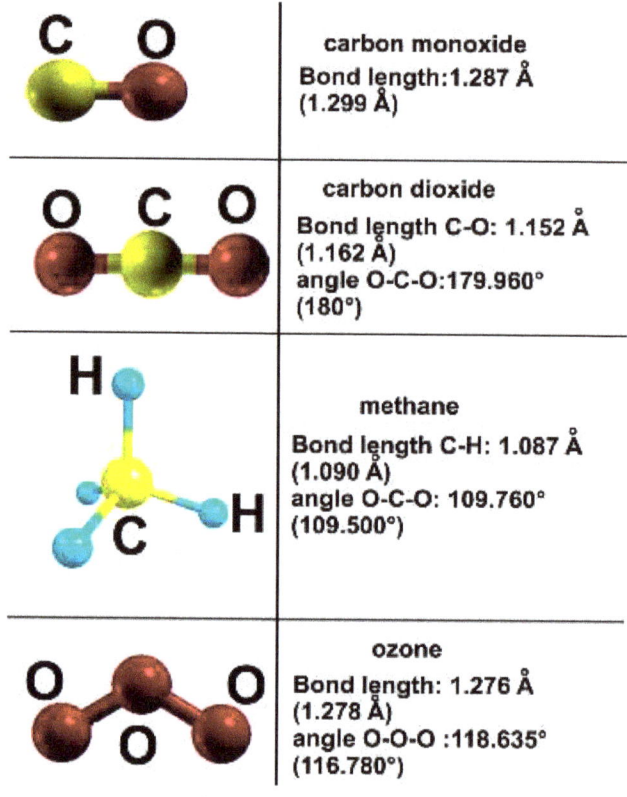

Figure 22. We compare our results, obtained by minimizing the molecule energy, and the experimental values are given in parenthesis.

4. Discussion

We performed computational simulations to investigate the adsorption of polluting molecules on graphene-semifullerene (C_{30}) surfaces, considering two C_{30} geometries: hexagonal and pentagonal base. We found it possible to dope all surfaces with the metals Li, Ti, and Pt, which we used as catalysts in the adsorption of the different polluting molecules. We consider as pollutant molecules CO, CO_2, CH_4, and O_3.

We obtained the semifullerene adsorbs on the graphene surface with adsorption energies of -14.97 eV and -16.41 eV, respectively, for pentagonal and hexagonal bases. The adsorption occurs on a six-vacancy cluster in a graphene layer. Besides, the catalysts adsorb on the graphene-C_{30} surface with a pentagonal base with adsorption energies of -4.02 eV, -6.3 eV, and -8.4 eV for Li, Pt, and Ti, respectively. For the hexagonal base, the adsorption energies are -1.87 eV, -4.7 eV, and -5.43 eV, in the same order. Notice that in each case (P or H), Li shows the adsorption energy with the minor magnitude and Ti with the largest.

The carbon monoxide molecule adsorbs on the pentagonal-base (P) surface only when Ti is the dopant, with an adsorption energy of -3.6 eV, and this adsorption is without dissociation. Furthermore, CO adsorbs on the hexagonal-base (H) surface only with Pt as the dopant with an adsorption energy of -0.89 eV. Again, the adsorption is without dissociation.

The carbon dioxide molecule adsorbs on both surfaces but only with Li as the dopant, with adsorption energies of: -0.67 eV for the P surface and -0.54 eV for the H surface. The adsorption of the CO_2 molecule is without dissociation.

The methane molecule did not adsorb on any surface.

Finally, we found that both surfaces always adsorb the ozone molecule. When Ti or Pt are dopants, we found that the O_3 molecule always dissociates into an oxygen molecule and an oxygen atom. For the P surface, the adsorption energies are -6.3953 and -0.8521 eV for the Ti and Pt doped surfaces, respectively. Furthermore, the adsorption energies for the Ti and Pt doped H surface are -0.82 eV and -1.43 eV, respectively. In the case of Li, the O_3 molecule adsorbs without dissociation. The adsorption energy is -1.777 eV for the P surface, and the adsorption energy is -2.119 eV for the H surface.

At 300 K, the P surface would not act as a suitable sensor in any case. The H surface would be a sensor for O_3 with Ti as the dopant ($\tau = 9.97$ s) and for CO_2 with Li as a dopant ($\tau = 0.13$ s).

Author Contributions: Conceptualization, methodology, validation, formal analysis, investigation, M.C., J.S.A., and L.F.M.; resources, L.F.M.; data curation, M.C. and J.M.R.-d.-A.; writing—original draft preparation, writing—review and editing, M.C., J.M.R.-d.-A., J.S.A., and L.F.M.; project administration, funding acquisition, L.F.M. All authors have read and agreed to the published version of the manuscript.

Funding: This research was funded by Dirección General de Asuntos del Personal Académico de la Universidad Nacional Autónoma de México, grant number IN113220. The APC was funded by Tecnologico de Monterrey.

Institutional Review Board Statement: Not applicable.

Informed Consent Statement: Not applicable.

Data Availability Statement: Not applicable.

Acknowledgments: We thank Dirección General de Asuntos del Personal Académico de la Universidad Nacional Autónoma de México, partial financial support by Grant IN-111817. We also appreciate UNAM-Miztli-Super-Computing Center technical assistance by the project LANCAD-UNAM-DGTIC-030, and the Yoltla (UAM-Iztapalapa) computer facilities. M. C. acknowledges CONACYT for a scholarship during this research.

Conflicts of Interest: The authors declare no conflict of interest.

References

1. Kjellstrom, T.; Lodh, M.; McMichael, T.; Ranmuthugala, G.; Shrestha, R.; Kingsland, S. Air and Water Pollution: Burden and Strategies for Control. In *Disease Control Priorities in Developing Countries*; Jamison, D.T., Breman, J.G., Measham, A.R., Alleyne, G., Claeson, M., Evans, D.B., Jha, P., Mills, A., Musgrove, P., Eds.; World Bank: Washington, DC, USA, 2006; ISBN 978-0-8213-6179-5.
2. World Health Organization Don't Pollute My Future! The Impact of the Environment on Children's Health. Available online: https://www.who.int/publications-detail-redirect/WHO-FWC-IHE-17.01 (accessed on 12 April 2022).
3. Hussain, A.; Rehman, F.; Rafeeq, H.; Waqas, M.; Asghar, A.; Afsheen, N.; Rahdar, A.; Bilal, M.; Iqbal, H.M.N. In-Situ, Ex-Situ, and Nano-Remediation Strategies to Treat Polluted Soil, Water, and Air–A Review. *Chemosphere* **2022**, *289*, 133252. [CrossRef]
4. Osawa, E. Superaromaticity. *Kagaku* **1970**, *25*, 854–863.
5. Kroto, H.W.; Heath, J.R.; O'Brien, S.C.; Curl, R.F.; Smalley, R.E. C60: Buckminsterfullerene. *Nature* **1985**, *318*, 162–163. [CrossRef]
6. Sygula, A. Chemistry on a Half-Shell: Synthesis and Derivatization of Buckybowls. *Eur. J. Org. Chem.* **2011**, *2011*, 1611–1625. [CrossRef]
7. Astefanei, A.; Núñez, O.; Galceran, M.T. Characterisation and Determination of Fullerenes: A Critical Review. *Anal. Chim. Acta* **2015**, *882*, 1–21. [CrossRef]
8. Murayama, H.; Tomonoh, S.; Alford, J.M.; Karpuk, M.E. Fullerene Production in Tons and More: From Science to Industry. *Fuller. Nanotub. Carbon Nanostruct.* **2005**, *12*, 1–9. [CrossRef]
9. Rabideau, P.W.; Abdourazak, A.H.; Folsom, H.E.; Marcinow, Z.; Sygula, A.; Sygula, R. Buckybowls: Synthesis and Ab Initio Calculated Structure of the First Semibuckminsterfullerene. *J. Am. Chem. Soc.* **1994**, *116*, 7891–7892. [CrossRef]
10. Rabideau, P.W.; Sygula, A. Buckybowls: Polynuclear Aromatic Hydrocarbons Related to the Buckminsterfullerene Surface. *Acc. Chem. Res.* **1996**, *29*, 235–242. [CrossRef]
11. Prinzbach, H.; Weiler, A.; Landenberger, P.; Wahl, F.; Wörth, J.; Scott, L.T.; Gelmont, M.; Olevano, D. Gas-Phase Production and Photoelectron Spectroscopy of the Smallest Fullerene, C20. *Nature* **2000**, *407*, 60–63. [CrossRef]
12. Barth, W.E.; Lawton, R.G. Dibenzo[Ghi,Mno]Fluoranthene. *J. Am. Chem. Soc.* **1966**, *88*, 380–381. [CrossRef]
13. Lawton, R.G.; Barth, W.E. Synthesis of Corannulene. *J. Am. Chem. Soc.* **1971**, *93*, 1730–1745. [CrossRef]

14. Filatov, A.S.; Petrukhina, M.A. Probing the Binding Sites and Coordination Limits of Buckybowls in a Solvent-Free Environment: Experimental and Theoretical Assessment. *Coord. Chem. Rev.* **2010**, *254*, 2234–2246. [CrossRef]
15. Meyer, J.C.; Geim, A.K.; Katsnelson, M.I.; Novoselov, K.S.; Booth, T.J.; Roth, S. The Structure of Suspended Graphene Sheets. *Nature* **2007**, *446*, 60–63. [CrossRef] [PubMed]
16. Geim, A.K.; Novoselov, K.S. The Rise of Graphene. *Nat. Mater.* **2007**, *6*, 183–191. [CrossRef] [PubMed]
17. Boustani, I. New Quasi-Planar Surfaces of Bare Boron. *Surf. Sci.* **1997**, *370*, 355–363. [CrossRef]
18. Švec, M.; Merino, P.; Dappe, Y.J.; González, C.; Abad, E.; Jelínek, P.; Martín-Gago, J.A. Van Der Waals Interactions Mediating the Cohesion of Fullerenes on Graphene. *Phys. Rev. B* **2012**, *86*, 121407. [CrossRef]
19. Zhang, X.; Huang, Y.; Wang, Y.; Ma, Y.; Liu, Z.; Chen, Y. Synthesis and Characterization of a Graphene–C60 Hybrid Material. *Carbon* **2009**, *47*, 334–337. [CrossRef]
20. Pu, J.; Mo, Y.; Wan, S.; Wang, L. Fabrication of Novel Graphene–Fullerene Hybrid Lubricating Films Based on Self-Assembly for MEMS Applications. *Chem. Commun.* **2013**, *50*, 469–471. [CrossRef] [PubMed]
21. Kim, K.; Lee, T.H.; Santos, E.J.G.; Jo, P.S.; Salleo, A.; Nishi, Y.; Bao, Z. Structural and Electrical Investigation of C_{60} –Graphene Vertical Heterostructures. *ACS Nano* **2015**, *9*, 5922–5928. [CrossRef] [PubMed]
22. Zhang, T.; Sun, H.; Wang, F.; Zhang, W.; Tang, S.; Ma, J.; Gong, H.; Zhang, J. Adsorption of Phosgene Molecule on the Transition Metal-Doped Graphene: First Principles Calculations. *Appl. Surf. Sci.* **2017**, *425*, 340–350. [CrossRef]
23. Khodadadi, Z. Evaluation of H2S Sensing Characteristics of Metals–Doped Graphene and Metals-Decorated Graphene: Insights from DFT Study. *Phys. E Low-Dimens. Syst. Nanostructures* **2018**, *99*, 261–268. [CrossRef]
24. Promthong, N.; Tabtimsai, C.; Rakrai, W.; Wanno, B. Transition Metal-Doped Graphene Nanoflakes for CO and CO2 Storage and Sensing Applications: A DFT Study. *Struct. Chem.* **2020**, *31*, 2237–2247. [CrossRef]
25. Khan, A.A.; Ahmad, I.; Ahmad, R. Influence of Electric Field on CO2 Removal by P-Doped C60-Fullerene: A DFT Study. *Chem. Phys. Lett.* **2020**, *742*, 137155. [CrossRef]
26. Perdew, J.P.; Burke, K.; Ernzerhof, M. Generalized Gradient Approximation Made Simple. *Phys. Rev. Lett.* **1997**, *78*, 1396. [CrossRef]
27. Kleinman, L.; Bylander, D.M. Efficacious Form for Model Pseudopotentials. *Phys. Rev. Lett.* **1982**, *48*, 1425–1428. [CrossRef]
28. Giannozzi, P.; Andreussi, O.; Brumme, T.; Bunau, O.; Buongiorno Nardelli, M.; Calandra, M.; Car, R.; Cavazzoni, C.; Ceresoli, D.; Cococcioni, M.; et al. Advanced Capabilities for Materials Modelling with Quantum ESPRESSO. *J. Phys. Condens. Matter* **2017**, *29*, 465901. [CrossRef] [PubMed]
29. Monkhorst, H.J.; Pack, J.D. Special Points for Brillouin-Zone Integrations. *Phys. Rev. B* **1976**, *13*, 5188–5192. [CrossRef]
30. Lide, D.R. (Ed.) *CRC Handbook of Chemistry and Physics: A Ready-Reference Book of Chemical and Physical Data*, 81st ed.; CRC Press: Boca Raton, FL, USA, 2000; ISBN 978-0-8493-0481-1.
31. Eyring, H. The Activated Complex in Chemical Reactions. *J. Chem. Phys.* **1935**, *3*, 107–115. [CrossRef]
32. Popa, I.; Fernández, J.M.; Garcia-Manyes, S. Direct Quantification of the Attempt Frequency Determining the Mechanical Unfolding of Ubiquitin Protein. *J. Biol. Chem.* **2011**, *286*, 31072–31079. [CrossRef]

Article

Resolving the Mechanism of Acoustic Plasmon Instability in Graphene Doped by Alkali Metals

Leonardo Marušić [1,*], Ana Kalinić [2], Ivan Radović [2], Josip Jakovac [3], Zoran L. Mišković [4] and Vito Despoja [3,5]

1. Maritime Department, University of Zadar, M. Pavlinovića 1, 23000 Zadar, Croatia
2. Department of Atomic Physics, "VINČA" Institute of Nuclear Sciences—National Institute of the Republic of Serbia, University of Belgrade, P.O. Box 522, 11001 Belgrade, Serbia; ana.kalinic@vin.bg.ac.rs (A.K.); iradovic@vin.bg.ac.rs (I.R.)
3. Institut za Fiziku, Bijenička 46, 10000 Zagreb, Croatia; jjakovac@ifs.hr (J.J.); vito@phy.hr (V.D.)
4. Department of Applied Mathematics and Waterloo Institute for Nanotechnology, University of Waterloo, Waterloo, ON N2L 3G1, Canada; zmiskovi@uwaterloo.ca
5. Donostia International Physics Center (DIPC), P. Manuel de Lardizabal 4, 20018 San Sebastián, Spain
* Correspondence: lmarusic@unizd.hr

Abstract: Graphene doped by alkali atoms (AC_x) supports two heavily populated bands (π and σ) crossing the Fermi level, which enables the formation of two intense two-dimensional plasmons: the Dirac plasmon (DP) and the acoustic plasmon (AP). Although the mechanism of the formation of these plasmons in electrostatically biased graphene or at noble metal surfaces is well known, the mechanism of their formation in alkali-doped graphenes is still not completely understood. We shall demonstrate that two isoelectronic systems, KC_8 and CsC_8, support substantially different plasmonic spectra: the KC_8 supports a sharp DP and a well-defined AP, while the CsC_8 supports a broad DP and does not support an AP at all. We shall demonstrate that the AP in an AC_x is not, as previously believed, just a consequence of the interplay of the π and σ intraband transitions, but a very subtle interplay between these transitions and the background screening, caused by the out-of-plane interband $C(\pi) \to A(\sigma)$ transitions.

Keywords: plasmon; acoustic plasmon; graphene; graphene intercalation compounds; EELS

1. Introduction

Studying acoustic plasmons (APs) in single-layer [1–5], double-layer [3,6–9] and multi-layer [10–13] graphene or in metal/dielectric/graphene superstructures [14] is a very active field of research. Recently, an AP was detected in a graphene–dielectric–metal structure using near-field scattering microscopy [2]. That experiment demonstrated a very robust character of the AP with small propagation loss, which enables its efficient coupling to the electromagnetic field at infra-red frequencies. The graphene/graphene-nanoribbon superstructure also supports a strong AP which can be excited by light and provide strong field enhancement inside the nano-gap, allowing efficient biosensing [6]. In this work, we focus on the AP in graphene doped by alkali metals.

As alkali atoms are added to graphene deposited on various substrates, they intercalate between the graphene layer and the substrate and metalize, i.e., they arrange in a two-dimensional (2D) crystal lattice, which supports a partially filled parabolic band [15–21]. The alkali atoms then usually donate electrons to the graphene π band, lifting the Fermi level above the Dirac point and transforming the graphene from a gapless semiconductor into a metal. Therefore, the alkali adlayer results in the formation of two thin, i.e., quasi two-dimensional (q2D), plasmas able to support a variety of different plasmonic modes, which do not exist in pristine or electrostatically doped graphene. Our previous theoretical research of electronic excitations in graphene doped by alkali atoms (AC_x) showed that this system supports two kinds of intraband or 2D plasmons: the Dirac plasmon (DP) and the acoustic plasmon (AP). The intensity of these modes depends on the type of the dielectric substrate, as well as on electrostatic or natural doping [22–25]. Moreover, these theoretical studies suggest the possibility of simple manipulation of the DP and AP intensities (they can be switched 'on' and 'off' in a controlled way) [25], thereby opening new possibilities for their application in various fields, such as plasmonics, photonics, transformation optics, optoelectronics, light emitters, detectors and photovoltaic devices [26–36]. Additionally, these 'tunable' 2D plasmons could be very useful in the area of chemical or biological sensing [27,37–40].

In this paper, we focus on the AP, firstly because of its linear dispersion, but also because of the still rather unclear mechanism of its occurrence in the AC_x. The mechanism of formation of the AP on a metal surface was resolved a long time ago [41–43]. In that case, a partially filled surface state supports a 2D plasmon, which hybridizes with the ordinary surface plasmon, resulting in the transformation of the usual 2D plasmon with square-root 2D dispersion into a linear AP [41]. However, our previous studies of plasmons in AC_x systems show that the AP is very sensitive and strongly dependent on several parameters: the type of the alkali metal atom (A), the coverage (x) and the electrostatic doping. In other words, it is substantially unstable, and the aim of this research is to explore the mechanism of the (in)stability of the AP in the AC_x.

In order to do that, we investigate two model systems of doped graphene, KC_8 and CsC_8, which are isostructural, but the first system does support the AP, while the second one does not. Moreover, the DP in the KC_8 is very sharp, while in the CsC_8 it is quite broad, even though the occupancies of the graphene π band and the alkali atom σ band in these two systems are almost identical. We shall show that the AP in the AC_x is not, as previously believed, just a consequence of the hybridization between the intraband 2D plasmons in the π and σ bands, but rather a very subtle interplay between these plasmons and the background screening caused by the interlayer interband $C(\pi) \rightarrow A(\sigma)$ transitions. We shall also demonstrate that the electronic screening coming from the high-energy interband transitions significantly reduces the intensity of the AP in both systems.

We study electronic excitations in the AC_x using two complementary methods: the ab initio random phase approximation (RPA) approach [24] and a reduced model. In the reduced model, the electrons in the alkali atom layer are approximated by a parabolic band with parameters (the effective mass m_σ and the Fermi energy $E_{F\sigma}$) taken from the ab initio calculations. We refer to this approximation as the "massive" 2D electron gas (m2DEG). The graphene band structure is approximated by the conical π bands with the occupation corresponding to the occupation of the π band (Fermi energy $E_{F\pi}$) in the ab initio AC_8 sample. This approximation is also known as the 'massless' Dirac fermion (MDF) approximation [44,45]. The polarization effects coming from the high-energy interband transitions are included through the polarizability parameters α_m and α_g deduced from the ab initio calculations.

The paper is organized as follows. In Section 2, we present the geometry of the system, a derivation of the ab initio ground state and the RPA spectra of the electronic excitations $S(Q, \omega)$ in the AC_x deposited on a dielectric substrate. We also present a theoretical formulation of the reduced model. In Section 3, we show the band structures of the KC_8 and CsC_8, as well as the intensities of the electronic excitations in these systems. After that, using a detailed analysis, which combines the ab initio RPA method and the reduced model, we resolve the mechanism of the AP instability. The conclusions are presented in Section 4.

2. Theoretical Formulation

In this section, we derive the spectral function of the electronic excitations in the self-standing KC_8 and CsC_8. However, to demonstrate the robustness of these electronic excitations, we show how a dielectric substrate can be included in our expressions for the spectral function, to enable the calculation of the electronic excitation spectra for systems consisting of crystals (CsC_8 and KC_8) deposited on Al_2O_3 substrate described by a local dielectric function $\epsilon_S(\omega)$. Additionally, we make use of the plausible assumption [25] that the Al$_2$O$_3$ substrate does not affect the ground state electronic structure (Kohn–Sham (KS) wave functions and energies) of the AC$_8$ crystal, but it can influence its dielectric properties.

Figure 1 shows the geometry of the studied system. The coordinates are oriented so that the AC$_8$ crystal is positioned in the $x - y$ plane, the z direction is perpendicular to the crystal plane, the graphene layer occupies the $z = 0$ plane, the alkali atom layer occupies the $z = -d$ plane and the dielectric substrate occupies the $z < -h$ half-space.

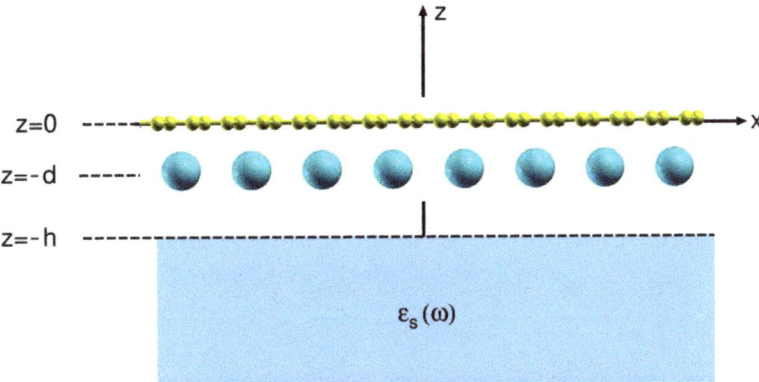

Figure 1. Schematic representation of the AC$_8$ crystal deposited on a dielectric substrate described by a local dielectric function $\epsilon_S(\omega)$. The substrate (blue) occupies the region $z < -h$, the graphene layer (yellow) is in the $z = 0$ plane and the alkali atom (turquoise) layer is in the $z = -d$ plane.

2.1. Calculation of the Surface Electronic Excitations Spectra

We shall briefly describe the method of calculation of the surface electronic excitation spectra at the arbitrary position $z > 0$, previously used in several studies of electronic excitations in 2D crystals [22,23,46–50].

We start from the 3D Fourier transform of the noninteracting (free) electron response function

$$\chi^0_{GG'}(Q,\omega) = \frac{2}{\Omega} \sum_{K \in SBZ} \sum_{n,m} \frac{f_n(K) - f_m(K+Q)}{\omega + i\eta + E_n(K) - E_m(K+Q)} \quad (1)$$

$$\times \rho_{nK,mK+Q}(G) \, \rho^*_{nK,mK+Q}(G'),$$

where $\Omega = S \times L$ is the normalization volume and $f_{nK} = [e^{(E_{nK}-E_F)/kT} + 1]^{-1}$ is the Fermi–Dirac distribution at temperature T. The matrix elements (or the charge vertices) are

$$\rho_{nK,mK+Q}(G) = \left\langle \Phi_{nK} \left| e^{-i(Q+G)\cdot r} \right| \Phi_{mK+Q} \right\rangle_{V'} \quad (2)$$

where $K = (K_x, K_y)$ is the 2D wave vector, Q is the momentum transfer vector parallel to the $x - y$ plane, $G = (G_\parallel, G_z)$ are the 3D reciprocal lattice vectors and $r = (\rho, z)$ is the 3D position vector. The integration is performed over the normalization volume Ω. The plane wave expansion of the wave function has the form

$$\Phi_{nK}(\rho, z) = \frac{1}{\sqrt{\Omega}} e^{iK\cdot\rho} \sum_G C_{nK}(G) e^{iG\cdot r},$$

where the coefficients C_{nK} are obtained by solving the KS equations within the local density approximation (LDA) self-consistently.

The Fourier expansion of the free electron response function in the z and z' coordinates is:

$$\chi^0_{\mathbf{G}_\| \mathbf{G}'_\|}(\mathbf{Q}, \omega, z, z') = \frac{1}{L} \sum_{G_z G'_z} \chi^0_{\mathbf{G}\mathbf{G}'}(\mathbf{Q}, \omega) e^{iG_z z - iG'_z z'}, \quad (3)$$

where we assume that our system is periodical in the z and z' direction as well, i.e., that it repeats periodically from supercell to supercell, and the supercells are q2D crystals separated by the distance L. We now need to determine the screened response function $\chi^0_{\mathbf{G}_\| \mathbf{G}'_\|}(\mathbf{Q}, \omega, z, z')$ of one supercell without including the polarization of the surrounding supercells. This spurious interaction with the replicas of the q2D crystal can be eliminated easily, by using the RPA Dyson equation

$$\chi_{\mathbf{G}_\| \mathbf{G}'_\|}(\mathbf{Q}, \omega, z, z') = \chi^0_{\mathbf{G}_\| \mathbf{G}'_\|}(\mathbf{Q}, \omega, z, z') + \sum_{\mathbf{G}_{\|1} \mathbf{G}_{\|2}} \int_{-L/2}^{L/2} dz_1 dz_2 \quad (4)$$
$$\times \chi^0_{\mathbf{G}_\| \mathbf{G}_{\|1}}(\mathbf{Q}, \omega, z, z_1) v^{2D}_{\mathbf{G}_{\|1} \mathbf{G}_{\|2}}(\mathbf{Q}, z_1, z_2) \chi_{\mathbf{G}_{\|2} \mathbf{G}'_\|}(\mathbf{Q}, \omega, z_2, z'),$$

where the matrix of the bare Coulomb interaction is

$$v^{2D}_{\mathbf{G}_{\|1} \mathbf{G}_{\|2}}(\mathbf{Q}, z, z') = v^{2D}_{\mathbf{G}_{\|1}}(\mathbf{Q}, z, z') \delta_{\mathbf{G}_{\|1} \mathbf{G}_{\|2}}, \quad (5)$$

and the 2D Fourier transform of the bare Coulomb interaction is

$$v^{2D}_{\mathbf{G}_{\|1}}(\mathbf{Q}, z, z') = \frac{2\pi}{|\mathbf{Q} + \mathbf{G}_{\|1}|} e^{-|\mathbf{Q}+\mathbf{G}_{\|1}||z-z'|}. \quad (6)$$

Since the integrations in (4) are performed from $-L/2$ to $L/2$, the interaction between the density fluctuations, via the Coulomb propagator $v^{2D}_{\mathbf{G}_{\|1} \mathbf{G}_{\|2}}(\mathbf{Q}, z, z')$, is limited to one supercell located at $-L/2 < z < L/2$. After inserting the Fourier expansion (3), and a similar one for χ, in RPA Dyson Equation (4), it again becomes a matrix equation

$$\chi_{\mathbf{G}\mathbf{G}'}(\mathbf{Q}, \omega) = \chi^0_{\mathbf{G}\mathbf{G}'}(\mathbf{Q}, \omega) + \sum_{\mathbf{G}_1 \mathbf{G}_2} \chi^0_{\mathbf{G}\mathbf{G}_1}(\mathbf{Q}, \omega) v^{2D}_{\mathbf{G}_1 \mathbf{G}_2}(\mathbf{Q}) \chi_{\mathbf{G}_2 \mathbf{G}'}(\mathbf{Q}, \omega), \quad (7)$$

where the matrix of the bare Coulomb interaction is

$$v^{2D}_{\mathbf{G}_1 \mathbf{G}_2}(\mathbf{Q}) = v^{3D}_{\mathbf{G}_1 \mathbf{G}_2}(\mathbf{Q}) - p_{G_{z1}} p_{G_{z2}} \frac{4\pi(1 - e^{-|\mathbf{Q}+\mathbf{G}_{\|1}|L})}{|\mathbf{Q}+\mathbf{G}_{\|1}|L} \quad (8)$$
$$\times \frac{|\mathbf{Q}+\mathbf{G}_{\|1}|^2 - G_{z1} G_{z2}}{(|\mathbf{Q}+\mathbf{G}_{\|1}|^2 + G_{z1}^2)(|\mathbf{Q}+\mathbf{G}_{\|1}|^2 + G_{z2}^2)} \delta_{\mathbf{G}_{\|1} \mathbf{G}_{\|2}},$$

with $v^{3D}_{\mathbf{G}_1 \mathbf{G}_2}(\mathbf{Q}) = \frac{4\pi}{|\mathbf{Q}+\mathbf{G}_1|^2} \delta_{\mathbf{G}_1 \mathbf{G}_2}$, $G_z = \frac{2\pi k}{L}$, $p_{G_z} = (-1)^k$, and $k \in \mathbb{Z}$.

The solution of Equation (7) has the form

$$\chi_{\mathbf{G}\mathbf{G}'}(\mathbf{Q}, \omega) = \sum_{\mathbf{G}_1} \mathcal{E}^{-1}_{\mathbf{G}\mathbf{G}_1}(\mathbf{Q}, \omega) \chi^0_{\mathbf{G}_1 \mathbf{G}'}(\mathbf{Q}, \omega), \quad (9)$$

where the dielectric matrix is

$$\mathcal{E}_{\mathbf{G}\mathbf{G}'}(\mathbf{Q}, \omega) = \delta_{\mathbf{G}\mathbf{G}'} - \sum_{\mathbf{G}_1} V^{2D}_{\mathbf{G}\mathbf{G}_1}(\mathbf{Q}) \chi^0_{\mathbf{G}_1 \mathbf{G}'}(\mathbf{Q}, \omega). \quad (10)$$

After solving Equation (7), the nonlocal screened response function in the z and z' direction becomes:

$$\chi_{\mathbf{G}_\| \mathbf{G}'_\|}(\mathbf{Q}, \omega, z, z') = \frac{1}{L} \sum_{G_z G'_z} \chi_{\mathbf{G}\mathbf{G}'}(\mathbf{Q}, \omega) e^{iG_z z - iG'_z z'}. \quad (11)$$

The propagator of the induced dynamically screened Coulomb interaction can be calculated from the response function (9) as [50]

$$W^{ind}_{\mathbf{G}_\|}(\mathbf{Q},\omega,z,z') = \int_{-L/2}^{L/2} dz_1 dz_2 v^{2D}_{\mathbf{G}_\|}(\mathbf{Q},z,z_1)\chi_{\mathbf{G}_\|0}(\mathbf{Q},\omega,z_1,z_2) v^{2D}_0(\mathbf{Q},z_2,z'), \quad (12)$$

where the index zero means that $\mathbf{G}'_\| = 0$. After using the expansion (11) and Equation (6), the integrations over z_1 and z_2 can be performed analytically, and the induced dynamically screened interaction at $z, z' > 0$ can be written as

$$W^{ind}_{\mathbf{G}_\|}(\mathbf{Q},\omega,z,z') = e^{-|\mathbf{Q}+\mathbf{G}_\|||z-Qz'|} D_{\mathbf{G}_\|}(\mathbf{Q},\omega) \quad (13)$$

where the propagator of the surface excitations is

$$D_{\mathbf{G}_\|}(\mathbf{Q},\omega) = \sum_{G_{z1} G_{z2}} \chi_{\mathbf{G}_\|,0,G_{z1},G_{z2}}(\mathbf{Q},\omega) F_{G_{z1}}(\mathbf{Q}+\mathbf{G}_\|) F^*_{G_{z2}}(\mathbf{Q}), \quad (14)$$

and the form factors F are

$$F_{G_z}(\mathbf{Q}) = \frac{4\pi p_{G_z}}{Q\sqrt{L}} \frac{\sinh(\frac{QL}{2})}{Q+iG_z}. \quad (15)$$

The spectral function, which defines the intensity of the energy loss by an external perturbation to the excitation of the $(\mathbf{Q},\omega)$ modes, can now be calculated as

$$S(\mathbf{Q},\omega) = -Im D_{\mathbf{G}_\|=0}(\mathbf{Q},\omega). \quad (16)$$

Up to this point, we derived the expressions for the self-standing q2D systems, but including a dielectric substrate polarization is now straightforward. It is obtained simply by replacing the bare Coulomb interaction (6) with the Coulomb interaction screened by the substrate

$$v^{2D}_{\mathbf{G}_\|}(\mathbf{Q},z,z') \to \tilde{v}^{2D}_{\mathbf{G}_\|}(\mathbf{Q},\omega,z,z') = v^{2D}_{\mathbf{G}_\|}(\mathbf{Q},z,z') + \frac{2\pi}{|\mathbf{Q}+\mathbf{G}_\||} D_S(\omega) e^{-|\mathbf{Q}+\mathbf{G}_\||(2h+z+z')}, \quad (17)$$

where the dielectric surface response function is

$$D_S(\omega) = \frac{1-\epsilon_S(\omega)}{1+\epsilon_S(\omega)}.$$

This also means that the matrix (8) used in matrix Dyson Equation (7) should be modified as

$$v^{2D}_{\mathbf{G}_1\mathbf{G}_2}(\mathbf{Q}) \to \tilde{v}^{2D}_{\mathbf{G}_1\mathbf{G}_2}(\mathbf{Q},\omega) =$$
$$v^{2D}_{\mathbf{G}_1\mathbf{G}_2}(\mathbf{Q}) + \frac{|\mathbf{Q}+\mathbf{G}_{\|1}|}{2\pi} e^{-2|\mathbf{Q}+\mathbf{G}_{\|1}|h} D_S(\omega) F_{G_{z1}}(|\mathbf{Q}+\mathbf{G}_{\|1}|) F^*_{G_{z2}}(|\mathbf{Q}+\mathbf{G}_{\|1}|)\delta_{\mathbf{G}_{\|1}\mathbf{G}_{\|2}}. \quad (18)$$

2.2. Calculation of the 2D Dynamical Polarizability Function $\alpha(\omega)$

In the long-wavelength or optical limit ($Q \to 0$), the in-plane dielectric function of a 2D crystal can be approximated as

$$\epsilon(Q,\omega) = 1 + 2\pi Q\alpha(\omega). \quad (19)$$

The 2D polarizability α can be divided into intraband and interband contributions

$$\alpha(\omega) = \alpha^{intra}(\omega) + \alpha^{inter}(\omega), \quad (20)$$

where

$$\alpha^{intra/inter}(\omega) = i\sigma^{intra/inter}_\mu(\omega)/\omega, \quad (21)$$

and $\mu = x$ or y. The intraband optical conductivity is [51]

$$\sigma_\mu^{\text{intra}}(\omega) = i\frac{e^2}{m}\frac{n_\mu}{\omega + i\eta_{\text{intra}}}, \qquad (22)$$

where the effective number of charge carriers is

$$n_\mu = -\frac{m}{Se^2}\sum_n \sum_{\mathbf{K}\in 1.SBZ} \frac{\partial f_{n\mathbf{K}}}{\partial E_{n\mathbf{K}}}\left|j_{n\mathbf{K},n\mathbf{K}}^\mu\right|^2. \qquad (23)$$

The interband optical conductivity is determined from the optical limit of the nonlocal interband conductivity

$$\sigma_\mu^{\text{inter}}(\omega) = L\sigma_\mu^{\text{inter}}(\omega, \mathbf{Q}\to 0).$$

The nonlocal interband conductivity is [51]

$$\sigma_\mu^{\text{inter}}(\mathbf{Q},\omega) = -\frac{i\hbar}{\Omega}\sum_{n\neq m}\sum_{\mathbf{K}\in 1.SBZ}\frac{\left|j_{n\mathbf{K},m\mathbf{K}+\mathbf{Q}}^\mu\right|^2}{E_{n\mathbf{K}} - E_{m\mathbf{K}+\mathbf{Q}}}\frac{f_{n\mathbf{K}} - f_{m\mathbf{K}+\mathbf{Q}}}{\hbar\omega + i\eta_{\text{inter}} + E_{n\mathbf{K}} - E_{m\mathbf{K}+\mathbf{Q}}}, \qquad (24)$$

where the current vertices are

$$j_{n\mathbf{K},m\mathbf{K}+\mathbf{Q}}^\mu = \int_\Omega d\mathbf{r}\, e^{-i\mathbf{Q}\cdot\mathbf{r}}\, j_{n\mathbf{K},m\mathbf{K}+\mathbf{Q}}^\mu(\mathbf{r}), \qquad (25)$$

and the current produced by the transitions between the Bloch states $\phi_{n\mathbf{K}}^* \to \phi_{m\mathbf{K}+\mathbf{Q}}$ is defined as

$$j_{n\mathbf{K},m\mathbf{K}+\mathbf{Q}}^\mu(\mathbf{r}) = \frac{e\hbar}{2im}\{\phi_{n\mathbf{K}}^*(\mathbf{r})\partial_\mu\phi_{m\mathbf{K}+\mathbf{Q}}(\mathbf{r}) - [\partial_\mu\phi_{n\mathbf{K}}^*(\mathbf{r})]\phi_{m\mathbf{K}+\mathbf{Q}}(\mathbf{r})\}.$$

Figure 2 shows the interband contribution to the dynamical polarizability $\alpha^{\text{inter}}(\omega)$ in the KC_8 (black), CsC_8 (orange) and doped graphene (red dashed). The graphene is doped so that the Fermi level is 1eV above the Dirac point, which corresponds to the doping of the π bands in the KC_8 and CsC_8. All three systems show qualitatively equal behavior; the peak at about $\omega = 2$ eV, indicating the onset for the $\pi \to \pi^*$ interband transitions, and the dip at $\omega \approx 4$ eV is a consequence of the high density of the $\pi \to \pi^*$ interband transitions at the M point of the Brillouin zone. Even though the π bands in all three systems are almost equally doped, we can see a substantial difference in the KC_8 and CsC_8 statical polarizabilities which are $\alpha^{\text{inter}}(\omega=0) = 2.2$ Å and $\alpha^{\text{inter}}(\omega=0) = 3.05$ Å, respectively. This difference probably comes from the difference in the intensities of the $C(\pi) \to A(\sigma)$ interlayer (interband) excitations in the two systems. These excitations are manifested as the peak at $\omega \approx 0.25$ eV for the CsC_8, which does not exist for the KC_8. Finally, we can see that the agreement between the dynamical polarizabilities in the KC_8 and the equivalently doped graphene is almost perfect. As we shall demonstrate later, this small deviation in the low-energy part of the dynamical polarizability is responsible for the disappearance of the AP in the CsC_8, but only if we take into account that these transitions represent a perpendicular polarization.

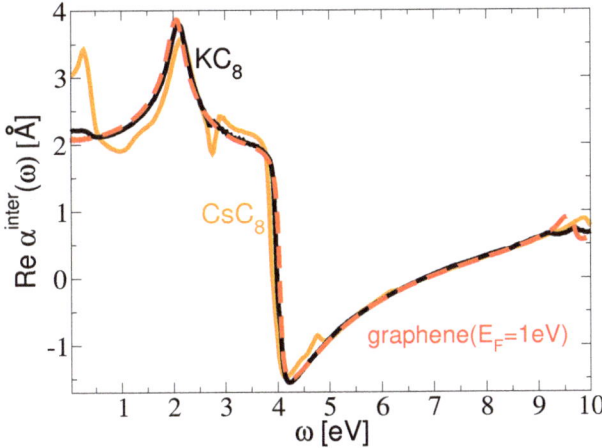

Figure 2. The interband contribution to the dynamical polarizability $\alpha^{\text{inter}}(\omega)$ for the KC$_8$ (black), CsC$_8$ (orange) and doped graphene (red dashed). The graphene is doped so that the Fermi energy is 1eV above Dirac point, which corresponds to the doping of the π bands in the KC$_8$ and CsC$_8$.

2.3. Calculation of the Substrate Dielectric Function

We assume that the dielectric media is vacuum (i.e., $\epsilon_0 = 1$), and that the substrate is aluminium oxide Al$_2$O$_3$ described by the macroscopic dielectric function $\epsilon_s(\omega)$. To calculate the $\epsilon_s(\omega)$, we start from the 3D Fourier transform of the independent electron response function

$$\chi^0_{\mathbf{GG'}}(\mathbf{q},\omega) = \frac{2}{\Omega} \sum_{\mathbf{k}\in 1.\text{BZ}} \sum_{n,m} \frac{f_n(\mathbf{k}) - f_m(\mathbf{k}+\mathbf{q})}{\omega + i\eta + E_n(\mathbf{k}) - E_m(\mathbf{k}+\mathbf{q})} \rho_{n\mathbf{k},m\mathbf{k}+\mathbf{q}}(\mathbf{G}) \rho^*_{n\mathbf{k},m\mathbf{k}+\mathbf{q}}(\mathbf{G'}), \quad (26)$$

where $\mathbf{k} \in 1.\text{BZ}$ indicates that the summation is performed within the first Brillouin zone. The charge vertices are defined as

$$\rho_{n\mathbf{k},m\mathbf{k}+\mathbf{q}}(\mathbf{G}) = \int_\Omega d\mathbf{r}\, \phi^*_{n\mathbf{k}}(\mathbf{r}) e^{-i(\mathbf{q}+\mathbf{G})\cdot\mathbf{r}} \phi_{m\mathbf{k}+\mathbf{q}}(\mathbf{r}), \quad (27)$$

where $\mathbf{k} = (k_x, k_y, k_z)$, $\mathbf{q} = (q_x, q_y, q_z)$ and $\mathbf{G} = (G_x, G_y, G_z)$ are the 3D wave vector, the transfer wave vector and the reciprocal lattice vector, respectively. The integration is performed over the normalization volume Ω. We use the response matrix (26) to determine the dielectric matrix as

$$\mathcal{E}_{\mathbf{GG'}}(\mathbf{q},\omega) = \delta_{\mathbf{GG'}} - \sum_{\mathbf{G_1}} v_{\mathbf{GG_1}}(\mathbf{q}) \chi^0_{\mathbf{G_1 G'}}(\mathbf{q},\omega), \quad (28)$$

where the bare Coulomb interaction is $v_{\mathbf{GG'}}(\mathbf{q}) = \frac{4\pi}{|\mathbf{q}+\mathbf{G}|^2} \delta_{\mathbf{GG'}}$. Finally, the macroscopic dielectric function is determined by inverting the dielectric matrix

$$\epsilon_s(\omega) = \epsilon_1(\omega) + i\epsilon_2(\omega) = 1/\mathcal{E}^{-1}_{\mathbf{G}=0\mathbf{G'}=0}(\mathbf{q} \approx 0, \omega). \quad (29)$$

2.4. Reduced Model

Analytical modeling of the energy loss spectra is achieved by representing each of the systems, CsC$_8$ and KC$_8$, by a two-layer structure consisting of a single sheet of doped graphene and an m2DEG, placed in vacuum at a distance d apart, as shown in Figure 3. In the reduced model, the substrate is neglected, i.e., we assume $\epsilon_S(\omega) = 1$, and the energy loss function $-\Im\{1/\epsilon(\mathbf{Q},\omega)\}$ is then obtained from the effective 2D dielectric permittivity for this two-layer structure, in the RPA given by [44,52]

$$\epsilon(\mathbf{Q},\omega) = \frac{1}{2}\left[1 + \coth(Qd) - 2v_Q\chi_g^0\right] - \frac{1}{2}\frac{\cosech^2(Qd)}{1+\coth(Qd) - 2v_Q\chi_m^0}, \quad (30)$$

where $v_Q = 2\pi/Q$ is the Coulomb interaction, while $\chi_g^0(Q,\omega)$ and $\chi_m^0(Q,\omega)$ are the response functions of the noninteracting electrons in the graphene and the m2DEG layers, respectively.

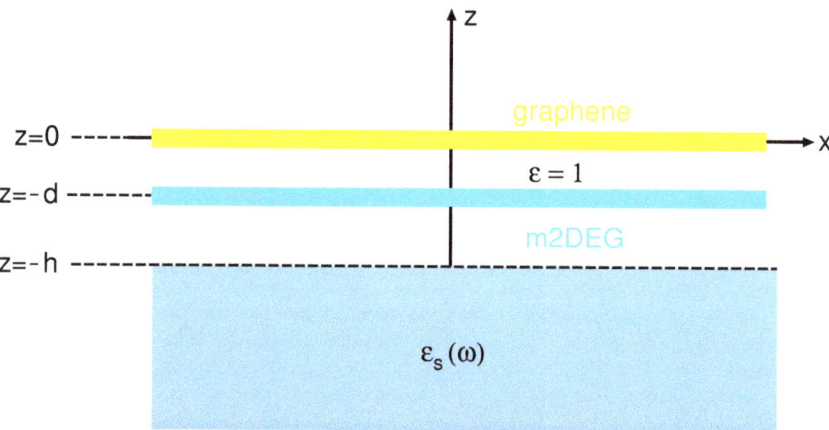

Figure 3. Schematic representation of the reduced model. The alkali atom layer is approximated by 'massive' 2D electron gas (parabolic σ band), and the graphene layer is described by the 'massless' Dirac fermion (MDF) approximation (conical π band).

For the doped graphene, we follow the method proposed by Gjerding et al. [53], and write the response function as $\chi_g^0(Q,\omega) = \chi_{\text{Dirac}}(Q,\omega) - \alpha_g Q^2$, where χ_{Dirac} is the response function given in Refs. [54,55], which describes both intraband and low-energy interband electron transitions within the π electron bands approximated by the Dirac cones with the Fermi energy $E_{F\pi}$, while α_g is the phenomenological parameter providing the correction due to the high-energy interband transitions. For the m2DEG, we similarly write $\chi_m^0(Q,\omega) = \chi_{\text{2DEG}}(Q,\omega) - \alpha_m Q^2$, where χ_{2DEG} is the polarization function given in Ref. [56], describing the intraband transitions in the 2DEG which occupies a single parabolic energy band with the effective mass m_σ and the Fermi energy $E_{F\sigma}$, while α_m is a phenomenological parameter taking into account interband transitions in the m2DEG. The expressions used for both response functions, $\chi_g^0(Q,\omega)$ and $\chi_m^0(Q,\omega)$, are formulated for zero temperature, but they are corrected by the Mermin procedure to take into account a finite damping parameter η in both graphene and m2DEG layers [53]. We use $\eta = 40$ meV as in the ab initio calculations, as well as $\eta = 65$ meV to take into account the additional smearing at room temperature.

We note that the relevant parameters for the KC$_8$ ($d = 2.92$ Å, $E_{F\pi} = 1.01$ eV, $E_{F\sigma} = 0.9$ eV, $m_\sigma = 0.92$) and CsC$_8$ ($d = 3.13$ Å, $E_{F\pi} = 1.03$ eV, $E_{F\sigma} = 1.03$ eV, $m_\sigma = 0.72$) are obtained from the electronic band structure calculation for these two systems, shown in Figure 4a,b, respectively. The parameters α are obtained from the static limit ($\omega \to 0$) of the ab initio results for the corresponding dynamic polarizability functions $\alpha^{\text{inter}}(\omega)$ in the optical limit. In particular, the value $\alpha_g \approx 1.3$ Å for high-energy interband transitions is deduced from the data for the intrinsic graphene [53], and by adding the value for the low-energy $\pi \to \pi^*$ interband transitions, estimated from the Dirac model in the optical limit at zero temperature [57] as $(4\pi E_{Fg})^{-1} \approx 1.15$ Å, we obtain that the total contribution of the interband transitions in the doped graphene is around 2.45 Å. This is close to the result $\alpha^{inter}(\omega = 0) = 2.2$ Å shown in Figure 2 for the KC$_8$, which indicates that $\alpha_m \approx 0$ for that system. On the other hand, the result $\alpha^{inter}(\omega = 0) = 3.05$ Å shown in Figure 2 for CsC$_8$ indicates that $\alpha_m \approx 0.6$ Å for that system.

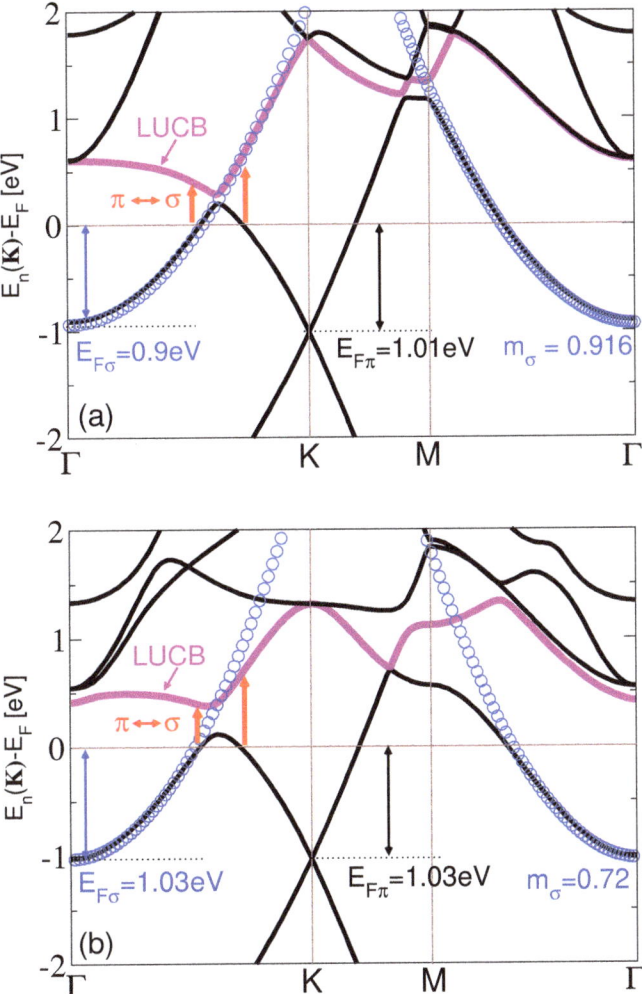

Figure 4. The electronic band structure in (**a**) KC$_8$ and in (**b**) CsC$_8$. The magenta line denotes the lowest unoccupied band (LUCB). The blue circles represent the parabolic fit $E(K) = E_{F\sigma} + \frac{\hbar^2 K^2}{2m_\sigma}$ of the alkali atom σ band.

2.5. Computational Details

The KC$_8$, CsC$_8$ and graphene KS wave functions ϕ_{nK} and energies E_{nK} are determined using the plane-wave self-consistent field DFT code (PWSCF) within the QUANTUM ESPRESSO (QE) package [58]. The core–electron interaction is approximated by the norm-conserving pseudopotentials [59,60]. For the KC$_8$ and CsC$_8$, the exchange correlation (XC) potential is approximated by the Perdew–Burke–Ernzerhof (PBE) generalized gradient approximation (GGA) functional [61]. The ground state electronic densities are calculated using the $8 \times 8 \times 1$ Monkhorst–Pack K-mesh [62] and the plane-wave cut-off energy is 60Ry. For both AC$_8$ crystals, we used the hexagonal Bravais lattice, where $a = 4.922$ Å and the separation between the AC$_8$ layers is $L = 2.5a$. The atomic and the unit cell relaxations were performed until maximum force below 0.001 Ry/a.u. was obtained. After performing the structural optimization, the obtained separations between the K and Cs layers and the graphene layer are d = 2.92 Å and d = 3.13 Å respectively. The graphene XC potential is approximated by the Perdew–Zunger (PZ) LDA [63]. The ground state electronic density is determined using $21 \times 21 \times 1$ K-point mesh, and for the plane-wave cut-off energy we choose 50 Ry. For the graphene unit-cell constant we use the experimental value $a_g = 2.45$ Å [64], while for the superlattice unit-cell constant we take $L = 5a_g$.

The AC$_8$ response functions (1) and conductivities (22)–(25) are evaluated from the wave functions $\Phi_{n\mathbf{K}}(\mathbf{r})$ and energies $E_n(\mathbf{K})$ calculated for the $201 \times 201 \times 1$ Monkhorst–Pack K-point mesh. The band summation (n, m) is performed over 60 bands, the damping parameters are $\eta = \eta_{intra} = \eta_{inter} = 10$ meV and the temperature is $T = 25$ meV. Due to the large spatial dispersivity of the dielectric response in the perpendicular (z) direction, the crystal local field effects are taken into account only in the z direction and neglected in the $x - y$ plane, i.e., we set the $\mathbf{G}_{\parallel}$ to zero. To calculate the matrix $\chi_{G_z, G_z'}$ we set the energy cut-off to 10Ry, which corresponds to a 23×23 matrix. Graphene conductivity (22)–(25) is calculated by performing the summation over $601 \times 601 \times 1$ K-point mesh, the band summations (n, m) are performed over 20 bands, the damping parameters are $\eta_{intra} = 10$ meV and $\eta_{inter} = 25$ meV and the temperature is $T = 25$ meV. The conductivity of the doped graphene is calculated using the rigid band approximation, i.e., the occupation parameter (the Fermi energy relative to the Dirac point) is adjusted a posteriori.

The ground state electronic density of the bulk Al$_2$O$_3$ is calculated using $9 \times 9 \times 3$ K-mesh, the plane-wave cut-off energy is 50Ry and the Bravais lattice is hexagonal (12 Al and 18 O atoms in the unit cell) with the lattice constants $a = 4.76$ Å and $c = 12.99$ Å. The response function (26) of the Al$_2$O$_3$ is calculated using the $21 \times 21 \times 7$ k-point mesh and the band summations (n, m) are performed over 120 bands. The damping parameter is $\eta = 100$ meV and the temperature is $T = 10$ meV. For the optically small wave vectors $\mathbf{q} \approx 0$, the crystal local field effects are negligible, so the crystal local field effect cut-off energy is set to zero. Using this approach, the Al$_2$O$_3$ dielectric function is estimated to be $\epsilon_S(\omega) \approx \epsilon_S(\omega = 0) = 2.9$ for $\omega < 2$ eV, which is in good agreement with the experimental value 3.4 [65]. However, in the calculations we used the full dynamical $\epsilon_S(\omega)$.

3. Results and Discussion

Figure 4 shows (a) KC$_8$ and (b) CsC$_8$ band structures, respectively, along the high symmetry $\Gamma \to K \to M \to \Gamma$ directions. The blue circles represent the parabolic fit $E(K) = E_{F\sigma} + \frac{\hbar^2 K^2}{2m_\sigma}$ for the alkali atom σ bands. In both figures, we can see that the graphene π band and the alkali atom σ band are significantly doped by the electrons, so the Fermi level is around 1eV above the Dirac point and the σ band bottom. It is also evident that the σ bands behave almost as an ideal 2D free electron gas. They are parabolic (described by effective masses $m_\sigma = 0.916$ and $m_\sigma = 0.72$, respectively) up to the Fermi energy, especially in Figure 4a where the σ band is parabolic almost through the entire Brillouin zone. Finally, the most important feature is the obvious similarity between the two band structures, which both fulfill conditions for the occurrence of the AP (two bands of different effective masses crossing the Fermi energy) [22,41]. However, as we shall see, the spectra of the low-energy electronic excitations in these two systems are quite different. In both systems, we can identify the lowest band that remains above the Fermi lever for all wave-vectors, and we call it the lowest unoccupied band (LUCB), even though for some wave-vectors there is another unoccupied band below that one, since this one turns out to be of particular importance, as explained below.

Figure 5 shows the intensities of the electronic excitations in the (a) KC$_8$ and (b) CsC$_8$ deposited on the dielectric Al$_2$O$_3$ surface. Since the graphene π bands are abundantly doped by the electrons, both systems support a strong DP. However, although the occupancies of both π bands are almost identical ($E_{F\pi} = 1.01$ eV for the KC$_8$ and 1.03 eV for the CsC$_8$), the intensities of the DP are quite different. We can see that the DP in the CsC$_8$

is broader and much more efficiently damped by the interband $\pi \to \pi^*$ electron hole excitations than the DP in the KC_8. Moreover, we can see that the DP in the KC_8 is very sharp and it extends deep into the interband $\pi \to \pi^*$ continuum while the DP in the CsC_8 decays immediately upon entering the interband $\pi \to \pi^*$ continuum. The most interesting difference between the two systems is that, even though the occupancies of their σ bands are very similar ($E_{F\sigma} = 0.9$ eV in the KC_8 and 1.03 eV in the CsC_8), the KC_8 does support the AP while the CsC_8 does not. Moreover, we can see that the π plasmon (denoted by the blue symbol π) in the KC_8 is well defined, and it appears to be a very intensive optically active plasmon (its intensity does not vanish for $Q \to 0$), while the π plasmon in the CsC_8 is less intensive, more diffuse and not in optically active mode. It is very interesting that these two very similar band structures lead to very different excitation spectra.

To demonstrate that these effects are not driven by the dielectric substrate, in Figure 5 we also show the intensities of the electronic excitations in the self-standing (c) KC_8 and (d) CsC_8. The magenta dots in Figure 5a,b denote the DP and AP dispersion relations in the self-standing samples, for comparison. It is worth noting that the insulator surface does not change the qualitative behavior of the electronic modes. The only quantitative differences are: the dielectric screening slightly reduces the DP energy and slightly reduces the intensities of the AP and π plasmons. However, the AP in the KC_8 is still clearly visible. Since the substrate does not affect the qualitative behavior of the excitation spectra, it will be omitted from further consideration. Therefore, the fact that one system does and the other does not support the AP, even though the bands crossing the Fermi energy in the two systems are almost identical, suggests that the mechanism responsible for this AP instability is probably in the screening coming from the interband excitations beyond the Fermi energy.

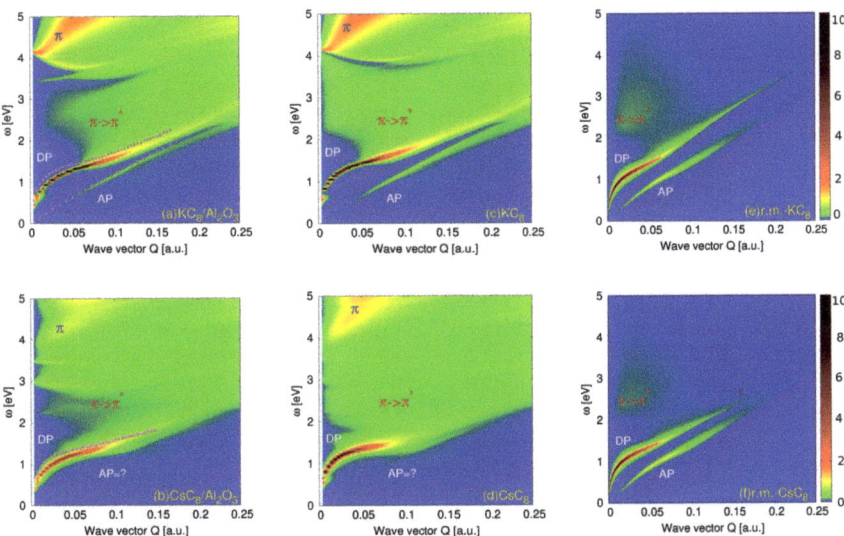

Figure 5. The ab initio spectra of the electronic excitations $S(\mathbf{Q}, \omega)$ in the (a) KC_8 and (b) CsC_8 deposited on dielectric Al_2O_3 surface with $h = 5.92$ Å and $h = 6.13$ Å respectively (i.e., the separation between the dielectric surface and the alkali atom layer is chosen to be 3 Å). The ab initio spectra of the electronic excitations in the self-standing (c) KC_8 and (d) CsC_8. The magenta dots in (**a**,**b**) denote the DP and AP dispersion relations in the self-standing samples, i.e., the DP and AP in (**c**,**d**). The energy loss function $-\Im\{1/\varepsilon(\mathbf{Q},\omega)\}$ in the self-standing (e) KC_8 and (f) CsC_8 was obtained using the reduced model.

To investigate this mechanism, we take advantage of the reduced model which clearly distinguishes between the different interband contributions to the dynamical response in the two systems. As discussed in Section 2.4, the statical polarizabilities due to the high-energy interband excitations in the metallic subsystem (characterized by the parameter α_m) are $\alpha_m \approx 0$ in the KC$_8$ and 0.6 in the CsC$_8$. This difference suggests that this polarization mechanism may be responsible for suppressing the AP. Figure 5e,f show the energy loss function $-\Im\{1/\epsilon(\mathbf{Q},\omega)\}$ in the self-standing KC$_8$ and CsC$_8$, respectively, obtained using the reduced model. We can see that the agreement between the ab initio and the reduced model intensities for the KC$_8$ for $\omega < 3$ eV and $Q < 0.14$ a.u. is almost perfect. Beyond these values, the plasmons in the reduced model appear at higher energies, which is reasonable since the reduced model neglects the dynamical effects of the high-energy interband transitions. For example, one can notice the absence of the π plasmon in the reduced model. In the case of the CsC$_8$, the agreement is no longer so good, and the most important difference is that the reduced model (at this level of approximation) is obviously not able to reproduce the disappearance of the AP. The stronger metallic interband screening in the CsC$_8$ is still not sufficient to suppress the AP. Moreover, even the implementation of the ab initio dynamic polarizability $\alpha^{\text{inter}}(\omega)$ in the reduced model (taking into account that the low-energy interband excitations in the χ_{Dirac} and χ_{2DEG} need to be extracted) fails to cause the disappearance of the AP. This means that the mechanism of the AP disappearance is more complex and obviously the explanation requires more accurate treatment of the spatial dispersivity of the interband dynamical response, beyond the 2D model. Therefore, we shall again exploit the ab initio method, which incorporates the effects of the local crystal field (the spatial dispersivity of the dynamical response in the z direction), where the interband screening will be modified by changing the number of valence bands participating in the interband screening.

Resolving the Mechanism of the AP Instability

To explore how the screening coming from the interband excitations beyond the Fermi energy influences the AP, we omit one or more unoccupied bands from the calculation, to determine exactly how each band influences the excitation spectra. As we can see from the band structures shown in Figure 4a,b, in both systems, CsC$_8$ and KC$_8$, there is one band crossing the Fermi level, i.e., for some wave-vectors that band is the highest occupied valence band (HOVB), while for other wave-vectors that same band is the lowest unoccupied band. Therefore, to avoid confusion, we use the expression lowest unoccupied band for the next one, i.e., the first band above the Fermi level at the Γ point (magenta line in Figure 4a,b), since that one remains above the Fermi level for all wave-vectors. That particular band is the one that has to be omitted from the calculation to achieve a qualitative difference with respect to the complete calculation (the one taking into account all bands). However, the difference becomes much more significant if we omit more bands. To keep track of the bands omitted from the calculation, we introduce the integer n which denotes the number of omitted bands. For example, $n = 0$ denotes that no bands are omitted, $n = 1$ denotes that only the lowest unoccupied band is omitted, $n = 2$ denotes that the first two unoccupied bands are omitted, etc.

Figure 6a,b show how the AP peak changes as we omit the unoccupied bands from the calculation, for $Q = 0.1$ a.u. and for the CsC$_8$ and KC$_8$, respectively. The thick black lines show the actual spectra, i.e., the ones obtained without omitting any bands ($n = 0$), and we can see that for the selected wave-vector the AP in the KC$_8$ is clearly present (though not very strong), while in the CsC$_8$ it does not exist. On the other hand, the red line shows that the AP peak exists in both systems if we omit the LUCB from the calculation ($n = 1$). In the CsC$_8$, that peak is very weak but it exists, while in the KC$_8$ it is, surprisingly, lower than the actual ($n = 0$) peak. The other lines show that the intensity of the peak increases as we omit more and more bands, in both systems, and it is the strongest when all the unoccupied bands are omitted ($n = \infty$). The same can be seen in Figure 6c, which shows how the intensity of the peak changes with n. It is important to point out that we also attempted to keep the lowest unoccupied band in the calculation while omitting any number of the other unoccupied bands, above that one, and that did not lead to the occurrence of the AP in the CsC_8. This means that omitting the lowest unoccupied band ($n = 1$) is crucial for the occurrence of the AP, while omitting the higher bands as well ($n > 1$) only enhances its intensity. This leads to the conclusion that the mechanism responsible for the disappearance of the AP in the CsC_8 is the small difference in the 'out-of-plane' polarization coming from the interband (interlayer) transitions between the graphene π band and the lowest unoccupied alkali atom σ band (denoted by red arrows in Figure 4). This difference is also manifested as the small peak at $\omega \approx 0.25$ eV in the CsC_8 dynamical polarizability $\alpha^{inter}(\omega)$ shown in Figure 2, which is missing for the KC_8. However, even if we include this difference in the dynamical polarizability in the reduced model, that still fails to reproduce the disappearance of the AP. This is because the reduced model is inherently a 2D model, i.e., it allows only the 'in-plane' polarization, while the $\pi \to \sigma$ transitions induce the 'out-of-plane' polarization.

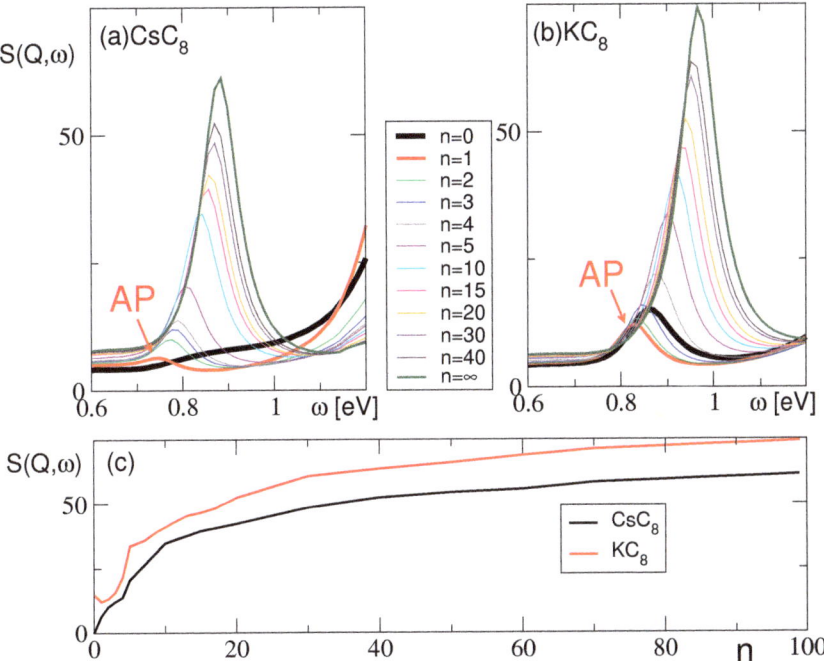

Figure 6. AP peak for $Q = 0.1$ a.u. when various numbers of the unoccupied bands are omitted from the calculation (n) for the (**a**) CsC_8 and (**b**) KC_8. (**c**) Intensity of the AP peak as a function of n for the CsC_8 (black line) and KC_8 (red line).

Another interesting point is that the difference in the intensities of the $\pi \to \sigma$ transitions is obviously correlated with the strength of the hybridization of the π and σ bands at the crossing point. We can see that the π and σ bands in the KC_8 intersect without any distortion while in the CsC_8 we notice a significant avoided crossing. In other words, the π and σ in the CsC_8 hybridize significantly while this is not the case in the KC_8. This leads to one very unusual conclusion; the intensity of the AP, which was to date believed to be a consequence of various long-range screening mechanisms, can be significantly affected by the short-range electronic correlations occurring between the atomic orbitals (i.e., by the chemical bonding between the alkaline atoms and graphene).

4. Conclusions

We analyzed the origin of AP instability in graphene doped by alkali metals on two prototype systems, KC_8 and CsC_8. Even though the band structures of these systems are almost identical, we proved that the hybridization between the $C(\pi)$ and the $K(\sigma)$ or $Cs(\sigma)$ bands at the crossing point causes a significant modification of the dynamic polarizability in the perpendicular direction, which has a substantial influence on the low-energy excitation spectra in these systems. We demonstrated that the net perpendicular screening in the CsC_8 causes the disappearance of the AP and significant weakening of the DP. We also demonstrated that the electronic screening coming from the high-energy interband transitions (beyond the HOVB–LUCB interval) significantly reduces the intensity of the AP in both systems, KC_8 and CsC_8. This illustrates the importance of the nature of the chemical bonding between the alkaline atoms and graphene, as well as the importance of perpendicular dispersivity of the dynamical response in theoretical simulations of technologically interesting low-energy plasmons in chemically doped graphenes.

Author Contributions: Conceptualization, V.D., L.M., I.R. and Z.L.M.; methodology, V.D. and Z.L.M.; software, V.D., J.J., I.R. and A.K.; investigation, V.D., L.M., J.J., A.K., I.R. and Z.L.M., visualization, V.D., L.M., J.J., A.K., I.R. and Z.L.M., writing—original draft preparation, V.D., L.M. and I.R.; writing—review and editing, V.D., L.M., Z.L.M. and I.R.; supervision V.D. All authors have read and agreed to the published version of the manuscript.

Funding: This research is funded by the Croatian Science Foundation (Grant No. IP-2020-02-5556), the European Regional Development Fund for the 'QuantiXLie Centre of Excellence' (Grant No. KK.01.1.1.01.0004), the Ministry of Education, Science and Technological Development of the Republic of Serbia, the Serbia–Croatia bilateral project (Grant No. 337-00-205/2019-09/28), and the Natural Sciences and Engineering Research Council of Canada (Grant No. 2016-03689).

Institutional Review Board Statement: Not applicable.

Informed Consent Statement: Not applicable.

Data Availability Statement: Not Applicable.

Acknowledgments: A.K., I.R. and V.D. would like to acknowledge networking support from COST Action CA19118 EsSENce, supported by the COST Association (European Cooperation in Science and Technology). Computational resources were provided by the Donostia International Physics Center (DIPC) computing center.

Conflicts of Interest: The authors declare no conflict of interest.

References

1. Li, J.; Lin, Z.; Miao, G.; Zhong, W.; Xue, S.; Li, Y.; Tao, Z.; Wang, W.; Guo, J.; Zhu, X. Geometric Effect of High-Resolution Electron Energy Loss Spectroscopy on the Identification of Plasmons: An Example of Graphene. *Surf. Sci.* **2022**, *721*, 122067. [CrossRef]
2. Sergey, G.M.; Lee, I.; Lee, S.; Ha, H.; Heiden, J.T.; Yoo, D.; Kim, T.; Low, T.; Lee, Y.H.; Oh, S.; et al. Real-space imaging of acoustic plasmons in large-area graphene grown by chemical vapor deposition. *Nat. Commun.* **2021**, *12*, 938.
3. Fateev, D.V.; Polischuk, O.V.; Mashinsky, K.V.; Moiseenko, I.M.; Morozov, M.Y.; Popov, V.V. Terahertz Lasing with Weak Plasmon Modes in Periodic Graphene Structures. *Phys. Rev. Appl.* **2021**, *15*, 034043. [CrossRef]
4. Kim, P.D.T.; Nguyen, M.V. Exchange-correlation effects and layer-thickness affect plasmon modes in gapped graphene-GaAs double-layer systems. *Eur. Phys. J. B* **2021**, *94*, 14.
5. Chiu, C.; Chung, Y.; Yang, C.; Liu, C.; Lin, C. Coulomb decay rates in monolayer doped graphene. *RSC Adv.* **2020**, *10*, 2337–2346. [CrossRef]
6. Wen, C.; Luo, J.; Xu, W.; Zhu, Z.; Qin, S.; Zhang, J. Enhanced Molecular Infrared Spectroscopy Employing Bilayer Graphene Acoustic Plasmon Resonator. *Biosensors* **2021**, *11*, 431. [CrossRef] [PubMed]
7. Mohammadi, Y. Tunable plasmon modes in doped AA-stacked bilayer graphene. *Superlattices Microstruct.* **2021**, *156*, 106955. [CrossRef]
8. Men, N.V.; Khanh, N.Q.; Phuong, D.T.K. Plasmon modes in double-layer gapped graphene. *Phys. E Low-Dimens. Syst. Nanostruct.* **2020**, *118*, 113859. [CrossRef]
9. Men, N.V.; Phuong, D.T.K. Plasmon modes in double-layer gapped graphene at zero temperature. *Phys. Lett. A* **2020**, *384*, 126221. [CrossRef]

10. Nguyen, V.-M. Temperature and inhomogeneity combination effects on collective excitations in three-layer graphene structures. *Phys. E Low-Dimens. Syst. Nanostruct.* **2022**, *140*, 115201. [CrossRef]
11. Dong-Thi, K.P.; Nguyen, V.M. Plasmonic Excitations in 4-MLG Structures: Background Dielectric Inhomogeneity Effects. *J. Low Temp. Phys.* **2022**, *206*, 51. [CrossRef]
12. Kim, P.D.T.; Van, M.N. Plasmon modes in N-layer graphene structures at zero temperature. *J. Low Temp. Phys.* **2020**, *201*, 311.
13. Nguyen, V.M. Plasmon modes in N-layer gapped graphene. *Phys. B Condens. Matter* **2020**, *578*, 411876.
14. Li, H.; Zhang, Y.; Xiao, H.; Qin, M.; Xia, S.; Wang, L. Investigation of acoustic plasmons in vertically stacked metal/dielectric/graphene heterostructures for multiband coherent perfect absorption. *Opt. Express* **2020**, *28*, 37577. [CrossRef] [PubMed]
15. Kumar, A.; Reddy, A.L.M.; Mukherjee, A.; Dubey, M.; Zhan, X.; Singh, N.; Ci, L.; Billups, W.E.; Nagurny, J.; Mital, G.; et al. Direct Synthesis of Lithium-Intercalated Graphene for Electrochemical Energy Storage Application. *ACS Nano* **2011**, *5*, 4345. [CrossRef] [PubMed]
16. Yang, S.L.; Sobota, J.A.; Howard, C.A.; Pickard, C.J.; Hashimoto, M.; Lu, D.H.; Mo, S.K.; Kirchmann, P.S.; Shen, Z.X. Superconducting graphene sheets in CaC6 enabled by phonon-mediated interband interactions. *Nat. Commun.* **2014**, *5*, 3493. [CrossRef]
17. Caffrey, N.M.; Johansson, L.I.; Xia, C.; Armiento, R.; Abrikosov, I.A.; Jacobi, C. Structural and electronic properties of Li-intercalated graphene on SiC(0001). *Phys. Rev. B* **2016**, *93* 195421. [CrossRef]
18. Ichinokura, S.; Sugawara, K.; Takayama, A.; Takahashi, T.; Hasegawa, S. Superconducting Calcium-Intercalated Bilayer Graphene. *ACS Nano* **2016**, *10*, 2761. [CrossRef]
19. Li, K.; Feng, X.; Zhang, W.; Ou, Y.; Chen, L.; He, K.; Wang, L.; Guo, L.; Liu, G.; Xue, Q.; et al. Superconducting Calcium-Intercalated Bilayer Graphene. *Appl. Phys. Lett.* **2013**, *103*, 062601. [CrossRef]
20. Pervan, P.; Lazić, P.; Petrović, M.; Rakić, I.Š.; Pletikosić, I.; Kralj, M.; Milun, M.; Valla, T. Li adsorption versus graphene intercalation on Ir(111): From quenching to restoration of the Ir surface state. *Phys. Rev. B* **2015**, *92*, 245415. [CrossRef]
21. Petrović, M.; Rakić, I.Š.; Runte, S.; Busse, C.; Sadowski, J.T.; Lazić, P.; Pletikosić, I.; Pan, Z.-H.; Milun, M.; Pervan, P.; et al. The mechanism of caesium intercalation of graphene. *Nat. Commun.* **2013**, *4*, 2772. [CrossRef] [PubMed]
22. Marušić, L.; Despoja, V. Prediction of measurable two-dimensional plasmons in Li-intercalated graphene LiC2. *Phys. Rev. B* **2017**, *95*, 201408(R). [CrossRef]
23. Despoja, V.; Marušić, L. UV-active plasmons in alkali and alkaline-earth intercalated graphene. *Phys. Rev. B* **2018**, *97*, 205426. [CrossRef]
24. Despoja, V.; Novko, D.; Lončarić, I.; Golenić, N.; Marušić, L.; Silkin, V.M. Strong acoustic plasmons in chemically doped graphene induced by a nearby metal surface. *Phys. Rev. B* **2019**, *100*, 195401. [CrossRef]
25. Despoja, V.; Jakovac, J.; Golenić, N.; Marušić, L. Bias-controlled plasmon switching in lithium-doped graphene on dielectric model Al2O3 substrate. *NPJ 2D Mater. Appl.* **2020**, *4*, 19. [CrossRef]
26. Jablan, M.; Buljan, H.; Soljačić, M. Plasmonics in graphene at infrared frequencies. *Phys. Rev. B* **2009**, *80*, 245435. [CrossRef]
27. Bonaccorso, F.; Sun, Z.; Hasan, T.; Ferrari, A.C. Graphene photonics and optoelectronics. *Nat. Photonics* **2010**, *4*, 611–622. [CrossRef]
28. Vakil, A.; Engheta, N. Transformation Optics Using Graphene. *Science* **2011**, *332*, 1291–1294. [CrossRef]
29. Hua-Qiang, W.; Chang-Yang, L.; Hong-Ming, L.; He, Q. Graphene applications in electronic and optoelectronic devices and circuits. *Chin. Phys. B* **2013**, *22*, 098106.
30. Pospischil, A.; Furchi, M.M.; Mueller, T. Solar-energy conversion and light emission in an atomic monolayer p–n diode. *Nat. Nanotechnol.* **2014**, *9*, 257–261. [CrossRef]
31. Ross, J.S.; Klement, P.; Jones, A.M.; Ghimire, N.J.; Yan, J.; Mandrus, D.G.; Taniguchi, T.; Watanabe, K.; Kitamura, K.; Yao, W.; et al. Electrically tunable excitonic light-emitting diodes based on monolayer WSe2 p-n junctions. *Nat. Nanotechnol.* **2014**, *9*, 268–272. [CrossRef] [PubMed]
32. Koppens, F.H.L.; Mueller, T.; Avouris, P.; Ferrari, A.C.; Vitiello, M.S.; Polini, M. Photodetectors based on graphene, other two-dimensional materials and hybrid systems. *Nat. Nanotechnol.* **2014**, *9*, 780–793. [CrossRef] [PubMed]
33. Jo, S.; Ubrig, N.; Berger, H.; Kuzmenko, A.B.; Morpurgo, A.F. Mono- and Bilayer WS2 Light-Emitting Transistors. *Nano Lett.* **2014**, *14*, 2019–2025. [CrossRef] [PubMed]
34. Lopez-Sanchez, O.; Llado, E.A.; Koman, V.; Morral, A.F.i.; Radenovic, A.; Kis, A. NEXT Light Generation and Harvesting in a van der Waals Heterostructure. *ACS Nano* **2014**, *8*, 3042. [CrossRef] [PubMed]
35. Lee, C.-H.; Lee, G.-H.; Zande, A.M.V.; Chen, W.; Li, Y.; Han, M.; Cui, X.; Arefe, G.; Nuckolls, C.; Heinz, T.F.; et al. Atomically thin p-n junctions with van der Waals heterointerfaces. *Nat. Nanotechnol.* **2014**, *9*, 676–681. [CrossRef] [PubMed]
36. Baugher, B.W.H.; Churchill, H.O.H.; Yang, Y.; Jarillo-Herrero, P. Optoelectronic Devices Based on Electrically Tunable P–n Diodes in a Monolayer Dichalcogenide. *Nat. Nanotechnol.* **2014**, *9*, 262–267. [CrossRef] [PubMed]
37. Low, T.; Avouris, P. Graphene Plasmonics for Terahertz to Mid-Infrared Applications. *ACS Nano* **2014**, *8*, 1086–1101. [CrossRef]
38. Britnell, L.; Ribeiro, R.M.; Eckmann, A.; Jalil, R.; Belle, B.D.; Mishchenko, A.; Kim, Y.J.; Gorbachev, R.V.; Georgiou, T.; Morozov, S.V.; et al. Absorption enhancement of thin layer black phosphorous in the mid-infrared with an all-dielectric metasurface. *Science* **2013**, *340*, 1311. [CrossRef]
39. Zhu, C.; Du, D.; Lin, Y. Graphene and graphene-like 2D materials for optical biosensing and bioimaging: A review. *2D Mater.* **2015**, *2*, 032004. [CrossRef]

40. Zhu, A.Y.; Cubukcu, E. Graphene nanophotonic sensors. *2D Mater.* **2015**, *2*, 032005. [CrossRef]
41. Pitarke, J.M.; Nazarov, V.U.; Silkin, V.M.; Chulkov, E.V.; Zaremba, E.; Echenique, P.M. Theory of acoustic surface plasmons. *Phys. Rev. B* **2004**, *70*, 205403. [CrossRef]
42. Diaconescu, B.; Pohl, K.; Vattuone, L.; Savio, L.; Hofmann, P.; Silkin, V.M.; Pitarke, J.M.; Chulkov, E.V.; Echenique, P.M.; Farías, D.; et al. Low-energy acoustic plasmons at metal surfaces. *Nature* **2007**, *448*, 57–59. [CrossRef] [PubMed]
43. Silkin, V.M.; Pitarke, J.M.; Chulkov, E.V.; Diaconescu, B.; Pohl, K.; Vattuone, L.; Savio, L.; Hofmann, P.; Farías, D.; Rocca, M.; et al. Band structure effects on the Be(0001) acoustic surface plasmon energy dispersion. *Phys. Stat. Sol.* **2008**, *205*, 1307–1311. [CrossRef]
44. Despoja, V.; Djordjević, T.; Karbunar, L.; Radović, I.; Mišković, Z.L. Ab initio study of the electron energy loss function in a graphene-sapphire-graphene composite system. *Phys. Rev. B* **2017**, *96*, 075433. [CrossRef]
45. Falkovsky, L.A.; Pershoguba, S.S. Optical far-infrared properties of a graphene monolayer and multilayer. *Phys. Rev. B* **2007**, *76*, 153410. [CrossRef]
46. Despoja, V.; Mowbray, D.J.; Vlahović, D.; Marušić, L. TDDFT study of time-dependent and static screening in graphene. *Phys. Rev. B* **2012**, *86*, 195429. [CrossRef]
47. Despoja, V.; Dekanić, K.; Šunjić, M.; Marušić, L. Ab initio study of energy loss and wake potential in the vicinity of a graphene monolayer. *Phys. Rev. B* **2012**, *86*, 165419. [CrossRef]
48. Despoja, V.; Novko, D.; Dekanić, K.; Šunjić, M.; Marušić, L. Two-dimensional and π plasmon spectra in pristine and doped graphene *Phys. Rev. B* **2013**, *87*, 075447. [CrossRef]
49. Novko, D.; Despoja, V.; Šunjić, M. Changing character of electronic transitions in graphene: From single-particle excitations to plasmons. *Phys. Rev. B* **2015**, *91*, 195407. [CrossRef]
50. Despoja, V.; Rukelj, Z.; Marušić, L. Ab initio study of electronic excitations and the dielectric function in molybdenum disulfide monolayer. *Phys. Rev. B* **2016**, *94*, 165446. [CrossRef]
51. Novko, D.; Šunjić, M.; Despoja, V. Optical absorption and conductivity in quasi-two-dimensional crystals from first principles: Application to graphene. *Phys. Rev. B* **2016**, *93*, 125413. [CrossRef]
52. Kalinic, A.; Radovic, I.; Karbunar, L.; Despoja, V.; Miskovic, Z.L. Wake effect in interactions of ions with graphene-sapphire-graphene composite system. *Phys. E Low-Dimens. Syst. Nanostruct.* **2021**, *126*, 114447. [CrossRef]
53. Gjerding, M.N.; Cavalcante, L.S.R.; Chaves, A.; Thygesen, K.S. Enhancing and Controlling Plasmons in Janus MoSSe–Graphene Based van der Waals Heterostructures. *J. Phys. Chem. C* **2021**, *124*, 11609. [CrossRef]
54. Wunsch, B.; Stauber, T.; Sols, F.; Guinea, F. Dynamical polarization of graphene at finite doping. *New J. Phys.* **2006**, *8*, 318. [CrossRef]
55. Hwang, E.H.; Sarma, S.D. Dielectric function, screening, and plasmons in two-dimensional graphene. *Phys. Rev. B* **2007**, *75*, 205418. [CrossRef]
56. Nersisyan, H.B.; Das, A.K. Interaction of fast charged projectiles with two-dimensional electron gas: Interaction and collisional-damping effects. *Phys. Rev. E* **2009**, *80*, 016402. [CrossRef]
57. Falkovsky, L.A. Optical properties of graphene. *J. Phys. Conf. Ser.* **2008**, *129*, 012004. [CrossRef]
58. Giannozzi, P.; Baroni, S.; Bonini, N.; Calandra, M.; Car, R.; Cavazzoni, C.; Ceresoli, D.; Chiarotti, G.L.; Cococcioni, M.; Dabo, I.; et al. Quantum espresso: A Modul. Open-Source Softw. Proj. Quantum Simulations Materials. *J. Phys. Condens. Matter* **2009**, *21*, 395502. [CrossRef]
59. Troullier, N.; Martins, J.L. Efficient pseudopotentials for plane-wave calculations. *Phys. Rev. B* **1991**, *43*, 1993. [CrossRef]
60. Hamann, D.R. Optimized norm-conserving Vanderbilt pseudopotentials *Phys. Rev. B* **2013**, *88*, 085117. [CrossRef]
61. Perdew, J.P.; Burke, K.; Ernzerhof, M. Generalized Gradient Approximation Made Simple. *Phys. Rev. Lett.* **1996**, *77*, 3865 [CrossRef] [PubMed]
62. Monkhorst, H.J.; Pack, J.D. Special points for Brillouin-zone integrations. *Phys. Rev. B* **1976**, *13*, 5188. [CrossRef]
63. Perdew, J.P.; Zunger, A. Self-interaction correction to density-functional approximations for many-electron systems. *Phys. Rev. B* **1981**, *23*, 5048. [CrossRef]
64. Saito, R.; Dresselhaus, G.; Dresselhaus, M.S. *Physical Properties of Carbon Nanotubes*; Imperial College Press: London, UK, 1998.
65. Arakawa, E.T.; Williams, M.W. Optical properties of aluminum oxide in the vacuum ultraviolet. *J. Phys. Chem. Solids* **1968**, *29*, 735–744. [CrossRef]

Article

Two-Photon–Near Infrared-II Antimicrobial Graphene-Nanoagent for Ultraviolet–Near Infrared Imaging and Photoinactivation

Wen-Shuo Kuo [1,2,3,†], Yen-Sung Lin [4,5,†], Ping-Ching Wu [6], Chia-Yuan Chang [7], Jiu-Yao Wang [3], Pei-Chi Chen [3], Miao-Hsi Hsieh [3], Hui-Fang Kao [8], Sheng-Han Lin [9,*] and Chan-Chi Chang [10,*]

[1] School of Chemistry and Materials Science, Nanjing University of Information Science and Technology, Nanjing 210044, China; wskuo88@gmail.com
[2] State Key Laboratory for Chemistry and Molecular Engineering of Medicinal Resources, Guangxi Normal University, Guilin 541004, China
[3] Center for Allergy, Immunology and Microbiome (AIM), China Medical University Children's Hospital, China Medical University, Taichung 404, Taiwan; a122@mail.ncku.edu.tw (J.-Y.W.); simple48686@gmail.com (P.-C.C.); karinadrift@gmail.com (M.-H.H.)
[4] Division of Pulmonary and Critical Care Medicine, An Nan Hospital, China Medical University, Tainan 709, Taiwan; chestlin@gmail.com
[5] Department of Nursing, Chung Hwa University of Medical Technology, Tainan 717, Taiwan
[6] Department of Biomedical Engineering, National Cheng Kung University, Tainan 701, Taiwan; wbcxyz@bme.ncku.edu.tw
[7] Department of Mechanical Engineering, National Cheng Kung University, Tainan 701, Taiwan; cychang0829@gs.ncku.edu.tw
[8] Department of Nursing, National Tainan Junior College of Nursing, Tainan 700, Taiwan; kaohuif@gmail.com
[9] Department of Anesthesiology, E-Da Hospital, Kaohsiung 824, Taiwan
[10] Department of Otolaryngology, National Cheng Kung University Hospital, College of Medicine, National Cheng Kung University, Tainan 70456, Taiwan
* Correspondence: ed111667@edah.org.tw (S.-H.L.); 109a0015@gs.ncku.edu.tw (C.-C.C.)
† These authors contributed equally to this work.

Abstract: Nitrogen doping and amino group functionalization through chemical modification lead to strong electron donation. Applying these processes to a large π-conjugated system of graphene quantum dot (GQD)-based materials as electron donors increases the charge transfer efficiency of nitrogen-doped amino acid-functionalized GQDs (amino-N-GQDs), resulting in enhanced two-photon absorption, post-two-photon excitation (TPE) stability, TPE cross-sections, and two-photon luminescence through the radiative pathway when the lifetime decreases and the quantum yield increases. Additionally, it leads to the generation of reactive oxygen species through two-photon photodynamic therapy (PDT). The sorted amino-N-GQDs prepared in this study exhibited excitation-wavelength-independent two-photon luminescence in the near-infrared region through TPE in the near-infrared-II region. The increase in size resulted in size-dependent photochemical and electrochemical efficacy, increased photoluminescence quantum yield, and efficient two-photon PDT. Therefore, the sorted amino-N-GQDs can be applicable as two-photon contrast probes to track and localize analytes in in-depth two-photon imaging executed in a biological environment along with two-photon PDT to eliminate infectious or multidrug-resistant microbes.

Keywords: sorted-graphene quantum dot; excitation-wavelength-independent photoluminescence; two-photon photoinactivation; near-infrared-II two-photon bioimaging; multi-drug resistant microbe

1. Introduction

Graphene quantum dot (GQD)-based materials with π–π configurations and surface groups have a high surface area, large diameter, and excellent surface grafting. These materials may cause intrinsic–state and defect-state emission to achieve photoluminescence

(PL). Intrinsic–state emission is induced by the quantum size effect, zigzag edge sites, or the recombination of localized electron–hole pairs; by contrast, defect-state emission originates from defect effects (energy traps) [1,2]. The PL emission of a material determines its suitability for imaging and photochemistry [3]. GQDs can be bonded with nitrogen atoms (N-GQDs) to alter their chemical composition and modulate their band gap, enhancing their photochemical properties and facilitating the fabrication of tunable luminescence in bioimaging and photodynamic (or photoinactivation) applications [4,5]. In addition, primary amine molecules (also known as amino group–functionalized molecules) can be chemically modified to cause strong electron donation and significantly alter the electronic properties of nitrogen–doped amino acid–functionalized GQD (amino-N-GQD) materials, increasing electrochemical and photochemical activities [6].

Combining multiphoton and near-infrared (NIR) excitation is an effective approach for investigating photoexcitation. This approach involves less absorption and a shorter photoexcitation period than other types of excitation. Moreover, this approach involves ultra-low energy consumption. These attributes enable the deep penetration of biological specimens and effective observation [7]. This study used a novel inverted optical microscopy system with a femtosecond Ti-sapphire laser with a repetition rate of 80 MHz and optical parametric oscillators (Mai Tai Spectra-Physics, Santa Clara, CA, USA; Scheme S1). The derived amino-N-GQDs with quantum confinement in the sp^2 domain and intrinsic–state and defect–state emissions can cause an excitation wavelength–dependent PL phenomenon. Therefore, the amino-N-GQDs were sieved through membranes with pores of various sizes to ensure that they exhibited homogeneous atom doping functionalities, which enabled the investigation of electronic and intrinsic properties related to optical behavior under quantum confinement effects [8,9]. This phenomenon results in excitation wavelength-independent two-photon luminescence (EWI-TPL) emission at a two-photon excitation (TPE) wavelength of 960 nm in the NIR-II region [7,10]. X-ray photoelectron spectroscopy (XPS) revealed the enlargement of the sorted amino-N-GQDs, which increased the number of C–N groups and pyridinic-, amino-, and pyrrolic-N functionalities. This increase could induce a radiative recombination of localized electron-hole pairs, resulting in significant two-photon properties, including favorable two-photon absorption (TPA), high TPL emission, excellent absolute TPE cross-sections, a short lifetime, a high ratio of radiative to nonradiative decay rates, and high post-TPE stability. In addition, the mean lateral size increased, which resulted in a high PL quantum yield (QY) and high efficiency in photodynamic therapy (PDT, or photoinactivation) under TPE [excitation wavelength (Ex): 960 nm; ultralow energy: 222.7 nJ pixel^{-1}; photoexcitation period: 100–170 scans; total effective exposure time: 0.65–1.11 s]. The results indicated that the sorted amino-N-GQDs are promising as two-photon contrast probes for tracking and localizing analytes in detail in two-photon imaging (TPI) of a biological environment in two-photon PDT; they can be used to eliminate infectious microbes effectively.

2. Results

Amino-N-GQDs were synthesized from graphene oxide sheets through ultrasonic shearing according to the modified Hummers' method (Figure S1, Table S1, and Scheme S2a) [11]. XPS indicated that the as-prepared amino-N-GQDs with homogeneous O and N distributors exhibiting high crystallinity and uniformity were sieved through membranes with pores of various sizes (Figures S2a–d, S3 and S4). Low-magnification and high-resolution transmission electron microscopy (HR-TEM; inset images in Figure S2a–d) were used to characterize the amino-N-GQDs. The mean lateral sizes of the sorted dots were set to 9.1 ± 0.2 nm (amino-N-GQD 9.1), 9.9 ± 0.2 nm (amino-N-GQD 9.9), 11.1 ± 0.3 nm (amino-N-GQD 11.1), and 12.0 ± 0.3 nm (amino-N-GQD 12.0). Further characterization results indicated that the sorted amino-N-GQDs were successfully prepared (Figures S3–S5).

Zigzag edge sites, localized electron–hole pairs, and quantum effects have been used to induce intrinsic-state emission in GQD-based materials; however, defect effects (energy traps) have been used to trigger defect-state emission [2,6,7]. To demonstrate such effects, Figure 1a displays the sorted amino-N-GQD dispersions, various levels of PL emission (gray-level images), dots with slight variations in size, and wavelengths at 630 nm encompassing the NIR-I window. The x–y focal point and z-axis resolution (full width at half maximum, FWHM) of the laser system were set to approximately 0.45 and 0.90 µm, respectively (Figure 1b). Satisfactory TPA in the NIR-II window was measured in subsequent experiments using the custom femtosecond Ti-sapphire laser optical system displayed in Scheme S1 (for details on the system, please refer to the Materials and Methods section), with an extension of approximately 960 nm (Figure 1c). Applying the most effective excitation wavelength can significantly enhance the two-photon properties of the materials used for bioimaging with TPEs [12]. Figure 2a shows the TPL spectra of the increase in the size of the sorted amino-N-GQDs, indicating red-shifted peaks of amino-N-GQD 9.1, amino-N-GQD 9.9, amino-N-GQD 11.1, and amino-N-GQD 12.0 at approximately 719, 772, 810, and 862 nm, respectively, in the NIR region under TPE (power: 222.7 nJ pixel^{-1}; scans: 20 or 170; total effective exposure time: approximately 0.13 or 1.11 s; Ex: 960 nm). The emission peaks determined via PL spectrophotometry for amino-N-GQD 9.1, amino-N-GQD 9.9, amino-N-GQD 11.1, and amino-N-GQD 12.0 were observed at approximately 719, 772, 810, and 862 nm, respectively, and they exhibited EWI-PL features (Figure S6). XPS revealed that the electron redistribution increased as the number of carbonyl groups increased (Figure S3), which decreased the energy gaps, resulting in TPL red shifts [13]. The quadratic dependence of the TPL increases with TPE power during this process [14]. Figure 2b demonstrates the existence of a two-photon process with an exponent of 2.00 ± 0.02 for sorted dots and conventional fluorophore (e.g., rhodamine B and fluorescein; Figure 2b).

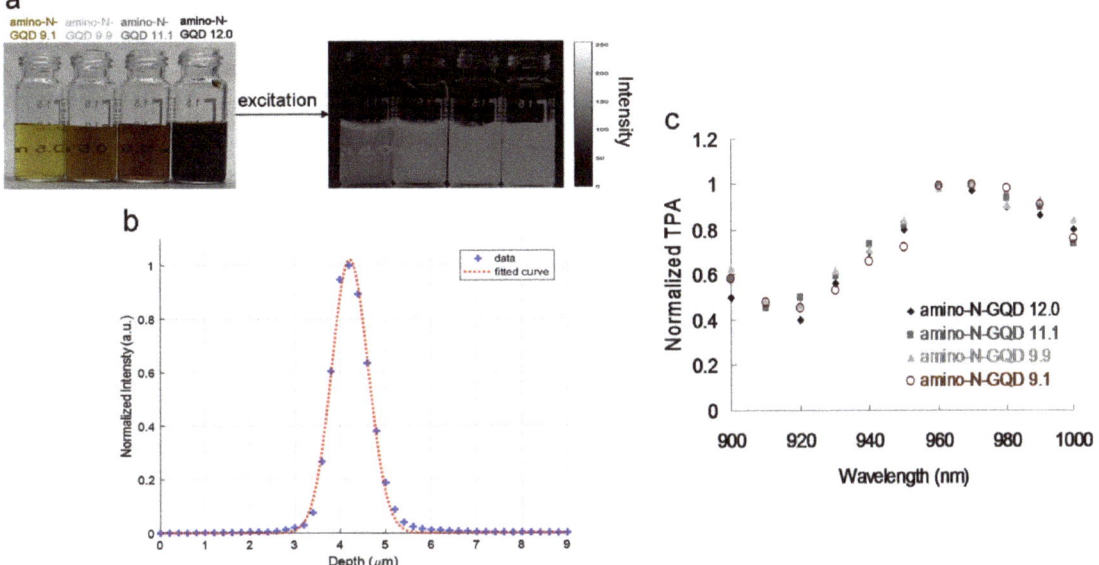

Figure 1. (a) Photographs of materials without and with 630 nm (gray-level) light excitation. (b) z-axis scan of thin gold film for measuring the second harmonic generation signal at various positions. The laser system's z-axis resolution (full width at half maximum, FWHM) was 0.90 µm (fit using the Gaussian function). (c) Relative TPA spectra of the sorted amino-N-GQDs. TPE signals were obtained at 900–1000 nm and at 127.3 nJ pixel^{-1}. Delivered dose: 0.75 µg mL^{-1} material.

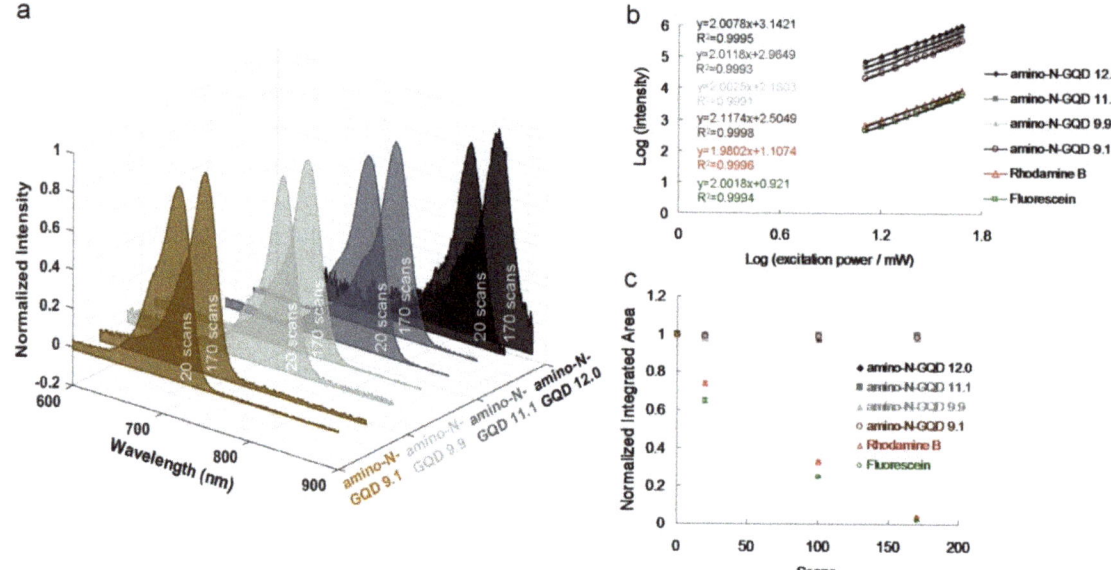

Figure 2. (**a**) Relative TPL spectra of materials at a TPE power of 222.7 nJ pixel^{-1} [20 and 170 scans (total effective exposure times: ~0.13 and 1.11 s), respectively; cut off = 900 nm, determined using cascading filters]. (**b**) TPL intensity dependence on the excitation power (logarithm) of materials and fluorophores; the slope is approximately 2.00 ± 0.02. TPE power = 1272.8–5091.2 nJ pixel^{-1}; $R^2 > 0.999$. (**c**) Two-photon stability of the amino-N-GQDs, rhodamine B, and fluorescein at a TPE power of 222.7 nJ pixel^{-1} with 20, 100, and 170 scans. The normalized integrated area was calculated by dividing the emission intensities of the integrated area after photoexcitation by those of the newly prepared material without photoexcitation. Delivered dose: 0.75 μg mL^{-1} material. Data are presented as means ± standard deviations ($n = 6$).

The sorted amino-N-GQDs with homogeneous O and N functionalities can be used to investigate the intrinsic electronic properties related to optical behavior with quantum confinement, leading to EWI-TPL under TPE. The sorted amino-N-GQDs exhibited two-photon stability, which could be attributed to the limited photobleaching due to the post-TPE TPL intensity of the dots (Figure 2c). Rhodamine B and fluorescein's fluorescence demonstrated low robustness against photobleaching upon TPE exposure (power: 222.7 nJ pixel^{-1}; scans: 20, 100, or 170; total effective exposure time: approximately 0.13, 0.65, or 1.11 s). Ultraviolet photoelectron spectroscopy revealed n–state levels that were maintained at approximately the same energetic positions (6.6–6.8 eV; Figure S7), regardless of the size determined through TEM and Raman spectroscopy (Figures S2 and S5), confirming the highest occupied orbital level of the sorted dots. The quantum confinement resulting from the particle size regulates the wavelengths of radiative transitions. The EWI-TPL emissions from the sorted amino-N-GQDs indicate the absence of trap states between the n-state and π^* energy levels. A change in particle size did not cause any disturbance at the n-state level. The EWI-TPL of the sorted dots could be attributed to $\pi^* \rightarrow n$ recombination, which triggers electron transition and phonon scattering. Measurements revealed that the absolute fluorescence QY [15] of the materials ranged from 0.39 (for amino-N-GQD 9.1) to 0.48 (for amino-N-GQD 12.0); these values are higher compared to those documented in other studies [16,17]. Desirable yields were achieved because of the electron-donating species in the sorted amino-N-GQD structure. XPS revealed that many C–N configurations functioned as electron-donating species and increased the material QY through the inhibition of nonradiative transitions (Figure S3). However, the low QY was because of the large number of electron-withdrawing carbonyl functional groups acting as non-radiative trap centers

(Figure S3). Characterization of the sorted amino-N-GQDs revealed that the successfully prepared GQDs exhibit EWI-TPL characteristics. However, a large cross-section is typically preferred for monitoring molecular actions. The sorted amino-N-GQDs exhibited a large absolute TPE cross-section, ranging from 55,946 to 60,728 Goeppert-Mayer (GM) units (1 GM = 10^{-50} cm^4 s photon^{-1}), which was more than 2900 times the magnitude of fluorescein (~19.2 GM; Table 1). The absolute TPE cross-sections of the amino-N-GQD 9.1, amino-N-GQD 9.9, amino-N-GQD11.1, and amino-N-GQD12.0 were approximately 55,946, 57,332, 59,051, and 60,728 GM, respectively (Table 2; for detailed calculations, please refer to the Materials and Methods section). This difference indicates a high ratio of energy absorption to energy input in the biospecimens. This phenomenon is highly favorable for TPI [18]. Moreover, the TPI emissions of the sorted dots (Figure 3a–d) occurred on a surface through the two-photon process, as shown in Figure 2b (power: 222.7 nJ pixel^{-1}; scans: 20; total effective illumination: ~0.13 s; Ex: 960 nm; scan rate: 6.53 ms scan^{-1}; scan area: 200 × 200 µm^2; for details regarding the calculation, see the Materials and Methods).

Table 1. TPE cross-section of materials at an excitation wavelength of 960 nm. Delivered dose: 0.75 µg mL^{-1} material.

Reference	Integrated Emission Intensity (Counts)		Action Cross-Section ($\eta\sigma$)
Rhodamine B [a]	53.9		13.4
Analyte	Integrated Emission Intensity (Counts)	Absolute Quantum Yield (η)	Absolute Cross-Section (σ)
amino-N-GQD 9.1	87,754	0.39	55,946
amino-N-GQD 9.9	94,550	0.41	57,332
amino-N-GQD 11.1	106,887	0.45	59,051
amino-N-GQD 12.0	117,250	0.48	60,728
Fluorescein	61.1	0.79 [b]	19.2

[a] Rhodamine B was selected as a reference to determine the TPE cross-section. The relevant calculations are shown in the Materials and Methods section; [b] Forster, L.S.; Livingston, R. The absolute quantum yields of the fluorescence of chlorophyll solutions. *J. Chem. Phys.* 1952, *20*, 1315–1320.

Table 2. Two-photon properties of the sorted amino-N-GQDs. Delivered dose: 0.75 µg mL^{-1} material.

	Absolute QY	Absolute Cross Section of TPE (GM)	Lifetime (ns)	Radiative Decay Rate ($\times 10^8$ s^{-1})	Nonradiative Decay Rate ($\times 10^8$ s^{-1})	Ratio of Radiative to Nonradiative Decay Rates
amino-N-GQD 9.1	0.39	55,946	1.13	3.45	5.40	0.64
amino-N-GQD 9.9	0.41	57,332	1.09	5.41	5.30	0.69
amino-N-GQD 11.1	0.45	59,051	1.04	4.33	5.29	0.82
amino-N-GQD 12.0	0.48	60,728	0.93	5.16	5.59	0.92

Because the inverted optical microscopy system was not suitable for investigating in vivo assay processes, the biological environment was mimicked by embedding an *Escherichia coli* (*E. coli*; 3.98 ± 1.37 µm in length and 0.98 ± 0.34 µm in width, calculated from 400 counts of bacteria) strain in a collagen matrix [19]. The TPI action occurred at a specimen depth of 180 µm under TPE (power: 222.7 nJ pixel^{-1}; scans: 20; total effective illumination: ~0.13 s; Ex: 960 nm; Figure 3e,f). Bacteria were observed under TEM (Figure 4a), but they were undetectable in the TPL images (Figure 3(e-1)), similar to the collagen matrix (Figure 3(e-2)). Lipopolysaccharides (LPSs) are major components of the outer membrane of *E. coli*. The physiologically stable and biocompatible sorted-amino-N-GQDs (Table S2 and Figure S8; a selected concentration of 0.75 µg mL^{-1} material was used in sequential experiments conducted in the dark) were coated with anti-LPS antibody (Ab$_{LPS}$) through electrostatic interaction to increase efficiency and specificity (Scheme S2b). This resulted in the absorption of a substantial amount of sorted dot–Ab$_{LPS}$ on the surface of the bacteria. No exceptional morphology (Figure 4b) was observed on the surface of the bacteria. In contrast, when the GQD size was increased using amino-N-GQD 9.1, amino-N-GQD

9.9, amino-N-GQD11.1, and amino-N-GQD12.0, a high fluorescence QY and large cross-section was detected in the TPL images (Figure 3f). However, the bacteria treated with the photoexcited material–Ab$_{LPS}$ hybrid were severely damaged when the power was increased to 222.7 nJ pixel^{-1} with 100 or 170 scans (with a total effective illumination of ~0.65 or 1.11 s), which resulted in abnormal morphology, as observed through TEM (Figure 4c,d). TPL decreased after 100 scans (Figure 3g) and was undetectable after 170 scans (Figure 3h) at a depth of 180 μm. For unlabeled bacteria, two-photon autofluorescence (TPAF) was not observed for the intrinsic fluorophores under TPE at the same power (Figure 3i). By contrast, TEM images revealed limited attachment and nonspecific binding for the surface of the *E. coli* strain (without antibody coating) treated with the sorted dots (power: 222.7 nJ pixel^{-1}; scans: 20; Ex: 960 nm; Figure 4e). Subsequently, TPI revealed almost no TPL emission at 180 μm (Figure 3j). Therefore, the *E. coli* strain treated with the photoexcited sorted dots exhibited a normal morphology even after photoexcitation (power: 222.7 nJ pixel^{-1}; scans: 170; Ex: 960 nm; Figure 4f). Under the same conditions, a clear TPI without TPL emission was observed for bacteria without the antibody-coated materials (Figure 3k). However, the images captured at a depth of >180 μm contained spherical aberrations, which severely degraded the image quality. Such aberrations were caused by the mismatch between the refractive indices of the aqueous sample and the maximum z-depth of the optical laser system, as well as the influence of the set objective, detection efficiency, and maximum *z-depth* of the optical laser system [20]. Therefore, TPI was not detected at a depth of 200 μm for all the sorted dots (Figure S9). In this study, the maximum *z-depth* for the detection of TPL emission with the laser optical system was 180 μm. This can be attributed to the detection efficiency and set objective, which was set to the depth to obtain the optimal resolution for examining the amino-N-GQDs used as two-photon contrast probes, particularly the sorted amino-N-GQDs with a large lateral size.

The changes in bacterial cell walls and oxidation were examined. The deterioration of the surrounding biological surface substrates was attributed to the reactive oxygen species (ROS) observed through PDT under TPE. These changes could cause bacterial atrophy, morphological damage, and distortion (Figure 4c,d) because of amino-N-GQD desorption from the bacterial surface (Figure 3g,h). The LIVE/DEAD kit was used to investigate the green fluorescence of the living bacteria and an additional incubation time of 3 h was necessary to induce PDT action effectively, leading to the elimination of bacteria (amino-N-GQD 12.0 was used to conduct this experiment; viability > 99%; Figure S10a). The results indicated that the bacteria were almost completely undamaged by exposure to laser treatment (power: 222.7 nJ pixel^{-1}; scans: 170; total effective illumination: ~1.11 s; Ex: 960 nm). The bacteria treated with the photoexcited amino-N-GQD 12.0–Ab$_{LPS}$ hybrid without incubation were also nearly undamaged (viability > 99%; Figure S10b). After 3 h of additional incubation, the same panel was observed, and the results indicated that the dead bacteria were somewhat distinguishable (represented by red fluorescence in Figure S10c). Bacterial viability was then quantified for further antimicrobial testing, which revealed nearly complete elimination of the bacteria treated with the amino-N-GQD 12.0–Ab$_{LPS}$ hybrid (elimination > 99%; Figure S10d, corresponding to Figures 3h and 4d) and the strong antibacterial effect of the amino-N-GQDs in PDT. Thus, no other photochemical activity (e.g., photothermal effect) was observed after photoexcitation. In addition, the bacteria not treated with the antibody–coated materials exhibited almost no antimicrobial effect under similar conditions (Figure S10e–h, corresponding to Figures 3k and 4f).

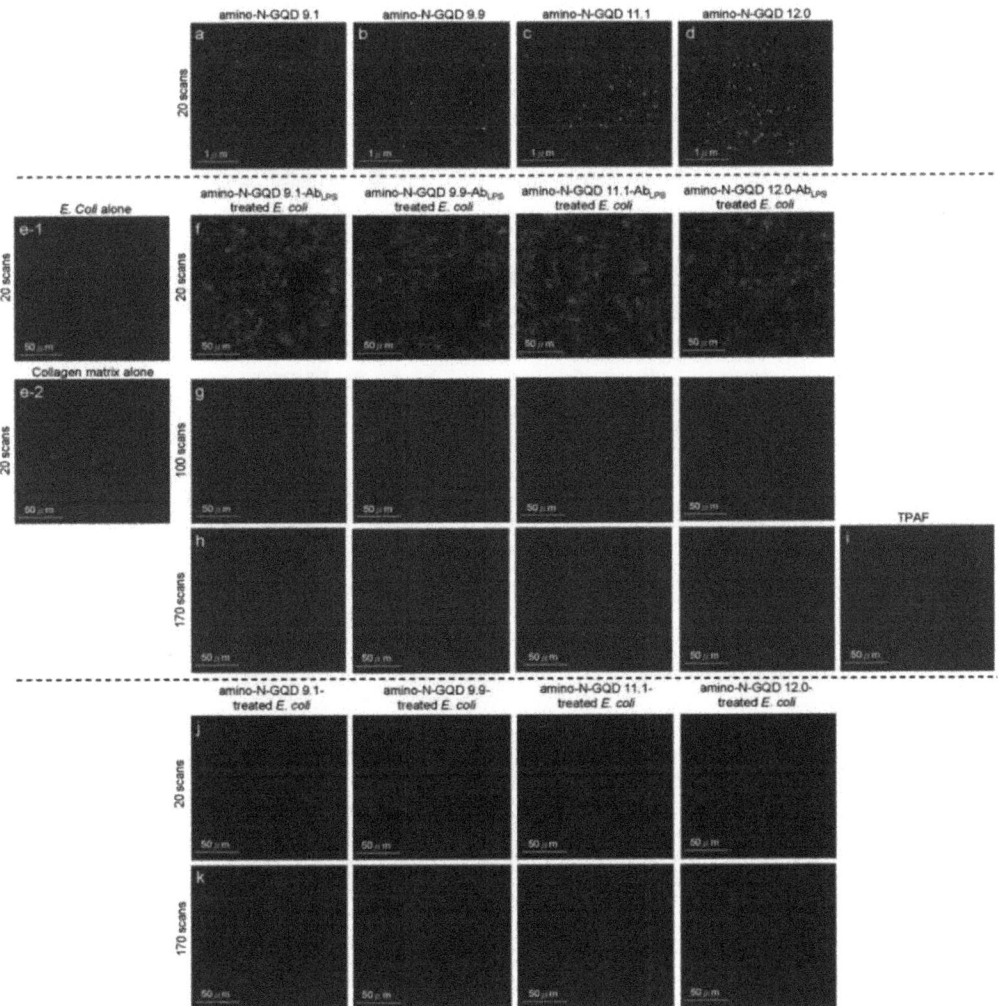

Figure 3. TPL images (gray-level) of the (**a**) amino-N-GQD 9.1, (**b**) amino-N-GQD 9.9, (**c**) amino-N-GQD 11.1, and (**d**) amino-N-GQD 12.0 at a TPE power of 222.7 nJ pixel^{-1} with 20 scans. (**e-1**) *E. coli* alone, (**e-2**) collagen matrix alone and bacteria subjected to sorted amino-N-GQD–Ab$_{LPS}$ treatment at a 180 μm depth (222.7 nJ pixel^{-1}) with (**f**) 20 scans and (**g**) 100 scans through TPE. (**h**) Images acquired after an additional 170 scans. (**i**) TPAF image of the unlabeled bacteria. TPL images of bacteria treated without the antibody-coated materials with (**j**) 20 scans and (**k**) 170 scans through photoexcitation under the same conditions. All images were acquired after 3 h of additional incubation to make the PDT action effectively. TPE wavelength: 960 nm. Delivered dose (OD$_{600}$): approximately 0.05 of *E. coli* or 0.75 μg mL^{-1} material–Ab$_{LPS}$.

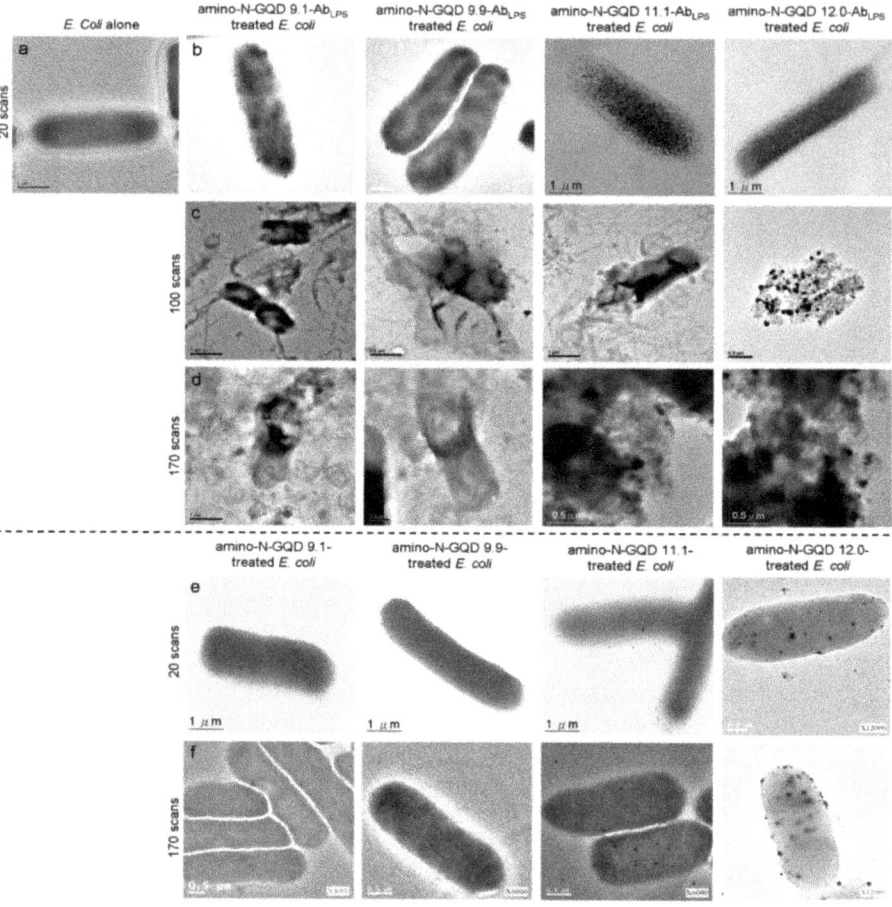

Figure 4. TEM images of the (**a**) bare *E. coli* (with 20 scans), (**b**–**d**) sorted amino-N-GQD–Ab$_{LPS}$-treated *E. coli* (with 20, 100, and 170 scans), and (**e**,**f**) sorted amino-N-GQD-treated *E. coli* (with 20 and 170 scans) under TPE (222.7 nJ pixel^{-1}). All images were acquired after 3 h of additional incubation. TPE wavelength: 960 nm. Delivered dose (OD$_{600}$): approximately 0.05 of *E. coli* or 0.75 μg mL^{-1} material–Ab$_{LPS}$.

ROS plays a crucial role in PDT by enabling the detection of superoxide anion radicals (O$_2^-$), hydrogen peroxide (H$_2$O$_2$), and singlet oxygen (^{1}O$_2$). In PDT, ROS are formed when molecular oxygen reacts with a photoexcited photosensitizer (PS) exposed to a suitable wavelength of light and energy. Photosensitized reactions involving oxygen are categorized as type I or II. A light-sensitized (excited) PS can directly react with a suitable substrate (unsaturated lipids, proteins, or nucleic acids) to produce unstable radicals through proton or electron transfer (type I reaction), leading to oxygenated products in the presence of oxygen, such as O$_2^-$, hydrogen peroxide (H$_2$O$_2$), or hydroxyl radicals (OH). Subsequently, it reacts with molecular oxygen to form ^{1}O$_2$ through energy transfer (type II reaction). ROS can induce DNA damage, inactivate enzymes, and oxidize amino acids, causing bacterial injury. However, a considerable amount of ^{1}O$_2$, O$_2^-$, and H$_2$O$_2$ was generated, and false-positive ROS signals were observed, which could be due to interactions among the sorted amino-N-GQDs, singlet oxygen sensor green (SOSG), trans-1-(2'-methoxyvinyl)pyrene (*t*-MVP), 2,3-bis (2-methoxy-4-nitro-5-sulfophenyl)-2H-tetrazolium-5-carboxanilide (XTT), glutathione (GSH), and 2,7-dichlorodihydrofluorescein diacetate (H_2DCFDA), which might compromise the results (Tables S3 and S4). The ROS generated by the bacteria treated with the sorted amino-N-GQD–Ab were monitored (Tables S5 and S6), and their signals were consistent with the $^1O_2$ phosphorescence signals emitted at 1270 nm from the sorted amino-N-GQDs (Figure 5a). The material without antibody coating (Tables S7 and S8) generated less ROS than the Ab_{LPS}–coated material. Furthermore, $^1O_2$ QY (Φ_Δ) values for amino-N-GQD 9.1, amino-N-GQD 9.9, amino-N-GQD 11.1, and amino-N-GQD 12.0 [21] were approximately 0.26, 0.28, 0.31, and 0.34, respectively. This study demonstrates the antimicrobial potential of the developed materials against *E. coli* in PDT. The bactericidal capability of the dots was investigated at a low dose of 0.75 μg mL^{-1} in the dark (TPE energy: 222.7 nJ pixel^{-1}; scans: 20; Ex: 960 nm). No significant difference in viability was observed between the panels (Figure S11a,b, corresponding to Figures 3f and 4b, respectively). After 100 scans, TPE still exhibited no bactericidal effect on the bacteria alone, and without TPE, the material in the panel exhibited considerable biocompatibility with the bacteria treated with the sorted dot–Ab hybrid (Figure S11c,d). However, under TPE, the sorted amino-N-GQDs exhibited excellent bactericidal capability (~89%, 93%, 98%, and 100% elimination for amino-N-GQD 9.1–Ab_{LPS}, amino-N-GQD 9.9–Ab_{LPS}, amino-N-GQD 11.1–Ab_{LPS}, and amino-N-GQD 12.0–Ab_{LPS}, respectively, amounting to an approximate 0.90–7.82 $\log_{10}$ reduction; Figure S10c,d, corresponding to Figures 3g and 4c). In contrast, the observed bacterial viability was higher for the materials without antibody coating (>98% viability) compared to the materials with the coating (Figure S11c,d). Although antimicrobial capabilities were still not apparent (~6%, 8%, 9%, and 11% elimination for amino-N-GQD 9.1, amino-N-GQD 9.9, amino-N-GQD 11.1, and amino-N-GQD 12.0 without the coating antibody, respectively), the sorted dots exhibited 100% antimicrobial efficacy against all *E. coli* strains treated with the sorted dot–Ab_{LPS} hybrid under TPE at 170 scans (Figure S11e,f, corresponding to Figures 3h and 4d). The surface protein, protein A, on the cell wall of a multidrug-resistant (MDR) strain of gram-positive methicillin-resistant *Staphylococcus aureus* (MRSA) was considered. Thus, the material was coated with $Ab_{protein\ A}$ to form a material–$Ab_{protein\ A}$ hybrid that eliminated MRSA (Figures S12 and S13), demonstrating a trend similar to that shown in Figure S11. These results were attributed to the sorted amino-N-GQDs functioning as a two-photon PS to generate ROS involved in PDT. These results also demonstrated the effectiveness of the antibody coating in enhancing the functions of the materials. Additionally, the trend of ROS generation in MRSA treated with the sorted dot–$Ab_{protein\ A}$ hybrid (Tables S9–S12) under TPE was similar to that of ROS generation in *E. coli* treated with the material–Ab_{LPS} hybrid (Tables S3–S8).

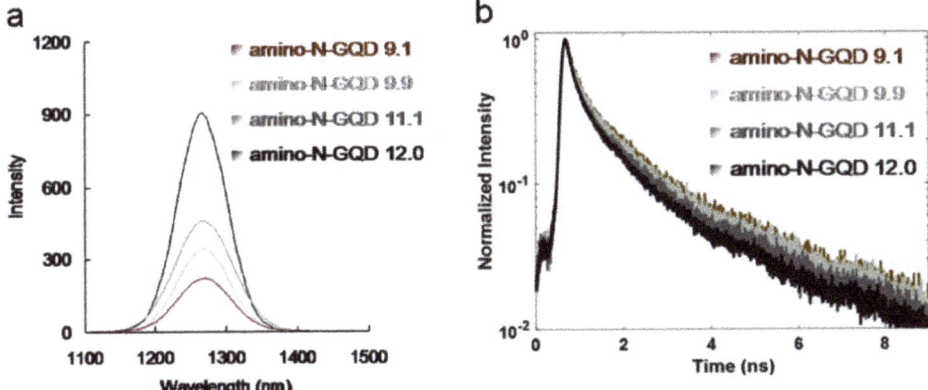

Figure 5. (a) Phosphorescence spectra of the sorted amino-N-GQDs (obtained at 1270 nm). (b) Decay profiles of the time-resolved room-temperature TPL material. Delivered dose: 0.75 μg mL^{-1} material.

Amino-N-GQDs exhibited remarkable quantum confinement, and their edge effects could be altered to increase their electrochemical, electrocatalytic, and photochemical activities [6,9]. Strong electron donation and large π-conjugated systems increased the charge transfer efficiency of the amino-N-GQDs [22], which resulted in favorable TPA, post-TPE stability, TPE cross-sections, and TPL. Additionally, they increased the ratio of radiative to non-radiative decay rates (amino-N-GQD 9.1: 0.64; amino-N-GQD 9.9: 0.69; amino-N-GQD 11.1: 0.82; and amino-N-GQD 12.0: 0.92; please refer to the Materials and Methods section for the calculation; Table 2). The results indicated that the material passed mainly through the radiative pathway as the fluorescence QY increased (amino-N-GQD 9.1: 0.39; amino-N-GQD 9.9: 0.41; amino-N-GQD 11.1: 0.45; and amino-N-GQD 12.0: 0.48) and the lifetime decreased (from 1.13 to 0.93 ns; Figure 5b, Tables 2 and 3). Radiative electron–hole pair recombination was observed and it was induced by N dopants and amino groups on the surface of the GQD-based material, which increased the intrinsic-state emission. However, for N dopants and amino groups, the presence of edge amine groups can increase the maximum occupied molecular orbital energy of the graphene flakes [23]. Thus, the narrowing of the orbital band gap, which increased the PL QY, could be attributed to the resonance between the delocalized π orbitals and the molecular orbital of the primary amine. XPS revealed that the C–O, C=O, and amide groups, which induced the nonradiative recombination of localized electron–hole pairs and prevented intrinsic–state emission [24], were favorable for small materials (Figures S3 and S4). The PL QY increased with increasing particle size. In addition, chemical modifications strongly affect the electronic properties of the amino-N-GQDs, enabling strong electron donation in primary amine molecules, which is also known as amino group functionalization. Singlet-triplet splitting of the amino-N-GQDs resulted in intersystem crossing and a high triplet-state yield. This splitting process was efficient, and it could compete with the process of internal conversion between multiple identical states, resulting in the creation of ROS for involvement in PDT [9,22]. As the number of edge sites increased, the number of C–N, pyridinic-, amino-, and pyrrolic-N groups increased (Figures S3 and S4). Similarly, as the size of the amino-N-GQDs (Figures S3 and S4) increased (Figures S2 and S5), their antibacterial ability and the number of ROS generated increased (Figure 5a and Tables S3–S8), leading to a highly efficient PDT process.

Table 3. Lifetime data and parameter generated using a time-correlated single-photon counting technique involving a triple-exponential fitting function while monitoring the emission under TPE. Delivered dose: 0.75 µg mL^{-1} material.

	3 Exp Fitting Model: (a0 × exp(a1x) + a2 × exp(a3x) + a4 × exp(a5x) + a6)						Lifetime 1	Lifetime 2	Lifetime 3	Average Lifetime (ns)
	a0	a1	a2	a3	a4	a5	a6			
amino-N-GQD 9.1	564.38	5.77	701.92	−0.97	187.90	−0.23	−0.53	0.17	1.03	4.39	1.13
amino-N-GQD 9.9	583.76	5.76	770.12	1.08	256.81	0.27	4.23	0.17	0.92	3.69	1.09
amino-N-GQD 11.1	668.75	6.00	821.54	1.07	255.08	0.27	3.85	0.17	0.94	3.64	1.04
amino-N-GQD 12.0	1747.39	6.20	1948.88	1.12	512.34	0.27	4.73	0.16	0.89	3.71	0.93

3. Materials and Methods

3.1. TEM Observation of the Negatively Stained Bacteria

Bacteria were picked from colonies and suspended in a 1% aqueous sodium phosphotungstate solution (Sigma Aldrich Co., St Louis, MO, USA) at pH 7.0. Droplets of the suspensions were allowed to dry on grids coated with the Formvar. Thereafter, the samples were subjected to TEM.

3.2. Molecular Weight of the Sorted-Amino-N-GQDs

The theoretical diameter of benzene is 0.243 nm with a molecular weight of 72 (ignoring the H atoms). According to Figure S2, the mean lateral sizes of the sorted-amino-N-GQDs were approximately 9.1 ± 0.2 nm (amino-N-GQD 9.1), 9.9 ± 0.2 nm (amino-N-

GQD 9.9), 11.1 ± 0.3 nm (amino-N-GQD 11.1), and 12.0 ± 0.3 nm (amino-N-GQD 12.0). For the sorted-amino-N-GQDs, assuming there was no leakage from a layer of material and ignoring the exposed functional groups, the benzene number and molecular weight could be approximately 1027 and 26,217 g mol^{-1}, 1261 and 32,027 g mol^{-1}, 1519 and 38,418 g mol^{-1}, 1801 and 45,390 g mol^{-1} (Table S14). The following measurement for the cross-section of TPE was performed using the estimated molecular weights.

3.3. Femtosecond Laser Optical System for the Measurements of TPA and TPL

A novel inverted optical microscopy system with a femtosecond Ti-sapphire laser [repetition rate: 80 MHz; Mai Tai with optical parametric oscillators, Spectra-Physics, Santa Clara, CA, USA] optical system: an inverted optical microscope (Zeiss, Oberkochen, Germany); an x–y galvanometer scanner (Cambridge, MA, USA); a triple-axis sample-positioning stage (Prior Scientific Instruments Ltd., London, UK); a z–axis piezoelectric nano-positioning stage (Mad City Labs, Madison, WI, USA); photomultiplier tubes (Hamamatsu, Shizuoka, Japan); a data acquisition card with a field-programmable gate array module (National Instruments, Austin, TX, USA) (Scheme S1).

All the Materials and Methods used in this study can be found in the Supplementary Materials.

4. Conclusions

Nitrogen doping and amino group functionalization, which result in strong electron donation, can be achieved through chemical modifications. Large π-conjugated systems of GQD-based materials acting as electron donors can be chemically manipulated with a low TPE energy in a short photoexcitation period to increase the charge transfer efficiency of the sorted amino-N-GQDs. This study used a novel femtosecond Ti-sapphire laser optical system (power: 222.7 nJ pixel^{-1}; scans: 100–170; total effective exposure time: ~0.65–1.11 s; excitation wavelength: 960 nm in the NIR-II region) for chemical modification. The sorted amino-N-GQDs exhibited increased TPA, post-TPE stability, TPE cross-sections, and TPL through the radiative pathway. The lifetime and quantum yield of the sorted amino-N-GQDs decreased and increased, respectively. Additionally, the sorted amino-N-GQDs exhibited EWI-PL in the NIR region and generated ROS after the TPE. Increasing the mean lateral size increased the number of C–N, pyridinic–N, amino–N, and pyrrolic–N functionalities, which induced the radiative recombination of localized electron–hole pairs and provided greater PL QY and efficient PDT action through TPE, enabling the sorted amino-N-GQDs to be applied in contrast probes to track and localize analytes in two-photon PDT.

Supplementary Materials: The following supporting information can be downloaded at: https://www.mdpi.com/article/10.3390/ijms23063230/s1.

Author Contributions: Conceptualization, data curation, formal analysis, project administration, investigation, methodology, funding acquisition, resources, software, supervision, validation, visualization, W.-S.K.; data curation, formal analysis, investigation, methodology, funding acquisition, resources, software, validation, visualization, P.-C.W., C.-Y.C. and J.-Y.W.; formal analysis, investigation, methodology, software, validation, visualization, P.-C.C., M.-H.H. and H.-F.K.; conceptualization, formal analysis, funding acquisition, investigation, methodology, project administration, resources, software, supervision, validation, S.-H.L., Y.-S.L. and C.-C.C. The manuscript was written through contributions of all authors. All authors have read and agreed to the published version of the manuscript.

Funding: This research was supported by Nanjing University of Information Science and Technology, China (2018r047); State Key Laboratory for Chemistry and Molecular Engineering of Medicinal Resources, Guangxi Normal University, China (CMEMR2021-B11); An Nan Hospital, China Medical University, Taiwan (ANHRF110-34); Allergy Immunology and Microbiome Center, China Medical University Children's Hospital, China Medical University, Taiwan (1JA8); E-Da Hospital, Taiwan; Ministry of Science and Technology, Taiwan (MOST 110-2221-E-006-013-MY3); Academia Sinica Healthy Longevity Grand Challenge Competition, Taiwan (AS-HLGC-110-07).

Institutional Review Board Statement: Not applicable.

Informed Consent Statement: Not applicable.

Data Availability Statement: Not applicable.

Conflicts of Interest: The authors declare no conflict of interest.

References

1. Gong, P.; Sun, L.; Wang, F.; Liu, X.; Yan, Z.; Wang, M.; Zhang, L.; Tian, Z.; Liu, Z.; You, L. Highly fluorescent N-doped carbon dots with two-photon emission for ultrasensitive detection of tumor marker and visual monitor anticancer drug loading and delivery. *Chem. Eng. J.* **2019**, *356*, 994–1002. [CrossRef]
2. Sun, Z.; Fang, S.; Hu, Y.H. 3D graphene materials: From understanding to design and synthesis control. *Chem. Rev.* **2020**, *120*, 10336–10453. [CrossRef] [PubMed]
3. Lee, J.; Wong, D.; Velasco, J., Jr.; Rodriguez-Nieva, J.F.; Kahn, S.; Tsai, H.Z.; Taniguchi, T.; Watanabe, K.; Zettl, A.; Wang, F.; et al. Imaging electrostatically confined Dirac fermions in graphene quantum dots. *Nat. Phys.* **2016**, *12*, 1032–1036. [CrossRef]
4. Sun, L.; Luo, Y.; Li, M.; Hu, G.; Xu, Y.; Tang, T.; Wen, J.; Li, X.; Wang, L. Role of pyridinic-N for nitrogen-doped graphene quantum dots in oxygen reaction reduction. *J. Colloid Interface Sci.* **2017**, *508*, 154–158. [CrossRef]
5. Li, M.; Wu, W.; Ren, W.; Cheng, H.M.; Tang, N.; Zhong, W.; Du, Y. Synthesis and upconversion luminescence of N-doped graphene quantum dots. *Appl. Phys. Lett.* **2012**, *101*, 103107. [CrossRef]
6. Torres, T. Graphene chemistry. *Chem. Soc. Rev.* **2012**, *46*, 4385–4386. [CrossRef] [PubMed]
7. Zhao, G.; Li, X.; Huang, M.; Zhen, Z.; Zhong, Y.; Chen, Q.; Zhao, X.; He, Y.; Hu, R.; Yang, T.; et al. The physics and chemistry of graphene-on-surfaces. *Chem. Soc. Rev.* **2017**, *46*, 4417–4449. [CrossRef]
8. Yildirim, M.; Sugihara, H.; So, P.T.C.; Sur, M. Functional imaging of visual cortical payers and subplate in awake mice with optimized three-photon microscopy. *Nat. Commun.* **2019**, *10*, 177. [CrossRef]
9. Wang, X.; Li, X.; Zhang, L.; Yoon, Y.; Weber, P.K.; Wang, H.; Guo, J.; Dai, H. N-doping of graphene through electrothermal reactions with ammonia. *Science* **2009**, *324*, 768–771. [CrossRef]
10. Kim, L.; Kim, S.; Jha, P.K.; Brar, V.W.; Atwater, H.A. Mid-infrared radiative emission from bright not plasmons in graphene. *Nat. Mater.* **2021**, *20*, 805–811. [CrossRef]
11. Hummers, W.S.; Offeman, R.E. Preparation of graphitic oxide. *J. Am. Chem. Soc.* **1958**, *80*, 1339. [CrossRef]
12. Zhao, W.; Li, Y.; Yang, S.; Chen, Y.; Zheng, J.; Liu, C.; Qing, Z.; Li, J.; Yang, R. Two-photon excitation/red emission, ratiometric fluorescent nanoprobe for intracellular pH imaging. *Anal. Chem.* **2016**, *88*, 4833–4840. [CrossRef] [PubMed]
13. Liu, J.; Li, D.; Zhang, K.; Yang, M.; Sun, H.; Yang, B. One-step hydrothermal synthesis of nitrogen-doped conjugated carbonized polymer dots with 31% efficient red emission for in vivo imaging. *Small* **2018**, *14*, e1703919. [CrossRef] [PubMed]
14. Horton, N.G.; Wang, K.; Kobat, D.; Clark, C.G.; Wise, F.W.; Schaffer, C.B.; Xu, C. In vivo three-photon microscopy of subcortical structures within an intact mouse brain. *Nat. Photonics* **2013**, *7*, 205–209. [CrossRef]
15. Würth, C.; Grabolle, M.; Pauli, J.; Spieles, M.; Resch-Genger, U. Relative and absolute determination of fluorescence quantum yields of transparent samples. *Nat. Protoc.* **2013**, *8*, 1535–1550. [CrossRef]
16. Li, B.; Wu, C.; Wang, M.; Charan, K.; Xu, C. An adaptive excitation source for high-speed multiphoton microscopy. *Nat. Methods* **2020**, *17*, 163–166. [CrossRef]
17. Zhang, F.; Liu, F.; Wang, C.; Xin, X.; Liu, J.; Guo, S.; Zhang, J. Effect of lateral size of graphene quantum dots on their properties and application. *ACS Appl. Mater. Interfaces* **2016**, *8*, 2104–2110. [CrossRef]
18. We, J.; Qiu, J.; Ren, L.; Zhang, K.; Wang, S.; Weeks, B. Size sorted multicolor fluorescence graphene oxide quantum dots obtained by differential velocity centrifugation. *Sci. Adv. Mater.* **2014**, *6*, 1052–1059. [CrossRef]
19. Roy, B.; Yuan, L.; Lee, Y.; Bharti, A.; Mitra, A.; Shivashankar, G.V. Fibroblast rejuvenation by mechanical reprogramming and redifferentiation. *Proc. Natl. Acad. Sci. USA* **2020**, *117*, 10131–10141. [CrossRef]
20. Richardson, D.S.; Lichtman, J.W. Clarifying tissue clearing. *Cell* **2015**, *162*, 246–257. [CrossRef]
21. Shi, L.; Hernandez, B.; Selke, M. Singlet oxygen generation from water-soluble quantum dot-organic dye nanocomposites. *J. Am. Chem. Soc.* **2006**, *128*, 6278–6279. [CrossRef] [PubMed]
22. Son, D.T.; Kwon, B.K.; Park, D.H.; Seo, W.S.; Yi, Y.; Angadi, B.; Lee, C.L.; Choi, K.W. Emissive ZnO-graphene quantum dots for white-light-emitting diodes. *Nat. Nanotechnol.* **2012**, *7*, 465–471. [CrossRef] [PubMed]
23. Tetsuka, H.; Asahi, R.; Nagoya, A.; Okamoto, K.; Tajima, I.; Ohta, R.; Okamoto, A. Optically tunable amino-functionalized graphene quantum dots. *Adv. Mater.* **2012**, *24*, 5333–5338. [CrossRef] [PubMed]
24. Bao, L.; Zhang, Z.L.; Tian, Z.Q.; Zhang, L.; Liu, C.; Lin, Y.; Qi, B.; Pang, D.W. Electrochemical tuning of luminescent carbon nanodots: From preparation to luminescence mechanism. *Adv. Mater.* **2011**, *23*, 5801–5806. [CrossRef]

Review

Graphene for Antimicrobial and Coating Application

Viritpon Srimaneepong [1], Hans Erling Skallevold [2], Zohaib Khurshid [3], Muhammad Sohail Zafar [4,5], Dinesh Rokaya [6,*] and Janak Sapkota [7,*]

1. Department of Prosthodontics, Faculty of Dentistry, Chulalongkorn University, Bangkok 10330, Thailand; viritpon.s@chula.ac.th
2. Department of Oral Biology, Faculty of Dentistry, University of Oslo, Sognsvannsveien 10, 0316 Oslo, Norway; herlings7b@msn.com
3. Department of Prosthodontics and Implantology, College of Dentistry, King Faisal University, Al-Hofuf 31982, Al Ahsa, Saudi Arabia; zsultan@kfu.edu.sa
4. Department of Restorative Dentistry, College of Dentistry, Taibah University, Al Madinah 41311, Al Munawwarah, Saudi Arabia; drsohail_78@hotmail.com
5. Department of Dental Materials, Islamic International Dental College, Riphah International University, Islamabad 44000, Pakistan
6. Department of Clinical Dentistry, Walailak University International College of Dentistry, Walailak University, Bangkok 10400, Thailand
7. Research Center, UPM Pulp Research and Innovations, 53200 Lappeenranta, Finland
* Correspondence: dinesh.ro@wu.ac.th (D.R.); sapkota.janak@outlook.com (J.S.); Tel.: +66-02-298-0244 (D.R.); +358-4-0729-1446 (J.S.)

Abstract: Graphene is a versatile compound with several outstanding properties, providing a combination of impressive surface area, high strength, thermal and electrical properties, with a wide array of functionalization possibilities. This review aims to present an introduction of graphene and presents a comprehensive up-to-date review of graphene as an antimicrobial and coating application in medicine and dentistry. Available articles on graphene for biomedical applications were reviewed from January 1957 to August 2020) using MEDLINE/PubMed, Web of Science, and ScienceDirect. The selected articles were included in this study. Extensive research on graphene in several fields exists. However, the available literature on graphene-based coatings in dentistry and medical implant technology is limited. Graphene exhibits high biocompatibility, corrosion prevention, antimicrobial properties to prevent the colonization of bacteria. Graphene coatings enhance adhesion of cells, osteogenic differentiation, and promote antibacterial activity to parts of titanium unaffected by the thermal treatment. Furthermore, the graphene layer can improve the surface properties of implants which can be used for biomedical applications. Hence, graphene and its derivatives may hold the key for the next revolution in dental and medical technology.

Keywords: graphene; coatings; bioactivity; tissue engineering; bone regeneration

1. Introduction

Graphene, having a sp^2 configuration, is made from a thin sheet of carbon atoms (Figure 1) [1–3]. The various forms of graphene include pure/pristine graphene, graphene oxide (GO) containing –COC–, –COOH, or –COH, reduced GO (rGO), and animated graphene oxide (AGO). Graphene materials have outstanding various properties with good mechanical strength, high surface area, elasticity, stiffness, excellent biocompatibility, superior electrical and thermal conductivity, and ease of functionalization [4–8]. Therefore, graphene is attractive in different fields including medicine and dentistry [3,9–11]. The review aims to present an introduction of graphene and presents a comprehensive up-to-date review of graphene as an antimicrobial and coating application in medicine and dentistry.

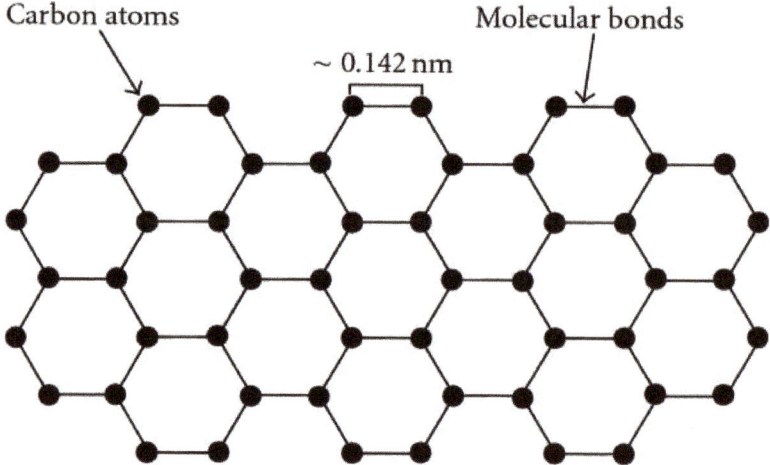

Figure 1. Structure of graphene [12].

2. Production of Graphene

Different grades of graphene can be prepared by various methods of production depending on the type of application. Such methods of production include mechanical exfoliation of graphite, epitaxial growth of graphene, liquid-phase exfoliation (LPE), chemical vapor deposition, and molecular assembly [3,4,13]. The most common methods are shown in Figure 2.

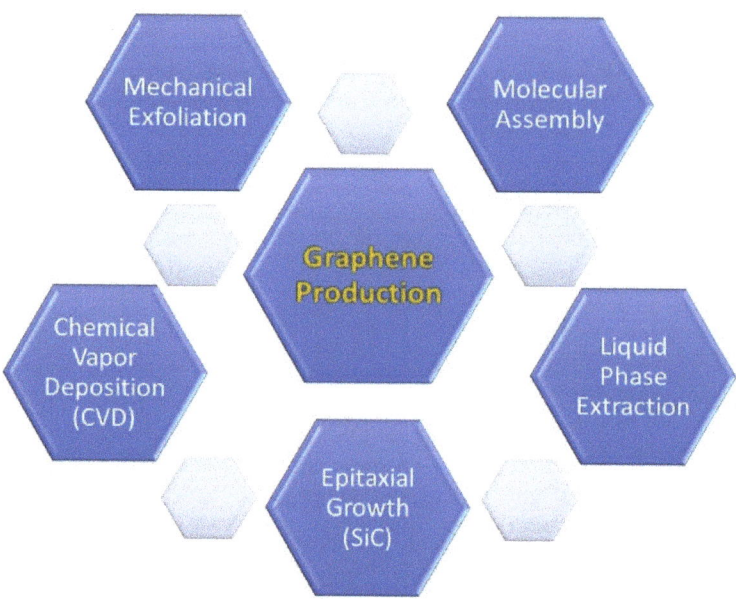

Figure 2. Various methods of production of graphene.

The simplest method of production of graphene is by mechanical exfoliation in which the graphite is subjected to tape exfoliation followed by transfer of graphene to a substrate [14,15]. Through this method, the greatest quality of graphene is produced, however, to scale-up the process is not possible [16]. The characteristics of graphene produced from various methods differ, as shown in Table 1. Graphene can grow epitaxially on SiC (silicon carbide) at high-temperature (1300–1800 °C) [17]. This method produces atomically smooth graphene nanostructures but may contain certain manufacturing defects. Furthermore, molecular assembly induces modulation of graphene using metal phthalocyanines [18] which is effective to improve the electronic properties [19,20] and the molecular ordering is critical to achieving potential shapes [19]. Liquid phase extraction (LPE) is important for the mass manufacture of graphene [21,22]. Common reported techniques of LPE include sonication [23], jet cavitation [24], micro-fluidization [25], and high-shear mixing [22]. Sonication can produce high concentrations of monolayer to few-layer graphene [23,26]. The factors responsible for the graphene exfoliation include the sonication process, shear forces, the dispersion medium, and the centrifugation process [27–29]. Graphene is also grown on non-metallic substrates such as SiO_2, h-BN, or quartz, using chemical vapor deposition, which allows direct deposition of high-quality graphene [30,31]. Chemical vapor deposition of graphene can result in 3D structures having low density, high surface area, and fast electron transport [32–34]. These properties are suitable for engineering, nanotechnology, and biomedical applications.

Table 1. Various methods of production of graphene and their properties [4].

Method	Crystallite Size (μm)	Sample Size (mm)	Charge Mobility ($cm^2\ V^{-1}\ s^{-1}$)
CVD processed graphene	>1000	~1000	10,000
Mechanical exfoliation of graphene	>1000	>1	$>2 \times 10^5$ and 10^6
Solution-processed graphene	~100	Infinite as a layer of graphene flakes	100
Epitaxial growth of graphene	50	100	10,000
Molecular assembly of graphene	<50	>1	NA

3. Structure and Properties of Graphene

A flat, 2D, sheet of graphene is single to multi-layered while graphene 3D structures can be produced to take various forms (flakes, foams, shells, and hierarchical structures) [32,35]. A graphene film may be comprised of a monolayer, bilayer, or multi-layer. Monolayer graphene is very thin (0.35 ± 0.01 nm) [36] and multilayer graphene has <10 layers [6], as reported by Raman scattering, scanning probe microscopy, and optical contrast [37]. The 2D graphene layers can have a pore size of less than a millimeter which can subsequently be incorporated into porous 3D graphene forms [38]. The 3D foam structures have a larger surface area, strength, are stiff, lightweight, and provide excellent electronic and thermal conductivity, and pathways for ionic transport.

The structure of GO and rGO and their process of production is shown in Figure 3 [39]. Generally, GO is manufactured by the oxidation of graphite from Hummers' method [40,41]. By thermal-, chemical-, and electrochemical reduction, GO yields rGO. GO and rGO have functional capabilities and wider applications beyond that of pristine graphene [3,42,43].

The AGO can be produced from the reduction and amination of graphene oxide via two-step liquid phase treatment with hydrobromic acid and ammonia solution in mild conditions [44]. The AGO is biocompatible, has electrical conductivity, and has the tendency to form wrinkled and corrugated graphene layers are observed in the AGO derivative compared to the pristine rGO. AGO can be used for biosensing, photovoltaic, catalysis application, and is used as a starting material for further chemical modifications.

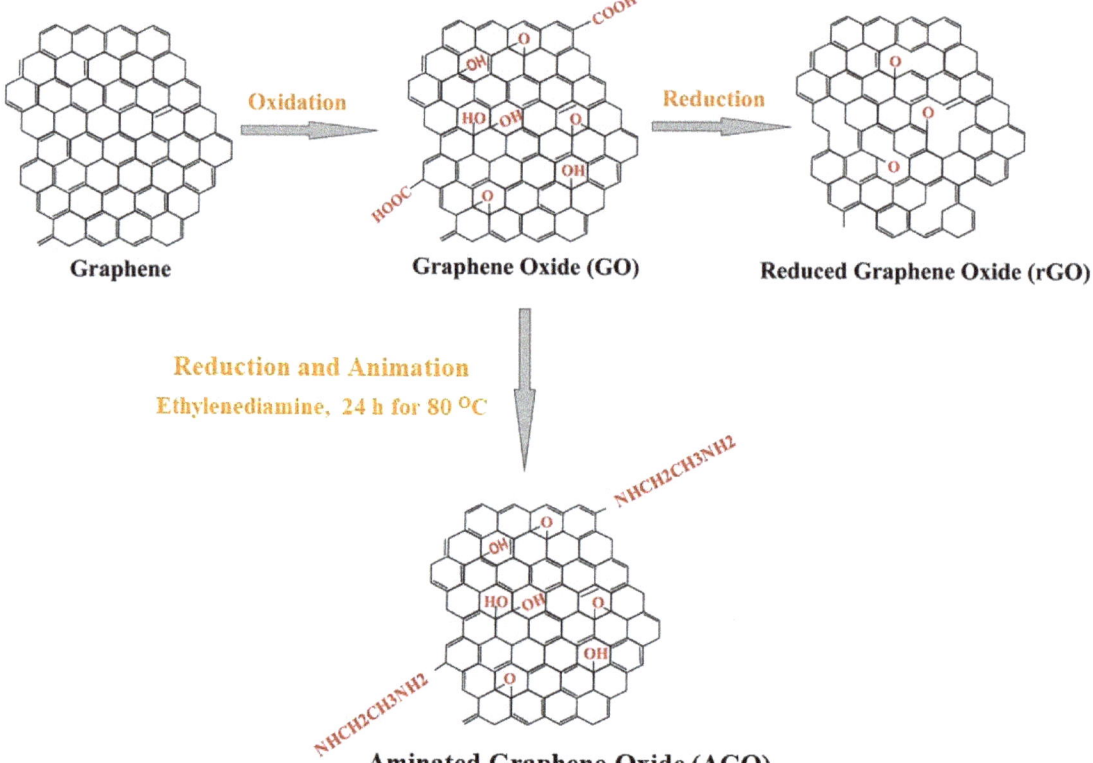

Figure 3. Production of graphene oxide (GO) reduced graphene oxide (rGO), and animated graphene oxide (AGO) from graphene [39].

Table 2 shows the essential properties of graphene, GO, and rGO [3,43]. Graphene has good electron mobility [45], increased surface area [46], good electrical conductivity [15], good thermal conductivity [47], high elastic modulus [48], strength and stiffness [48], and good wear and friction properties [4,7]. Large surface area and the ability to form nanocomposites, graphene-based materials have wide applications in regenerative medicine and drug delivery. High strength, wear-resistant, and low friction are useful in coatings and nanocomposites. Good electrical property is suitable for biosensors, semiconductors, and supercapacitors.

Table 2. Physical and mechanical properties of graphene [3,43].

Properties	Graphene	GO	rGO
Thermal conductivity	5000 W/m-K	2000 W/m-K	0.14–0.87 W/mK
Electrical conductivity	10^4 S/cm	10^{-1} S/cm	200–35,000 S/cm
Electrical resistance	10^{-6} Ω-cm	NA	NA
Tensile strength	130 GPa	120 GPa	NA
Elastic modulus	1 TPa	0.22 TPa	NA
Poisson's ratio	0.18	-	-

4. Characterization and Properties of Graphene

Several methods to study graphene's surface structure are of use, these include transmission electron microscopy, scanning electron microscope, energy dispersive spectroscopy, Raman spectroscopy, X-ray diffraction, and atomic force microscopy, [49–53]. A notable characteristic of GO includes a somewhat rough surface morphology, as observed using a scanning electron microscope [52,53]. Raman spectra of graphene-based materials exhibit a D- and G band at about 1320/cm and 1570/cm, respectively [54]. D bands signify the breathing mode of κ-point phonons with A_{1g} symmetry and the G band signifies the tangential stretching mode of the E_{2g} phonon of the carbon sp^2 atoms. The I_D/I_G ratio of around 0.84–0.97 has been reported [49].

The crystallinity and spacing of the interplane of graphene can be studied from XRD. The deflection, height, and 3D images are obtained at micron and nanoscale, by the XRD. AFM can reveal the surface structure and allow for the observation of features at the molecular and atomic levels. R_a (roughness average) can be also calculated from the AFM. XPS makes it possible to study the binding between C−O−C and C−C, and elemental composition [55].

5. Functionalization of Graphene

The development of nanocomposites has a long history. Although graphene has potential applications in engineering and biomedicine, its properties can be further improved via functionalization and doping due to its sp2 carbon atoms [56–58]. Graphene-based materials can be strengthened by various biopolymers (e.g., epoxy and polyketone) and metals (e.g., Zr, Ag, Cu, Zn, Au, Al, Ni, and Mg) [9,59–62], and nonmetals. As graphene is atomically thin, flat, and conducting material, it is suitable to produce energy storage devices [63]. At, present, the biomedical application of graphene nanocomposites is increasing [9,61,64,65]. The graphene nanocomposites have improved biocompatibility [66–68], surface properties [3,60], and mechanical properties [60] compared to pristine graphene. Graphene oxide, which is more amenable to chemical modification than pristine graphene. These properties permit applications involving protective and anticorrosion coatings [67,68], friction reduction [60], and antibacterial utilizations [69].

In graphene, n- or p-type doping Fermi level production is generally seen by physical or chemical bonds [70,71]. In graphene, chemical functionalization offers an obvious solution to the problems associated with graphene [72]. Electron-donating or -withdrawing groups can be bonded to the graphene network by synthetic chemistry methods, which could contribute to the bandgap widening and good dispersibility in common organic solvents.

The functionalization of graphene can be through covalent or non-covalent. Covalent bonds with graphene can occur using radical species, including nitrene, carbene, and aryl intermediates [72]. Conversely, modification of graphene occurs through noncovalent interactions, such as π–π interactions, van der Waals forces, ionic interactions, and hydrogen bonding, and result in major alteration of its structure and electronic properties [72]. The noncovalent interaction of graphene occurs with aromatic species, organic molecules, other carbon nanostructures, and inorganic species.

An aryl group can be grafted on the sp^2 carbon network of graphene using a diazonium salt and this has been widely applied to form covalently functionalized conducting or semiconducting materials [73,74]. A dinitrogen molecule is eliminated, and then, an electron is transferred from graphene to the diazonium salt to form an aryl radical. Thionine (Th) diazonium cation—covalently attached to the glassy carbon (GC) electrode via graphene nanosheets (GNs) (GC–GNs–Th)—has potential for application in sensors for detecting glucose and nitrite [74]. In addition, perfluorophenyl azides (PFPAs) can be covalently functionalized with graphene [75,76]. The functionalized graphene exhibits new chemical functionalities because the PFPAs groups impart solubility in both water and organic solvents [76].

Similarly, adsorption of aromatic molecules onto graphene, e.g., borazine ($B_3N_3H_6$), triazine ($C_3N_3H_3$), and benzene (C_6H_6) occurs through non-covalent bonds [77]. Park et al. [78] studied the influence of pyridine adsorption and the applied electric field on the band structure and metallicity of zigzag graphene nanoribbons (ZGNRs) using density functional theory. They found that adsorption of an electron-accepting organic molecule, such as pyridine, on ZGNRs should provide a simple and useful way to widen the band gap and can be used to turn the band structure of nanoscale electronic devices based on graphene applications.

Zhang et al. [79] developed a biosensor for the detection of microRNAs (miRNAs) based on graphene quantum dots (GQDs) and pyrene-functionalized molecular beacon probes (py-MBs). The pyrene unit served to shorten the distance between py-MBs and GQDs and to generate an increased fluorescence signal from dyes appended on the probes. When hybridized with the target miRNAs, the hairpin structure of py-MBs opened and formed more precise duplex structures.

Another important application of functionalized graphene is antimicrobial applications. Silver nanoparticles (AgNPs) be able to be decorated on the GO to make GO/Ag nanocomposite (Figure 4) [49,66]. This nanocomposite can be applied for coating and antimicrobial applications [49]. The ratio of D and G bands (I_D/I_G) of the GO/Ag nanocomposite may be elevated as a result of the disorder of the GO/Ag matrix [49,80].

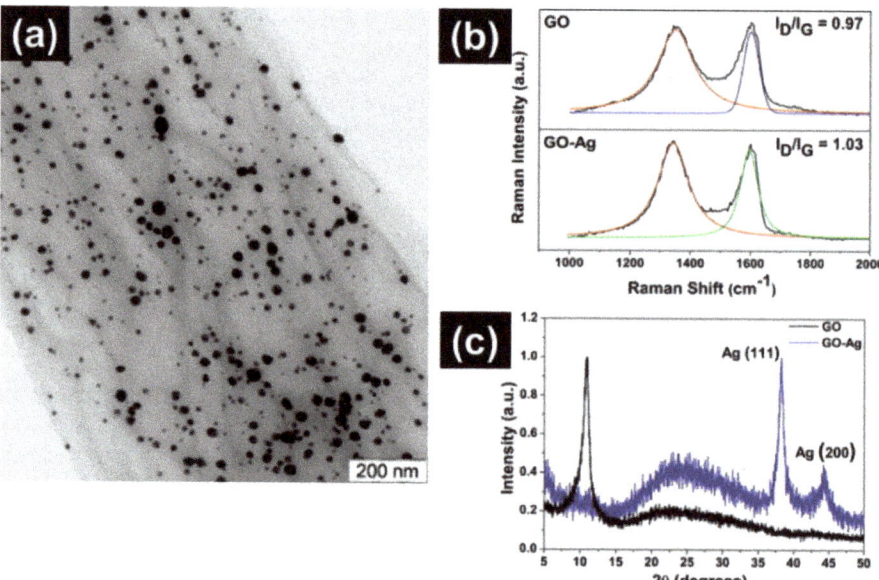

Figure 4. Characterization of graphene oxide (GO) nanocomposite formed from GO sheets decorated with Ag (GO/Ag). (**a**) Transmission electron microscopy (TEM) image, (**b**) Raman spectra, and (**c**) X-ray diffraction (XRD) [49].

Furthermore, Jeyaseelan et al. [81] developed the AGO for fluoride removal application, which was studied in terms of adsorption isotherms, kinetics (particle/intraparticle diffusion and pseudo-first/second-order models), and thermodynamic studies of AGO. The fluoride removal mechanism of AGO was found to be an electrostatic attraction.

Tissue engineering has emerged as an important approach to bone regeneration/ substitution [82]. Functionalized graphene and its derivates have been also used in bone regeneration and tissue engineering. Graphene can be combined with natural and synthetic biomaterials to enhance the osteogenic potential and mechanical properties of tissue engineering scaffolds [83–85]. Scaffolds play a central role in tissue engineering as structural support for specific cells and provide the templates to guide new tissue growth and construction [84]. Nishida et al. [86] coated collagen scaffolds with various concentrations of GO and evaluated the bioactivity, cell proliferation, and differentiation both in vivo and in vitro. The results showed that GO affected both cell proliferation and differentiation and improves the properties of collagen scaffolds. Subcutaneous implant tests showed that low concentrations of GO scaffold enhance cell in-growth and are highly biodegradable, whereas high concentrations of GO coating resulted in adverse biological effects. Consequently, scaffolds modified with a suitable concentration of GO are useful as a bioactive material for tissue engineering.

Similarly, Kang et al. [87] studied the covalent conjugation of GO flakes to 3D collagen scaffolds improves the mechanical properties of the scaffolds and promotes the osteogenic differentiation of human MSCs (hMSCs) cultured on the scaffolds. The covalent conjugation of GO flakes to collagen scaffolds increased the scaffold stiffness by 3-fold and did not cause cytotoxicity. hMSCs cultured on the GO/collagen scaffolds showed significantly enhanced osteogenic differentiation compared to cells cultured on non-modified collagen scaffolds. The enhanced osteogenic differentiation observed on the stiffer scaffolds was mediated by MSC mechanosensing because molecules that are involved in cell adhesion to stiff substrates were either up-regulated or activated. The 3D GO/collagen scaffolds could offer a powerful platform for stem cell research and orthopedic regenerative medicine.

Recently, graphene-based bioactive glass is studied as a potential drug/growth factor carrier, which includes the composition–structure–drug delivery relationship and the functional effect on the tissue-stimulation properties [82,88,89]. Wang et al. [88] designed a scaffold composed of mesoporous bioactive glasses (MBG) and GO and studied the composite porous scaffold that promotes local angiogenesis and bone healing. This in vitro study found that the MBG/GO scaffolds have better cytocompatibility and higher osteogenesis differentiation ability with rat bone marrow mesenchymal stem cells (rBMSCs) than the purely MBG scaffold. Moreover, MBG/GO scaffolds promote vascular ingrowth and, importantly, enhance bone repair at the defect site in a rat cranial defect model. The new bone was fully integrated not only with the periphery but also with the center of the scaffold. Hence, the MBG/GO scaffolds possess excellent osteogenic-angiogenic properties which will make them appealing candidates for repairing bone defects.

Finally, biodegradable composites have been used in various regeneration processes applications such as the regeneration of bones, cartilage, and soft tissues. Stepanova et al. [90] synthesized aminated graphene with oligomers of glutamic acid and their use for the preparation of composite materials based on poly(ε-caprolactone) for tissue regeneration applications. The poly(ε-caprolactone) films filled with modified aminated graphene were produced and characterized for their mechanical and biological properties. They found that grafting of glutamic acid oligomers from the surface of aminated graphene improved the distribution of the filler in the polymer matrix that, in turn, improved the mechanical properties of composite materials. In addition, the modification improved the biocompatibility of human MG-63 osteoblast-like cells.

6. Graphene Coating Applications

The potential application of graphene for various biomedical applications is promising [3,9], such as anticorrosion, antibacterial coatings, and friction reduction [67], as shown in Figure 5. Graphene is chemically inert, atomically smoothness and high durability make it an alternative candidate for implant coatings [91].

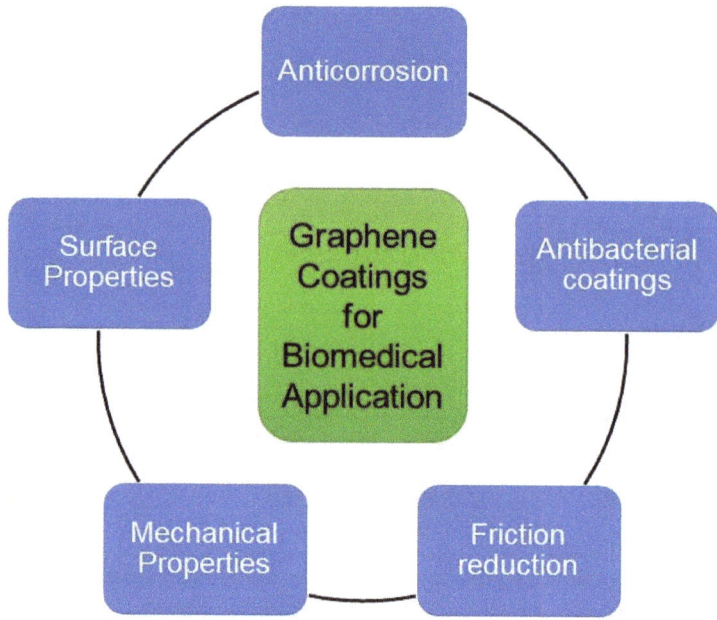

Figure 5. Biomedical applications of graphene-based coatings.

6.1. Anticorrosion Coating

There are various applications of metallic materials in medicine and dentistry, such as dental implants, orthopedic fixations, orthodontic, joint replacements, stents, endodontic files, and reamers [92]. However, the disadvantage of such biomaterials is the metal ions release, such as Ni, Ti, Ag, hence, coating of metallic materials plays an important role in such problems [92,93]. Although various coatings are being tried on metallic biomaterials, especially nitinol (NiTi), producing a successful coating has been always a challenge [94–107]. Notable disadvantages of polymer coating include toxicity of the component's roughness, porosity, and detachment of the coatings [108].

Even though graphene is an atom thick, it is inert and it is water-resistant and oxygen [4]. Hence, these properties combined with their durability and atomically stability has proven graphene to be useful as a corrosion barrier film [68,109–112]. Graphene can be directly grown on metallic surfaces (Mg, Zn, Ni, Al, etc.) to produce a protective coating [109,111,113]. Singh et al. [114] successfully developed an anti-corrosion graphene composite coating on Cu. In dentistry, graphene coatings can prevent corrosion of various metallic biomaterials such as archwires, files and reamers, and various metallic prostheses [65,68,109]. Furthermore, Hikku et al. [115] studied the anti-corrosion property of graphene and polyvinyl nanocomposite (GPVA) coating on the aluminum-2219 alloy (Al-2219). The corrosion rate for the coated Al-2219 alloys was better (polyvinyl alcohol coated alloy: 2.57 mm/year and GPVA coated alloy: 3.85×10^{-4} mm/year), whereas for untreated alloy: 45.25 mm/year in 3.5% NaCl solution (Figure 6). Hence, the GPVA coated Al-2219 alloy showed the best corrosion resistance than the uncoated alloy.

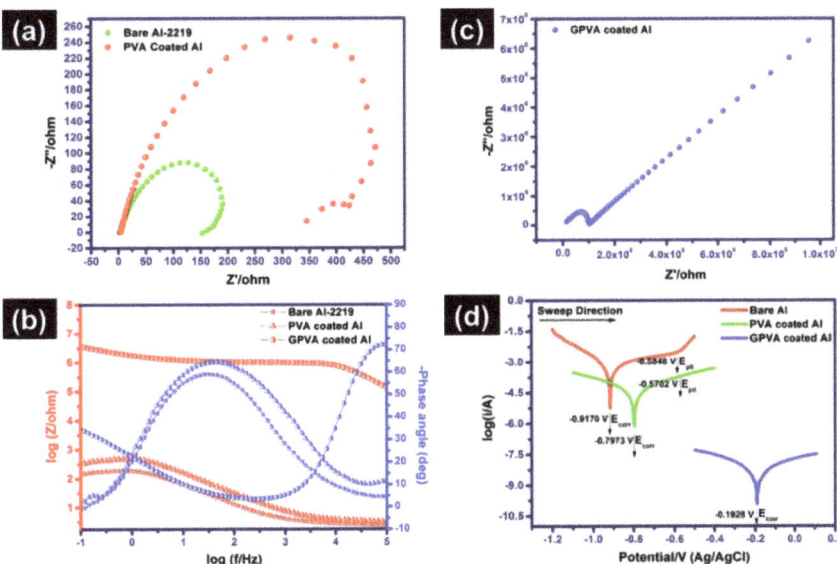

Figure 6. Anti-corrosion graphene blended with polyvinyl alcohol (GPVA) coatings: (**a**) Nyquist plot of bare Al-2219 and PVA coated Al-2219, (**b**) Nyquist plot of GPVA coated Al-2219, (**c**) Bode plot from the impedance analysis, and (**d**) Tafel plot of bare Al, PVA coated Al, and GPVA coated Al [115].

Graphene coatings can improve implant surface properties and reduce corrosion [91,116,117]. Podila et al. [91] produced graphene on Cu using chemical vapor deposition technique and transported it onto NiTi implant samples and studied the effects of the coatings on cell morphology and adhesion and they noted that the biological responses (cell adhesion and protein adsorption) were increased on the graphene-coated NiTi substrates, in comparison to the uncoated NiTi substrates. Thus, graphene-coated NiTi can be applied to the stent. Additionally, graphene could improve the osseointegration of Ti implants. In addition, Suo et al. [116] produced a homogeneous GO/chitosan/hydroxyapatite (GO/CS/HA) coating using electrophoretic deposition (EPD) on Ti. The GO/CS/HA coating's wettability and bonding strength were greater than the HA, GO/HA, and CS/HA coatings. Moreover, the GO/CS/HA coating significantly enhanced the cell–material interactions in vitro and osseointegration in vivo. Hence, the GO/CS/HA coatings on Ti can be a potential coating in implant dentistry.

Magnesium (Mg) can be used to make biodegradable implants; however, its major drawbacks of difficult-to-control corrosion. Catt et al. [110] produced a conducting polymer 3,4-ethylene dioxythiophene (PEDOT) and a GO coating for Mg implants to prevent corrosion. It was found that the significant reduction of Mg ions concentrations and pH of the media from the PEDOT/GO coating suggests a significant corrosion resistance. A positive finding was that of decreased hydrogen amounts. Three important factors were due to the passive layer preventing the ingress of a solution, film's negative charges, and development of a corrosion-resistant Mg-phosphate coat. Additionally, promising biocompatibility, in vitro, was observed as the coating did not show signs of toxicity to cultured neurons. Hence, the PEDOT/GO coating is successful in preventing Mg-based implants corrosion.

GO coating is also useful in tissue engineering and regenerative applications. Root fracture treatment, cementation of prostheses, pulp therapy, filling, repair, and regeneration of bone defects, may all indicate the use of bioactive cement. A bioactive cement typically releases calcium-ions (Ca^{2+}), increases the alkalinity in its surrounding environment, and induces cell differentiation and formation of mineralized tissue. However, the cement tends to possess poor mechanical properties, at risk of fracture due to poor strength and fracture toughness [118]. The mechanical properties are improved by the addition of graphene. A doubling of the strength of 58S bioactive glass was observed by the addition of 0.5 wt.% [119]. The addition of GO [119] and rGO [120] have also shown significant improvements in mechanical parameters, the latter (rGO 1 wt.%) resulted in a 200% increase in the fracture toughness of hydroxyapatite [119]. Additionally, the bone cement's bioactive properties are enhanced due to the addition of graphene. Several cell types, including bone marrow stem cells, periodontal ligament stem cells (PDLSCs), and dental pulp stem cells have shown spontaneous osteogenic differentiation as promoted by chemical vapor deposition-produced pristine graphene scaffolds and substrates [121,122]. Indeed, in vivo bone formation was exhibited by implanting GO-coated collagen scaffolds into tooth extraction sockets of beagle dogs. The GO-coated scaffolds showed increased bone formation and calcium absorption after 14 days, whereas the control scaffold was mostly filled with connective tissue [123]. Similarly, Zhou et al. [112] evaluated the bioactive effects of GO coated Ti substrate on PDLSCs and compared them to sodium titanate substrate. It was seen that the GO coated Ti substrate-induced PDLSCs exhibit suggestively higher alkaline phosphatase (ALP) activity, proliferation rate, and higher gene expression of osteogenesis markers, ALP, runt-related transcription factor 2 (Runx2), bone sialoprotein, and osteocalcin (OCN) compared to the sodium titanate substrate. Protein expressions of Runx2, bone sialoprotein, and OCN were additionally promoted by GO. Together, the findings suggest that GO and PDLSCs represent a favorable combination for regenerative medicine and dentistry.

6.2. Antibacterial Application

Bacteria and fungiform biofilms on the teeth surface, prostheses, or implant-anchored restorations [124]. If left untreated, the biofilm on dental implants may result in loss of the implant. It is challenging to produce implants with a high degree of osseointegration at the same time as inhibiting bacterial colonization [125–127]. The peri-implant diseases around implant result in implants failure due to supporting bone loss around the implant [128–132].

Various antimicrobial nanomaterials include polymers, nanoparticles such as gold nanoparticles (AuNPs), AgNPs, nanodiamond, and graphene-based materials [133–135]. Even though AgNPs show promising antibacterial properties, clinical applications of AgNPs are frequently impeded by their tendency to aggregate and consequent loss of antibacterial activity [133,136]. Additionally, the cytotoxicity of AgNPs towards human cells has been observed [137]. The amount of AgNPs should be minimal to avoid complications. However, AgNPs can be decorated onto GO to produce GO/Ag nanocomposite for increased antimicrobial activity [49,55]. AuNPs are used more for microbial identification rather than antimicrobial applications [138,139].

The graphene-based materials have powerful antimicrobial properties and inhibit bacterial colonization [69,125,140,141]. Agarwalla et al. [140] studied the graphene coating on Ti and their interaction with a biofilm of *Pseudomonas aeruginosa*, *Enterococcus faecalis*, *Streptococcus mutants*, and *Candida albicans*. They observed that when repeated twice, it reduces the formation of biofilm due to the hydrophobicity of graphene. These all findings show that coating Ti with graphene is useful for biofilm prevention on implants.

Graphene coatings enhance the adhesion of cells and osteogenic differentiation. Gu et al. [142] studied the osteoinductive and antibacterial effects of graphene sheets modified Ti implants. Chemical vapor deposition growth of graphene sheets by thermal treatment at 160 °C for 2 h and transferring to Ti discs. It was found that the graphene coatings on Ti enhanced adhesion of cells, osteogenic differentiation, and exhibited antibacterial properties. Similarly, another study also found similar results, i.e., osteogenic differentiation of mesenchymal stem cells using graphene [143]. Hence, graphene is capable to enhance the surface properties of NiTi-based implants.

Similarly, functionalized GO can improve the antimicrobial property, as demonstrated by the GO/Ag nanocomposite (Figure 4) [49,55]. Graphene nanocomposite has excellent antibacterial action against *Escherichia coli* and *Staphylococcus aureus* [55]. Zhao et al. [43] fabricated gelatin-functionalized GO (Gogel) surface coatings on NiTi substrates. The

Gogel's biocompatibility and antimicrobial properties were investigated, and it exhibited the highest rate of mouse osteoblastic adhesion, proliferation, as well as differentiation of cells compared to GO coated NiTi. Moreover, they reported that *E. coli* was suppressed on the surfaces of Gogel and GO. Following incubation on Gogel and GO, the integrity of the *E. coli* cell membrane was compromised and showed a low live/dead ratio. Therefore, GO-based coatings have both a high degree of biocompatibility and antimicrobial activity.

Chen et al. [144] studied the interaction of GO with four phytopathogens (two bacteria and two fungi). The studied bacteria were *Xanthomonas campestris pv. undulosa* and *pseudomonas* and studied fungus were *Fusarium oxysporum* and *Fusarium graminearum* (Figure 7). It was found that GO killed nearly 90% of the bacteria and repressed 80% macroconidia germination along with partial cell swelling and lysis at 500 µg mL^{-1}. They mentioned that GO sheets intertwined the bacterial and fungal spores resulting in the local perturbation of their cell membrane, decreasing the bacterial membrane potential, and resulting in the leakage of electrolytes of fungal spores causing cell lysis (Figure 8).

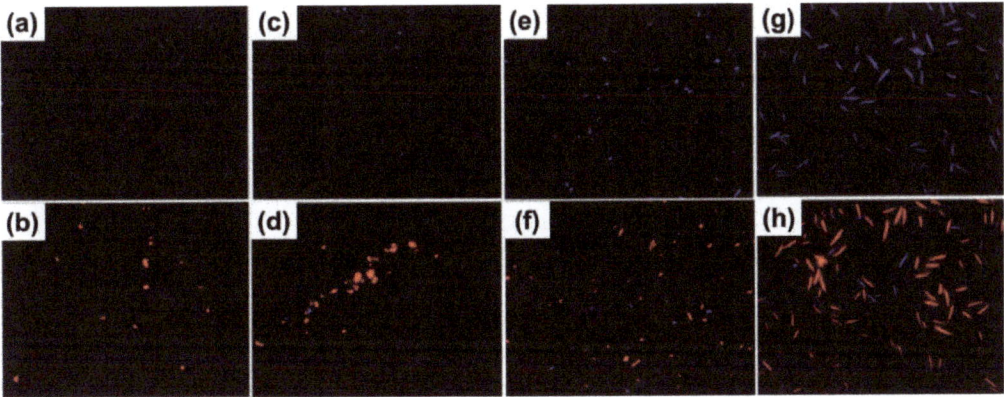

Figure 7. Fluorescence microscopy images of cells following exposure to graphene oxide (500 µg mL^{-1}): (**a**) *X. campestris pv. undulosa*, (**c**) *P. syringae*, (**e**) *F. oxysporum*, and (**g**) *F. graminearum* and images following staining of cells with propidium iodide and fluorescence stain (**b,d,f,h**) [144].

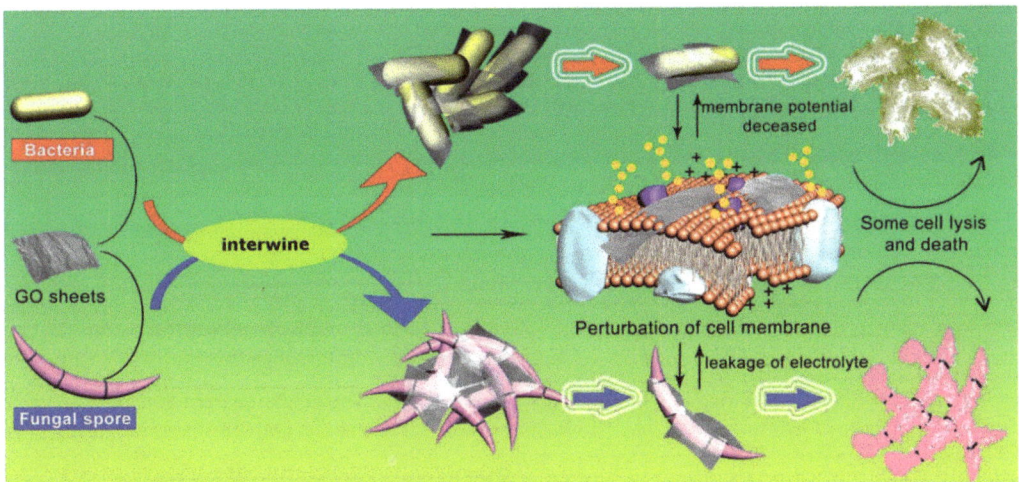

Figure 8. Antibacterial mechanism of graphene oxide against pathogens and fungal spores [144].

7. Conclusions

The available literature shows that graphene-based coatings can improve the bioactivity of biomaterials, provide microbial- and corrosion-protection of implants, both in vitro and in vivo. Peri-implant infections causing peri-implantitis are among the most common reasons for implant loss and may be prevented by the coating of antimicrobial graphene. These additive properties of graphene can be modified by methods of functionalization. Graphene exhibits high biocompatibility, corrosion prevention, and antimicrobial properties to prevent the colonization of bacteria. Graphene coatings enhance adhesion of cells, osteogenic differentiation, and exhibit antibacterial activity to parts of Ti unaffected by the thermal treatment. Graphene-based materials are promising and may hold the key for the next material-based revolution for antimicrobial and coatings applications in dental and medical technology. More research is urged before clinical utilization will be a widespread reality.

Author Contributions: Conceptualization, V.S., H.E.S. and D.R.; methodology, V.S., H.E.S., J.S. and D.R.; software, D.R.; validation, V.S. and D.R.; formal analysis, H.E.S. and D.R.; investigation, H.E.S. and D.R.; resources, H.E.S. and D.R.; data curation, H.E.S. and D.R.; writing—original draft preparation, H.E.S. and D.R.; writing—review and editing, V.S., J.S., Z.K., M.S.Z. and D.R.; visualization, V.S., Z.K., M.S.Z., J.S. and D.R.; supervision, V.S. and J.S.; project administration, V.S.; funding acquisition, V.S. All authors have read and agreed to the published version of the manuscript.

Funding: This research received no external funding.

Institutional Review Board Statement: Not applicable.

Informed Consent Statement: Not applicable.

Data Availability Statement: Not applicable.

Conflicts of Interest: The authors declare no conflict of interest.

References

1. Novoselov, K.S.; Geim, A.K.; Morozov, S.V.; Jiang, D.; Zhang, Y.; Dubonos, S.V.; Grigorieva, I.V.; Firsov, A.A. Electric field effect in atomically thin carbon films. *Science* **2004**, *306*, 666–669. [CrossRef]
2. Katsnelson, M.I. Graphene: Carbon in two dimensions. *Mater. Today* **2007**, *10*, 20–27. [CrossRef]
3. Kuill, T.; Bhadra, S.; Yao, D.; Kim, N.H.; Bose, S.; Lee, J.H. Recent advances in graphene-based polymer composites. *Prog. Polym. Sci.* **2010**, *35*, 1350–1375. [CrossRef]
4. Novoselov, K.S.; Fal'ko, V.I.; Colombo, L.; Gellert, P.R.; Schwab, M.G.; Kim, K. A roadmap for graphene. *Nature* **2012**, *490*, 192–200. [CrossRef] [PubMed]
5. Si, Y.; Samulski, E.T. Synthesis of water-soluble graphene. *Nano Lett.* **2008**, *8*, 1679–1682. [CrossRef] [PubMed]
6. Geim, A.K.; Novoselov, K.S. The rise of graphene. *Nat. Mater.* **2007**, *6*, 183–191. [CrossRef]
7. Dubey, N.; Bentini, R.; Islam, I.; Cao, T.; Castro Neto, A.H.; Rosa, V. Graphene: A versatile carbon-based material for bone tissue engineering. *Stem. Cells Int.* **2015**, *2015*, 804213. [CrossRef]
8. Rosa, V.; Zhang, Z.; Grande, R.H.; Nor, J.E. Dental pulp tissue engineering in full-length human root canals. *J. Dent. Res.* **2013**, *92*, 970–975. [CrossRef] [PubMed]
9. Stankovich, S.; Dikin, D.A.; Dommett, G.H.B.; Kohlhaas, K.M.; Zimney, E.J.; Stach, E.A.; Piner, R.D.; Nguyen, S.T.; Ruoff, R.S. Graphene-based composite materials. *Nature* **2006**, *442*, 282–286. [CrossRef]
10. Mao, H.Y.; Laurent, S.; Chen, W.; Akhavan, O.; Imani, M.; Ashkarran, A.A.; Mahmoudi, M. Graphene: Promises, facts, opportunities, and challenges in nanomedicine. *Chem. Rev.* **2013**, *113*, 3407–3424. [CrossRef]
11. Shen, H.; Zhang, L.; Liu, M.; Zhang, Z. Biomedical applications of graphene. *Theranostics* **2012**, *2*, 283–294. [CrossRef] [PubMed]
12. Roberts, M.; Clemons, C.; Wilber, J.; Young, G.; Buldum, A.; Quinn, D. Continuum plate theory and atomistic modeling to find the flexural rigidity of a graphene sheet interacting with a substrate. *J. Nanotechnol.* **2010**, *2010*, 868492. [CrossRef]
13. Al-Sherbini, A.-S.; Bakr, M.; Ghoneim, I.; Saad, M. Exfoliation of graphene sheets via high energy wet milling of graphite in 2-ethylhexanol and kerosene. *J. Adv. Res.* **2017**, *8*, 209–215. [CrossRef]
14. Geim, A.K. Nobel lecture: Random walk to graphene. *Rev. Mod. Phys.* **2011**, *83*, 851–862. [CrossRef]
15. Papageorgiou, D.G.; Kinloch, I.A.; Young, R.J. Mechanical properties of graphene and graphene-based nanocomposites. *Prog. Mater. Sci.* **2017**, *90*, 75–127. [CrossRef]
16. Guy, O.J.; Walker, K.-A.D. Chapter 4—Graphene functionalization for biosensor applications. In *Silicon Carbide Biotechnology*, 2nd ed.; Saddow, S.E., Ed.; Elsevier: Amsterdam, The Netherlands, 2016; pp. 85–141.

17. Michon, A.; Vézian, S.; Ouerghi, A.; Zielinski, M.; Chassagne, T.; Portail, M. Direct growth of few-layer graphene on 6h-sic and 3c-sic/si via propane chemical vapor deposition. *Appl. Phys. Lett.* **2010**, *97*, 171909. [CrossRef]
18. Wang, X.; Xu, J.-B.; Xie, W.; Du, J. Quantitative analysis of graphene doping by organic molecular charge transfer. *J. Phys. Chem. C* **2011**, *115*, 7596–7602. [CrossRef]
19. Järvinen, P.; Hämäläinen, S.K.; Banerjee, K.; Häkkinen, P.; Ijäs, M.; Harju, A.; Liljeroth, P. Molecular self-assembly on graphene on SiO_2 and h-bn substrates. *Nano Lett.* **2013**, *13*, 3199–3204. [CrossRef] [PubMed]
20. Wang, Q.H.; Hersam, M.C. Room-temperature molecular-resolution characterization of self-assembled organic monolayers on epitaxial graphene. *Nat. Chem.* **2009**, *1*, 206–211. [CrossRef] [PubMed]
21. Xu, Y.; Cao, H.; Xue, Y.; Li, B.; Cai, W. Liquid-phase exfoliation of graphene: An overview on exfoliation media, techniques, and challenges. *Nanomaterials* **2018**, *8*, 942. [CrossRef] [PubMed]
22. Paton, K.R.; Varrla, E.; Backes, C.; Smith, R.J.; Khan, U.; O'Neill, A.; Boland, C.; Lotya, M.; Istrate, O.M.; King, P.; et al. Scalable production of large quantities of defect-free few-layer graphene by shear exfoliation in liquids. *Nat. Mater.* **2014**, *13*, 624–630. [CrossRef]
23. Ciesielski, A.; Samori, P. Graphene via sonication assisted liquid-phase exfoliation. *Chem. Soc. Rev.* **2014**, *43*, 381–398. [CrossRef]
24. Shen, Z.; Li, J.; Yi, M.; Zhang, X.; Ma, S. Preparation of graphene by jet cavitation. *Nanotechnology* **2011**, *22*, 365306. [CrossRef]
25. Karagiannidis, P.G.; Hodge, S.A.; Lombardi, L.; Tomarchio, F.; Decorde, N.; Milana, S.; Goykhman, I.; Su, Y.; Mesite, S.V.; Johnstone, D.N.; et al. Microfluidization of graphite and formulation of graphene-based conductive inks. *ACS Nano* **2017**, *11*, 2742–2755. [CrossRef] [PubMed]
26. Lotya, M.; King, P.J.; Khan, U.; De, S.; Coleman, J.N. High-concentration, surfactant-stabilized graphene dispersions. *ACS Nano* **2010**, *4*, 3155–3162. [CrossRef]
27. Han, J.T.; Jang, J.I.; Kim, H.; Hwang, J.Y.; Yoo, H.K.; Woo, J.S.; Choi, S.; Kim, H.Y.; Jeong, H.J.; Jeong, S.Y.; et al. Extremely efficient liquid exfoliation and dispersion of layered materials by unusual acoustic cavitation. *Sci. Rep.* **2014**, *4*, 5133. [CrossRef] [PubMed]
28. Pavlova, A.S.; Obraztsova, E.A.; Belkin, A.V.; Monat, C.; Rojo-Romeo, P.; Obraztsova, E.D. Liquid-phase exfoliation of flaky graphite. *J. Nanophotonics* **2016**, *10*, 012525. [CrossRef]
29. Lin, Z.; Karthik, P.S.; Hada, M.; Nishikawa, T.; Hayashi, Y. Simple technique of exfoliation and dispersion of multilayer graphene from natural graphite by ozone-assisted sonication. *Nanomaterials* **2017**, *7*, 125. [CrossRef]
30. Wang, H.; Yu, G. Direct cvd graphene growth on semiconductors and dielectrics for transfer-free device fabrication. *Adv. Mater.* **2016**, *28*, 4956–4975. [CrossRef]
31. Teng, P.Y.; Lu, C.C.; Akiyama-Hasegawa, K.; Lin, Y.C.; Yeh, C.H.; Suenaga, K.; Chiu, P.W. Remote catalyzation for direct formation of graphene layers on oxides. *Nano Lett.* **2012**, *12*, 1379–1384. [CrossRef] [PubMed]
32. Chen, K.; Shi, L.; Zhang, Y.; Liu, Z. Scalable chemical-vapor-deposition growth of three-dimensional graphene materials towards energy-related applications. *Chem. Soc. Rev.* **2018**, *47*, 3018–3036. [CrossRef] [PubMed]
33. Chen, K.; Chai, Z.; Li, C.; Shi, L.; Liu, M.; Xie, Q.; Zhang, Y.; Xu, D.; Manivannan, A.; Liu, Z. Catalyst-free growth of three-dimensional graphene flakes and graphene/g-c3n4 composite for hydrocarbon oxidation. *ACS Nano* **2016**, *10*, 3665–3673. [CrossRef]
34. Wang, H.; Sun, K.; Tao, F.; Stacchiola, D.J.; Hu, Y.H. 3D honeycomb-like structured graphene and its high efficiency as a counter-electrode catalyst for dye-sensitized solar cells. *Angew. Chem. Int. Ed. Engl.* **2013**, *52*, 9210–9214. [CrossRef]
35. Rokaya, D.; Srimaneepong, V.; Thunyakitpisal, P.; Qin, J.; Rosa, V.; Sapkota, J. Potential applications of graphene-based nanomaterials in biomedical, dental, and implant applications. In *Advances in Dental Implantology Using Nanomaterials and Allied Technology Applications*; Chaughule, R.S., Dashaputra, R., Eds.; Springer International Publishing: Cham, Switzerland, 2021; pp. 77–105.
36. Gupta, A.; Chen, G.; Joshi, P.; Tadigadapa, S.; Eklund, P.C. Raman scattering from high-frequency phonons in supported n-graphene layer films. *Nano Lett.* **2006**, *6*, 2667–2673. [CrossRef]
37. Shearer, C.J.; Slattery, A.D.; Stapleton, A.J.; Shapter, J.G.; Gibson, C.T. Accurate thickness measurement of graphene. *Nanotechnology* **2016**, *27*, 125704. [CrossRef] [PubMed]
38. Lu, L.; De Hosson, J.T.M.; Peia, Y. Three-dimensional micron-porous graphene foams for lightweight current collectors of lithium-sulfur batteries. *Carbon* **2019**, *144*, 713–723. [CrossRef]
39. Priyadarsini, S.; Mohanty, S.; Mukherjee, S.; Basu, S.; Mishra, M. Graphene and graphene oxide as nanomaterials for medicine and biology application. *J. Nanostruct. Chem.* **2018**, *8*, 123–137. [CrossRef]
40. Hummers, W.S., Jr.; Offeman, R.E. Preparation of graphitic oxide. *J. Am. Chem. Soc.* **1958**, *80*, 1339. [CrossRef]
41. Kovtyukhova, N.I.; Ollivier, P.J.; Martin, B.R.; Mallouk, T.E.; Chizhik, S.A.; Buzaneva, E.V.; Gorchinskiy, A.D. Layer-by-layer assembly of ultrathin composite films from micron-sized graphite oxide sheets and polycations. *Chem. Mater.* **1999**, *11*, 771–778. [CrossRef]
42. Sang Tran, T.; Dutta, N.K.; Roy Choudhury, N. Graphene-based inks for printing of planar micro-supercapacitors: A review. *Materials* **2019**, *12*, 978. [CrossRef]
43. Zhao, C.; Pandit, S.; Fu, Y.; Mijakovic, I.; Jesorka, A.; Liu, J. Graphene oxide-based coatings on nitinol for biomedical implant applications: Effectively promote mammalian cell growth but kill bacteria. *RSC Adv.* **2016**, *6*, 38124–38134. [CrossRef]

44. Rabchinskii, M.K.; Ryzhkov, S.A.; Kirilenko, D.A.; Ulin, N.V.; Baidakova, M.V.; Shnitov, V.V.; Pavlov, S.I.; Chumakov, R.G.; Stolyarova, D.Y.; Besedina, N.A.; et al. From graphene oxide towards aminated graphene: Facile synthesis, its structure, and electronic properties. *Sci. Rep.* **2020**, *10*, 6902. [CrossRef] [PubMed]
45. Novoselov, K.S.; Geim, A.K.; Morozov, S.V.; Jiang, D.; Katsnelson, M.I.; Grigorieva, I.V.; Dubonos, S.V.; Firsov, A.A. Two-dimensional gas of massless dirac fermions in graphene. *Nature* **2005**, *438*, 197–200. [CrossRef] [PubMed]
46. Zhu, Y.; Murali, S.; Cai, W.; Li, X.; Suk, J.W.; Potts, J.R.; Ruoff, R.S. Graphene and graphene oxide: Synthesis, properties, and applications. *Adv. Mater.* **2010**, *22*, 3906–3924. [CrossRef]
47. Balandin, A.A.; Ghosh, S.; Bao, W.; Calizo, I.; Teweldebrhan, D.; Miao, F.; Lau, C.N. Superior thermal conductivity of single-layer graphene. *Nano Lett.* **2008**, *8*, 902–907. [CrossRef] [PubMed]
48. Lee, C.; Wei, X.; Kysar, J.W.; Hone, J. Measurement of the elastic properties and intrinsic strength of monolayer graphene. *Science* **2008**, *321*, 385–388. [CrossRef]
49. de Faria, A.F.; Martinez, D.S.T.; Meira, S.M.M.; de Moraes, A.C.M.; Brandelli, A.; Filho, A.G.S.; Alves, O.L. Anti-adhesion and antibacterial activity of silver nanoparticles supported on graphene oxide sheets. *Colloids Surf. B* **2014**, *113*, 115–124. [CrossRef]
50. Graf, D.; Molitor, F.; Ensslin, K.; Stampfer, C.; Jungen, A.; Hierold, C.; Wirtz, L. Spatially resolved raman spectroscopy of single- and few-layer graphene. *Nano Lett.* **2007**, *7*, 238–242. [CrossRef] [PubMed]
51. Lin, J.; Wang, L.; Chen, G. Modification of graphene platelets and their tribological properties as a lubricant additive. *Tribol. Lett.* **2011**, *41*, 209–215. [CrossRef]
52. Park, J.H.; Park, J.M. Electrophoretic deposition of graphene oxide on mild carbon steel for anti-corrosion application. *Surf. Coat. Technol.* **2014**, *254*, 167–174. [CrossRef]
53. Hao, J.; Ji, L.; Wu, K.; Yang, N. Electrochemistry of zno@reduced graphene oxides. *Carbon* **2018**, *130*, 480–486. [CrossRef]
54. Chen, J.; Zheng, X.; Wang, H.; Zheng, W. Graphene oxide-ag nanocomposite: In situ photochemical synthesis and application as a surface-enhanced raman scattering substrate. *Thin Solid Films* **2011**, *520*, 179–185. [CrossRef]
55. Shao, W.; Liu, X.L.; Min, H.; Dong, G.; Feng, Q.; Zuo, S. Preparation, characterization, and antibacterial activity of silver nanoparticle-decorated graphene oxide nanocomposite. *ACS Appl. Mater Interfaces* **2015**, *7*, 6966–6973. [CrossRef]
56. Banerjee, A.N. Graphene and its derivatives as biomedical materials: Future prospects and challenges. *Interface Focus* **2018**, *8*, 20170056. [CrossRef]
57. Usachov, D.; Vilkov, O.; Gruneis, A.; Haberer, D.; Fedorov, A.; Adamchuk, V.K.; Preobrajenski, A.B.; Dudin, P.; Barinov, A.; Oehzelt, M.; et al. Nitrogen-doped graphene: Efficient growth, structure, and electronic properties. *Nano Lett.* **2011**, *11*, 5401–5407. [CrossRef]
58. Hoik, L.; Keewook, P.; Ick, S.K. A review of doping modulation in graphene. *Synth. Met.* **2018**, *244*, 36–47.
59. Wang, L.; Yang, Z.; Cui, Y.; Wei, B.; Xu, S.; Sheng, J.; Wang, M.; Zhu, Y.; Fei, W. Graphene-copper composite with micro-layered grains and ultrahigh strength. *Sci. Rep.* **2017**, *7*, 41896. [CrossRef] [PubMed]
60. Rokaya, D.; Srimaneepong, V.; Qin, J.; Siraleartmukul, K.; Siriwongrungson, V. Graphene oxide/silver nanoparticles coating produced by electrophoretic deposition improved the mechanical and tribological properties of niti alloy for biomedical applications. *J. Nanosci. Nanotechnol.* **2018**, *18*, 3804–3810. [CrossRef]
61. Rokaya, D.; Srimaneepong, V.; Sapkota, J.; Qin, J.; Siraleartmukul, K.; Siriwongrungson, V. Polymeric materials and films in dentistry: An overview. *J. Adv. Res.* **2018**, *14*, 25–34. [CrossRef]
62. Ramanathan, T.; Abdala, A.A.; Stankovich, S.; Dikin, D.A.; Herrera-Alonso, M.; Piner, R.D.; Adamson, D.H.; Schniepp, H.C.; Chen, X.; Ruoff, R.S.; et al. Functionalized graphene sheets for polymer nanocomposites. *Nat. Nanotechnol.* **2008**, *3*, 327–331. [CrossRef]
63. Yoo, J.J.; Balakrishnan, K.; Huang, J.; Meunier, V.; Sumpter, B.G.; Srivastava, A.; Conway, M.; Reddy, A.L.; Yu, J.; Vajtai, R.; et al. Ultrathin planar graphene supercapacitors. *Nano Lett.* **2011**, *11*, 1423–1427. [CrossRef] [PubMed]
64. Zhang, C.; Fu, L.; Liu, N.; Liu, M.; Wang, Y.; Liu, Z. Synthesis of nitrogen-doped graphene using embedded carbon and nitrogen sources. *Adv. Mater.* **2011**, *23*, 1020–1024. [CrossRef]
65. Yang, Y.; Asiri, A.M.; Tang, Z.; Du, D.; Lin, Y. Graphene based materials for biomedical applications. *Mater. Today* **2013**, *16*, 365–373. [CrossRef]
66. Rokaya, D.; Srimaneepong, V.; Qin, J.; Thunyakitpisal, P.; Siraleartmukul, K. Surface adhesion properties and cytotoxicity of graphene oxide coatings and graphene oxide/silver nanocomposite coatings on biomedical niti alloy. *Sci. Adv. Mater.* **2019**, *11*, 1474–1487. [CrossRef]
67. Tong, Y.; Bohm, S.; Song, M. Graphene based materials and their composites as coatings. *Austin J. Nanomed. Nanotechnol.* **2013**, *1*, 1003.
68. Kirkland, N.T.; Schiller, T.; Medhekar, N.; Birbilis, N. Exploring graphene as a corrosion protection barrier. *Corros. Sci.* **2012**, *56*, 1–4. [CrossRef]
69. Nam, J.A.; Nahain, A.-A.; Kim, S.M.; In, I.; Park, S.Y. Successful stabilization of functionalized hybrid graphene for high-performance antimicrobial activity. *Acta. Biomater.* **2013**, *9*, 7996–8003. [CrossRef]
70. Johannsen, J.C.; Ulstrup, S.; Crepaldi, A.; Cilento, F.; Zacchigna, M.; Miwa, J.A.; Cacho, C.; Chapman, R.T.; Springate, E.; Fromm, F.; et al. Tunable carrier multiplication and cooling in graphene. *Nano Lett.* **2015**, *15*, 326–331. [CrossRef]
71. Wei, D.; Liu, Y.; Wang, Y.; Zhang, H.; Huang, L.; Yu, G. Synthesis of n-doped graphene by chemical vapor deposition and its electrical properties. *Nano Lett.* **2009**, *9*, 1752–1758. [CrossRef] [PubMed]

72. Yang, G.; Bao, D.; Liu, H.; Zhang, D.; Wang, N.; Li, H. Functionalization of graphene and applications of the derivatives. *J. Inorg. Organomet. Polym.* **2017**, *27*, 1129–1141. [CrossRef]
73. Sinitskii, A.; Dimiev, A.; Corley, D.A.; Fursina, A.A.; Kosynkin, D.V.; Tour, J.M. Kinetics of diazonium functionalization of chemically converted graphene nanoribbons. *ACS Nano* **2010**, *4*, 1949–1954. [CrossRef] [PubMed]
74. Hetemi, D.; Noël, V.; Pinson, J. Grafting of diazonium salts on surfaces: Application to biosensors. *Biosensors* **2020**, *10*, 4. [CrossRef]
75. Liu, L.-H.; Yan, M. Simple method for the covalent immobilization of graphene. *Nano Lett.* **2009**, *9*, 3375–3378. [CrossRef]
76. Liu, L.-H.; Yan, M. Functionalization of pristine graphene with perfluorophenyl azides. *J. Mater. Chem.* **2011**, *21*, 3273–3276. [CrossRef]
77. Chang, C.-H.; Fan, X.; Li, L.-J.; Kuo, J.-L. Band gap tuning of graphene by adsorption of aromatic molecules. *J. Phys. Chem. C* **2012**, *116*, 13788–13794. [CrossRef]
78. Park, H.; Lee, J.Y.; Shin, S. Tuning of the band structures of zigzag graphene nanoribbons by an electric field and adsorption of pyridine and bf3: A dft study. *J. Phys. Chem. C* **2012**, *116*, 20054–20061. [CrossRef]
79. Zhang, H.; Wang, Y.; Zhao, D.; Zeng, D.; Xia, J.; Aldalbahi, A.; Wang, C.; San, L.; Fan, C.; Zuo, X.; et al. Universal fluorescence biosensor platform based on graphene quantum dots and pyrene-functionalized molecular beacons for detection of micrornas. *ACS Appl. Mater. Interfaces* **2015**, *7*, 16152–16156. [CrossRef] [PubMed]
80. Hui, K.S.; Hui, K.N.; Dinh, D.A.; Tsang, C.H.; Cho, Y.R.; Zhou, W.; Hong, X.; Chun, H.H. Green synthesis of dimension- controlled silver nanoparticle-graphene oxide with in situ ultrasonication. *Acta Mater.* **2014**, *64*, 326–332. [CrossRef]
81. Jeyaseelan, A.; Ghfar, A.A.; Naushad, M.; Viswanathan, N. Design and synthesis of amine functionalized graphene oxide for enhanced fluoride removal. *J. Environ. Chem. Eng.* **2021**, *9*, 105384. [CrossRef]
82. Shadjou, N.; Hasanzadeh, M. Graphene and its nanostructure derivatives for use in bone tissue engineering: Recent advances. *J. Biomed. Mater. Res. Part A* **2016**, *104*, 1250–1275. [CrossRef]
83. Prasadh, S.; Suresh, S.; Wong, R. Osteogenic potential of graphene in bone tissue engineering scaffolds. *Materials* **2018**, *11*, 1430. [CrossRef] [PubMed]
84. Shadjou, N.; Hasanzadeh, M.; Khalilzadeh, B. Graphene based scaffolds on bone tissue engineering. *Bioengineered* **2018**, *9*, 38–47. [PubMed]
85. Cheng, J.; Liu, J.; Wu, B.; Liu, Z.; Li, M.; Wang, X.; Tang, P.; Wang, Z. Graphene and its derivatives for bone tissue engineering: In vitro and in vivo evaluation of graphene-based scaffolds, membranes and coatings. *Front. Bioeng. Biotechnol.* **2021**, *9*, 734688. [CrossRef]
86. Nishida, E.; Miyaji, H.; Takita, H.; Kanayama, I.; Tsuji, M.; Akasaka, T.; Sugaya, T.; Sakagami, R.; Kawanami, M. Graphene oxide coating facilitates the bioactivity of scaffold material for tissue engineering. *Jpn. J. Appl. Phys.* **2014**, *53*, 06JD04. [CrossRef]
87. Kang, S.; Park, J.B.; Lee, T.-J.; Ryu, S.; Bhang, S.H.; La, W.-G.; Noh, M.-K.; Hong, B.H.; Kim, B.-S. Covalent conjugation of mechanically stiff graphene oxide flakes to three-dimensional collagen scaffolds for osteogenic differentiation of human mesenchymal stem cells. *Carbon* **2015**, *83*, 162–172. [CrossRef]
88. Wang, W.; Liu, Y.; Yang, C.; Qi, X.; Li, S.; Liu, C.; Li, X. Mesoporous bioactive glass combined with graphene oxide scaffolds for bone repair. *Int. J. Biol. Sci.* **2019**, *15*, 2156–2169. [CrossRef] [PubMed]
89. Skallevold, H.E.; Rokaya, D.; Khurshid, Z.; Zafar, M.S. Bioactive glass applications in dentistry. *Int. J. Mol. Sci.* **2019**, *20*, 5960. [CrossRef] [PubMed]
90. Stepanova, M.; Solomakha, O.; Rabchinskii, M.; Averianov, I.; Gofman, I.; Nashchekina, Y.; Antonov, G.; Smirnov, A.; Ber, B.; Nashchekin, A.; et al. Aminated graphene-graft-oligo(glutamic acid)/poly(ε-caprolactone) composites: Preparation, characterization and biological evaluation. *Polymers* **2021**, *13*, 2628. [CrossRef]
91. Podila, R.; Moore, T.; Alexis, F.; Rao, A. Graphene coatings for biomedical implants. *J. Vis. Exp.* **2013**, *73*, e50276. [CrossRef]
92. Prasad, K.; Bazaka, O.; Chua, M.; Rochford, M.; Fedrick, L.; Spoor, J.; Symes, R.; Tieppo, M.; Collins, C.; Cao, A.; et al. Metallic biomaterials: Current challenges and opportunities. *Materials* **2017**, *10*, 884. [CrossRef]
93. McMahon, R.E.; Ma, J.; Verkhoturov, S.V.; Munoz-Pinto, D.; Karaman, I.; Rubitschek, F.; Maier, H.J.; Hahn, M.S. A comparative study of the cytotoxicity and corrosion resistance of nickel-titanium and titanium-niobium shape memory alloys. *Acta Biomater.* **2012**, *8*, 2863–2870. [CrossRef]
94. Goryczka, T.; Szaraniec, B. Characterization of polylactide layer deposited on ni-ti shape memory alloy. *J. Mater. Eng. Perform.* **2014**, *23*, 2682–2686. [CrossRef]
95. Williams, R.P.; Rinchuse, D.J.; Zullo, T.G. Perceptions of midline deviations among different facial types. *Am. J. Orthod. Dentofac. Orthop.* **2014**, *145*, 249–255. [CrossRef] [PubMed]
96. Li, P.; Wu, G.; Xu, R.; Wang, W.; Wu, S.; Yeung, K.W.K.; Chu, P.K. In vitro corrosion inhibition on biomedical shape memory alloy by plasma-polymerized allylamine film. *Mater. Lett.* **2012**, *89*, 51–54. [CrossRef]
97. Mazumder, M.M.; Mehta, J.L.; Mazumder, N.N.A.; Trigwell, S.; Sharma, R.; De, S. Encased Stent for Rapid Endothelialization for Preventing Restenosis. Patent 20040225346A1, 11 November 2004.
98. Schellhammer, F.; Walter, M.; Berlis, A.; Bloss, H.-G.; Wellens, E.; Schumacher, M. Polyethylene terephthalate and polyurethane coatings for endovascular stents: Preliminary results in canine experimental arteriovenous fistulas. *Radiology* **1999**, *211*, 169–175. [CrossRef]
99. Villermaux, F.; Tabrizian, M.; Yahia, L.H.; Czeremuszkin, G.; Piron, D.L. Corrosion resistance improvement of niti osteosynthesis staples by plasma polymerized tetrafluoroethylene coating. *Biomed. Mater. Eng.* **1996**, *6*, 241–254. [PubMed]

100. Tepe, G.; Schmehl, J.; Wendel, H.P.; Schaffner, S.; Heller, S.; Gianotti, M.; Claussen, C.D.; Duda, S.H. Reduced thrombogenicity of nitinol stents—in vitro evaluation of different surface modifications and coatings. *Biomaterials* **2006**, *27*, 643–650. [CrossRef]
101. Anjum, S.S.; Rao, J.; Nicholls, J.R. Polymer (ptfe) and shape memory alloy (niti) intercalated nano-biocomposites. *Mater. Sci. Eng.* **2012**, *40*, 012006. [CrossRef]
102. De Jesús, C.; Cruz, G.J.; Olayo, M.G.; Gómez, L.M.; López-Gracia, O.G. Coatings by plasmas of pyrrole on nitinol and stainless steel substrates. *Superf. Vacío* **2012**, *25*, 157–160.
103. Yang, M.-R.; Wu, S.K. Dc plasma-polymerized hexamethyldisilazane coatings of an equiatomic tini shape memory alloy. *Surf. Coat. Technol.* **2000**, *127*, 274–281. [CrossRef]
104. Bhattacharyya, A.; Dervishi, E.; Berry, B.; Viswanathan, T.; Bourdo, S.; Kim, H.; Sproles, R.; Hudson, M.K. Energy efficient graphite–polyurethane electrically conductive coatings for thermally actuated smart materials. *Smart Mater. Struct.* **2006**, *15*, S187. [CrossRef]
105. Lin, W.S.; Metz, M.J.; Pollini, A.; Ntounis, A.; Morton, D. Digital data acquisition for a cad/cam-fabricated titanium framework and zirconium oxide restorations for an implant-supported fixed complete dental prosthesis. *J. Prosthet. Dent.* **2014**, *112*, 1324–1329. [CrossRef]
106. Carroll, W.M.; Rochev, Y.; Clarke, B.; Burke, M.; Bradley, D.J.; Plumley, D.L. *Influence of Nitinol Wire Surface Preparation Procedures, on Cell Surface Interactions and Polymer Coating Adherence, Materials & Processes for Medical Devices Conference, Anaheim, CA, USA, 8–10 September 2003*; ASM International: Anaheim, CA, USA, 2004; pp. 63–68.
107. Rokaya, D.; Srimaneepong, V.; Qin, J. Modification of titanium alloys for dental applications. In *Metal, Metal Oxides and Metal Sulphides for Biomedical Applications*; Rajendran, S., Naushad, M., Durgalakshmi, D., Lichtfouse, E., Eds.; Springer International Publishing: Cham, Switzerland, 2021; pp. 51–82.
108. Raza, M.A.; Rehman, Z.U.; Ghauri, F.A.; Ahmad, A.; Ahmad, R.; Raffi, M. Corrosion study of electrophoretically deposited graphene oxide coatings on copper metal. *Thin Solid Films* **2016**, *620*, 150–159. [CrossRef]
109. Nayak, P.K.; Hsu, C.-J.; Wang, S.-C.; Sung, J.C.; Huang, J.-L. Graphene coated ni films: A protective coating. *Thin Solid Films* **2013**, *529*, 312–316. [CrossRef]
110. Catt, K.; Lia, H.; Cui, X.T. Poly (3,4-ethylenedioxythiophene) graphene oxide composite coatings for controlling magnesium implant corrosion. *Acta. Biomater.* **2017**, *48*, 530–540. [CrossRef] [PubMed]
111. Asgar, H.; Deen, K.M.; Rahman, Z.U.; Shah, U.H.; Raza, M.A.; Haider, W. Functionalized graphene oxide coating on ti6al4v alloy for improved biocompatibility and corrosion resistance. *Mater. Sci. Eng. C Mater. Biol. Appl.* **2019**, *94*, 920–928. [CrossRef]
112. Zhou, Q.; Yang, P.; Li, X.; Liu, H.; Ge, S. Bioactivity of periodontal ligament stem cells on sodium titanate coated with graphene oxide. *Sci. Rep.* **2016**, *6*, 19343. [CrossRef]
113. Catt, K.; Li, H.; Hoang, V.; Beard, R.; Cui, X.T. Self-powered therapeutic release from conducting polymer/graphene oxide films on magnesium. *Nanomedicine.* **2018**, *14*, 2495–2503. [CrossRef]
114. Singh, B.P.; Nayak, S.; Nanda, K.K.; Jena, B.K.; Bhattacharjee, S.; Besra, L. The production of a corrosion resistant graphene reinforced composite coating on copper by electrophoretic deposition. *Carbon* **2013**, *61*, 47–56. [CrossRef]
115. Hikku, G.S.; Jeyasubramanian, K.; Venugopal, A.; Ghosh, R. Corrosion resistance behaviour of graphene/polyvinyl alcohol nanocomposite coating for aluminium-2219 alloy. *J. Alloys Compd.* **2017**, *716*, 259–269. [CrossRef]
116. Suo, L.; Jiang, N.; Wang, Y.; Wang, P.; Chen, J.; Pei, X.; Wang, J.; Wan, Q. The enhancement of osseointegration using a graphene oxide/chitosan/hydroxyapatite composite coating on titanium fabricated by electrophoretic deposition. *J. Biomed. Mater. Res. B Appl. Biomater.* **2019**, *107*, 635–645. [CrossRef]
117. Li, K.; Wang, C.; Yan, J.; Zhang, Q.; Dang, B.; Wang, Z.; Yao, Y.; Lin, K.; Guo, Z.; Bi, L.; et al. Evaluation of the osteogenesis and osseointegration of titanium alloys coated with graphene: An in vivo study. *Sci. Rep.* **2018**, *8*, 1843. [CrossRef]
118. Rosa, V.; Rodríguez-Lozano, F.J.; Min, K. 22—Graphene to improve the physicomechanical properties and bioactivity of the cements. In *Advanced Dental Biomaterials*; Khurshid, Z., Najeeb, S., Zafar, M.S., Sefat, F., Eds.; Woodhead Publishing: Soston, UK, 2019; pp. 599–614.
119. Nair, M.; Nancy, D.; Krishnan, A.G.; Anjusree, G.S.; Vadukumpully, S.; Nair, S.V. Graphene oxide nanoflakes incorporated gelatin-hydroxyapatite scaffolds enhance osteogenic differentiation of human mesenchymal stem cells. *Nanotechnology* **2015**, *26*, 161001. [CrossRef] [PubMed]
120. Liu, Y.; Huang, J.; Li, H. Synthesis of hydroxyapatite-reduced graphite oxide nanocomposites for biomedical applications: Oriented nucleation and epitaxial growth of hydroxyapatite. *J. Mater. Chem. B* **2013**, *1*, 1826–1834. [CrossRef]
121. Xie, H.; Chua, M.; Islam, I.; Bentini, R.; Cao, T.; Viana-Gomes, J.C.; Castro Neto, A.H.; Rosa, V. Cvd-grown monolayer graphene induces osteogenic but not odontoblastic differentiation of dental pulp stem cells. *Dent. Mater.* **2017**, *33*, e13–e21. [CrossRef] [PubMed]
122. Xie, H.; Cao, T.; Gomes, J.V.; Castro Neto, A.H.; Rosa, V. Two and three-dimensional graphene substrates to magnify osteogenic differentiation of periodontal ligament stem cells. *Carbon* **2015**, *93*, 266–275. [CrossRef]
123. Nishida, E.; Miyaji, H.; Kato, A.; Takita, H.; Iwanaga, T.; Momose, T.; Ogawa, K.; Murakami, S.; Sugaya, T.; Kawanami, M. Graphene oxide scaffold accelerates cellular proliferative response and alveolar bone healing of tooth extraction socket. *Int. J. Nanomed.* **2016**, *11*, 2265–2277.
124. Høiby, N.; Ciofu, O.; Johansen, H.K.; Song, Z.-j.; Moser, C.; Jensen, P.Ø.; Molin, S.; Givskov, M.; Tolker-Nielsen, T.; Bjarnsholt, T. The clinical impact of bacterial biofilms. *Int. J. Oral Sci.* **2011**, *3*, 55–65. [CrossRef]

125. Dubey, N.; Ellepola, K.; Decroix, F.E.D.; Morin, J.L.P.; Castro Neto, A.H.; Seneviratne, C.J.; Rosa, V. Graphene onto medical grade titanium: An atom-thick multimodal coating that promotes osteoblast maturation and inhibits biofilm formation from distinct species. *Nanotoxicology* **2018**, *12*, 274–289. [CrossRef]
126. Raphel, J.; Holodniy, M.; Goodman, S.B.; Heilshorn, S.C. Multifunctional coatings to simultaneously promote osseointegration and prevent infection of orthopaedic implants. *Biomaterials* **2016**, *84*, 301–314. [CrossRef]
127. Gallo, J.; Holinka, M.; Moucha, C.S. Antibacterial surface treatment for orthopaedic implants. *Int. J. Mol. Sci.* **2014**, *15*, 13849–13880. [CrossRef]
128. Smeets, R.; Henningsen, A.; Jung, O.; Heiland, M.; Hammächer, C.; Stein, J.M. Definition, etiology, prevention and treatment of peri-implantitis—A review. *Head Face Med.* **2014**, *10*, 1–13. [CrossRef] [PubMed]
129. Monje, A.; Insua, A.; Wang, H.-L. Understanding peri-implantitis as a plaque-associated and site-specific entity: On the local predisposing factors. *J. Clin. Med.* **2019**, *8*, 279. [CrossRef]
130. Manor, Y.; Oubaid, S.; Mardinger, O.; Chaushu, G.; Nissan, J. Characteristics of early versus late implant failure: A retrospective study. *J. Oral Maxillofac. Surg.* **2009**, *67*, 2649–2652. [CrossRef] [PubMed]
131. Fragkioudakis, I.; Tseleki, G.; Doufexi, A.E.; Sakellari, D. Current concepts on the pathogenesis of peri-implantitis: A narrative review. *Eur. J. Dent.* **2021**, *15*, 379–387. [CrossRef]
132. Rokaya, D.; Srimaneepong, V.; Wisitrasameewon, W.; Humagain, M.; Thunyakitpisal, P. Peri-implantitis update: Risk indicators, diagnosis, and treatment. *Eur. J. Dent.* **2020**, *14*, 672–682. [CrossRef]
133. Szunerits, S.; Boukherroub, R. Antibacterial activity of graphene-based materials. *J. Mater. Chem. B* **2016**, *4*, 6892–6912. [CrossRef]
134. Yañez-Macías, R.; Muñoz-Bonilla, A.; De Jesús-Tellez, M.A.; Maldonado-Textle, H.; Guerrero-Sánchez, C.; Schubert, U.S.; Guerrero-Santos, R. Combinations of antimicrobial polymers with nanomaterials and bioactives to improve biocidal therapies. *Polymers* **2019**, *11*, 1789. [CrossRef] [PubMed]
135. Sierra-Fernandez, A.; De la Rosa-García, S.C.; Yañez-Macías, R.; Guerrero-Sanchez, C.; Gomez-Villalba, L.S.; Gómez-Cornelio, S.; Rabanal, M.E.; Schubert, U.S.; Fort, R.; Quintana, P. Sol–gel synthesis of Mg(OH)$_2$ and Ca(OH)$_2$ nanoparticles: A comparative study of their antifungal activity in partially quaternized p(dmaema) nanocomposite films. *J. Sol-Gel Sci. Technol.* **2019**, *89*, 310–321. [CrossRef]
136. Wong, K.K.Y.; Liu, X. Silver nanoparticles—The real "silver bullet" in clinical medicine? *MedChemComm* **2010**, *1*, 125–131. [CrossRef]
137. Kong, H.; Jang, J. Antibacterial properties of novel poly(methyl methacrylate) nanofiber containing silver nanoparticles. *Langmuir* **2008**, *24*, 2051–2056. [CrossRef]
138. Huang, C.C.; Chen, C.T.; Shiang, Y.C.; Lin, Z.H.; Chang, H.T. Synthesis of fluorescent carbohydrate-protected au nanodots for detection of concanavalin a and escherichia coli. *Anal. Chem.* **2009**, *81*, 875–882. [CrossRef]
139. Senapati, T.; Senapati, D.; Singh, A.K.; Fan, Z.; Kanchanapally, R.; Ray, P.C. Highly selective sers probe for Hg(II) detection using tryptophan-protected popcorn shaped gold nanoparticles. *Chem. Commun.* **2011**, *47*, 10326–10328. [CrossRef] [PubMed]
140. Agarwalla, S.V.; Ellepola, K.; Costa, M.; Fechine, G.J.M.; Morin, J.L.P.; Castro Neto, A.H.; Seneviratne, C.J.; Rosa, V. Hydrophobicity of graphene as a driving force for inhibiting biofilm formation of pathogenic bacteria and fungi. *Dent. Mater.* **2019**, *35*, 403–413. [CrossRef]
141. Pipattanachat, S.; Qin, J.; Rokaya, D.; Thanyasrisung, P.; Srimaneepong, V. Biofilm inhibition and bactericidal activity of niti alloy coated with graphene oxide/silver nanoparticles via electrophoretic deposition. *Sci. Rep.* **2021**, *11*, 14008. [CrossRef] [PubMed]
142. Gu, M.; Lv, L.; Du, F.; Niu, T.; Chen, T.; Xia, D.; Wang, S.; Zhao, X.; Liu, J.; Liu, Y.; et al. Effects of thermal treatment on the adhesion strength and osteoinductive activity of single-layer graphene sheets on titanium substrates. *Sci. Rep.* **2018**, *8*, 8141. [CrossRef]
143. Li, J.; Wang, G.; Geng, H.; Zhu, H.; Zhang, M.; Di, Z.; Liu, X.; Chu, P.K.; Wang, X. Cvd growth of graphene on niti alloy for enhanced biological activity. *ACS Appl. Mater. Interfaces* **2015**, *7*, 19876–19881. [CrossRef] [PubMed]
144. Chen, J.; Peng, H.; Wang, X.; Shao, F.; Yuan, Z.; Han, H. Graphene oxide exhibits broad-spectrum antimicrobial activity against bacterial phytopathogens and fungal conidia by intertwining and membrane perturbation. *Nanoscale* **2014**, *6*, 1879–1889. [CrossRef]

Review

Role of Defect Engineering and Surface Functionalization in the Design of Carbon Nanotube-Based Nitrogen Oxide Sensors

Manuel A. Valdés-Madrigal [1], Fernando Montejo-Alvaro [2], Amelia S. Cernas-Ruiz [3], Hugo Rojas-Chávez [4], Ramon Román-Doval [2], Heriberto Cruz-Martinez [2,*] and Dora I. Medina [5,*]

1. Instituto Tecnológico Superior de Ciudad Hidalgo, Tecnológico Nacional de México, Av. Ing. Carlos Rojas Gutiérrez 2120, Fracc. Valle de la Herradura, Ciudad Hidalgo 61100, Mexico; mavm8405@hotmail.com
2. Instituto Tecnológico Del Valle de Etla, Tecnológico Nacional de México, Abasolo S/N, Barrio Del Agua Buena, Santiago Suchilquitongo, Oaxaca 68230, Mexico; moaf1217@gmail.com (F.M.-A.); rrdoval.11@gmail.com (R.R.-D.)
3. Instituto Tecnológico del Istmo, Tecnológico Nacional de México, Panamericana 821, 2da., Juchitán de Zaragoza, Oaxaca 70000, Mexico; amelia.cr@istmo.tecnm.mx
4. Instituto Tecnológico de Tláhuac II, Tecnológico Nacional de México, Camino Real 625, Tláhuac, Ciudad de México 13508, Mexico; rojas_hugo@ittlahuac2.edu.mx
5. School of Engineering and Sciences, Tecnologico de Monterrey, Atizapan de Zaragoza 52926, Mexico
* Correspondence: heri1234@hotmail.com (H.C.-M.); dora.medina@tec.mx (D.I.M.)

Abstract: Nitrogen oxides (NO_x) are among the main atmospheric pollutants; therefore, it is important to monitor and detect their presence in the atmosphere. To this end, low-dimensional carbon structures have been widely used as NO_x sensors for their outstanding properties. In particular, carbon nanotubes (CNTs) have been widely used as toxic-gas sensors owing to their high specific surface area and excellent mechanical properties. Although pristine CNTs have shown promising performance for NO_x detection, several strategies have been developed such as surface functionalization and defect engineering to improve the NO_x sensing of pristine CNT-based sensors. Through these strategies, the sensing properties of modified CNTs toward NO_x gases have been substantially improved. Therefore, in this review, we have analyzed the defect engineering and surface functionalization strategies used in the last decade to modify the sensitivity and the selectivity of CNTs to NO_x. First, the different types of surface functionalization and defect engineering were reviewed. Thereafter, we analyzed experimental, theoretical, and coupled experimental–theoretical studies on CNTs modified through surface functionalization and defect engineering to improve the sensitivity and selectivity to NO_x. Finally, we presented the conclusions and the future directions of modified CNTs as NO_x sensors.

Keywords: experimental studies; DFT calculations; adsorption; selectivity; sensibility

1. Introduction

Novel technologies have undoubtedly allowed human civilization to reach a rapid development stage, which is mainly associated with the rapid industrialization of many countries. Approximately 75% of the global energy consumption used to achieve this was nonrenewable; that is, the energy requirements were supplied mainly from fossil fuels [1]. It is no overstatement to say that toxic emission constituents depend upon the incomplete combustion of hydrocarbons, which results in several by-products, such as O_x, CO_x, HO_x, SO_x, PO_x, RO_x, MO_x, and NO_x [2]. For this reason, although fossil fuels are limited in quantity, they have harmed the environment irreparably, despite governments implementing tax policies to discourage their use [3–5]. Accordingly, NO_x, among many other pollutants, is a component of our atmosphere that has considerably decreased the air quality around us. In this context, air pollution has direct and indirect effects on the human health, ecosystems, and climate, with consequent economic and social costs. For example, in the last two decades, health expenditures have increased due to air pollution [3].

Lamentably, air pollution constitutes a major problem in urban areas. In this sense, nitrogen oxides (NO_x) are among the primary air pollutants. Anthropogenic NO_x is formed during combustion processes at high temperatures during the operation of motor vehicles and various industrial activities [3,6]. Automotive exhaust is the main source for NO_x in urban areas. Several statistical epidemiological studies have associated air pollution with human health and mortality. For instance, air containing large amounts of NO_x can cause respiratory problems in the elderly, children, and patients with asthma [3]. Furthermore, NO_x has been recognized as an important factor in the deterioration of materials.

Air quality still affects the health of the population and perpetuates environmental degradation, e.g., the disruption of ecological balance and climate change. According to emissions projections from the World Health Organization, a massive increase in air pollution will lead to increased premature mortality caused by environmental degradation by 2050 [3]. Therefore, it is important to monitor air pollution by NO_x, in addition to other pollutants. Undoubtedly, accurate measurement of NO_x exposure in any given area, which is a demanding task, is required. From a theoretical and experimental perspective, this demonstrates the necessity of developing new sensors for NO_x detection. Therefore, the detection of toxic gases has become an important field of research.

It is not surprising that many nanomaterials have been proposed to detect such pollutants [7,8]. Even though toxic-gas sensors are conventionally designed and manufactured using semiconducting oxides (e.g., ZnO [9], SnO_2 [10], and Fe_2O_3 [11]), their use has been limited owing to poor sensibility and selectivity, as well as high operating temperatures [9–11]. In addition, it is worth highlighting that toxic-gas sensors, to be used in practice, should fulfil many requirements in terms of purposes and conditions of sensor operation. These are all connected with the aim to save energy, which is of key importance to have a remarkable increment in the toxic-gas sensors' life [12]. Along with this, the decrease in the power consumption—for gas detection technologies—should allow their fast integration into a wide range of common electronic devices associated to further improvements of modern life services, but it is still a challenging task [12]. To overcome these limitations, carbon nanostructures are currently the most promising materials to achieve such purposes. It is well known that carbon can form several different synthetic allotropes (e.g., fullerene, graphene, and nanotubes) [13], but it also exists as natural structures (e.g., diamond and graphite) that can be interconverted under specific conditions [14,15]. Among these materials, carbon nanotubes (CNTs) have attracted great attention in the design of NO_x sensors. In fact, the properties of CNTs have become active fields in modern research on new materials for toxic-gas sensors.

CNTs were first reported in the seminal work of Iijima in 1991 [16]. These nanomaterials are attractive for their interesting properties and possible applications as sensors for toxic gases. For instance, these exhibit fascinating properties, such as superior electrical conductivity [17,18], large surface area [19], excellent mechanical flexibility [20], high thermal/chemical stability [21,22], and high electron mobility [23]. Although pristine CNTs have shown promising performance for NOx detection [24–26], several strategies have been developed such as surface functionalization and defect engineering to improve the NOx sensing of pristine CNT-based sensors [27–37]. Through these strategies, the sensing properties of modified CNTs toward NO_x gases have been substantially improved [38–40]. Therefore, the modification of CNTs via surface functionalization and defect engineering is a relevant research field at both the theoretical and experimental levels for designing novel CNT-based NO_x sensors. Since CNTs are an interesting subject to be studied, a recent progress in gas sensors based on modified CNTs to detect NO_x has been revised in the literature [41,42], but those review articles are mainly focused on experimental findings. To date, there has not been a review article that analyzes the current approaches employing theoretical calculations and combining theoretical–experimental investigations. Therefore, the goal of this review is focused on recent advances (in the last decade) about modified CNTs as a promising material for sensing NO_x from both a theoretical and an experimental viewpoint, which allows a progress in the state-of-the-art. First, the types of surface functionalization and defect engineering are explained. After, the different modifications made to the CNTs from the experimental, theoretical, and combined theoretical–experimental studies are reviewed. Finally, we present the conclusions and the current challenges.

2. Surface Functionalization and Defect Engineering on CNTs

The CNTs have been widely studied and used due to their excellent mechanical, thermal, electromechanical, and chemical properties, as well as their high specific surface area [43,44]. They have been used in different fields such as catalysis, sensors, water treatment, electronics, and crop protection [45–50]. CNTs are cylindrical molecules that consist of rolled-up sheets of carbon hexagons that can be single (SWCNTs) or multiwall (MWCNTs); normally, the diameter varies from 0.8 to 2 nm and 5 to 20 nm, respectively, and their length reaches a few microns [51,52], which is generally synthesized by chemical vapor deposition (CVD), laser ablation, or electric arc [53,54]. CNTs are composed of strong sp^2 bonds that provide excellent strength [55], although these strong sp^2 bonds do not permit good chemical reactivity. Therefore, their application in some fields is limited (e.g., sensor and catalysis fields) [56], and several strategies have been developed to improve their chemical reactivity, such as surface functionalization and defect engineering (Figure 1). These strategies have substantially improved the reactivity of nanotubes to various gases [57,58]. Consequently, CNTs have become promising candidates for applications in the sensor field.

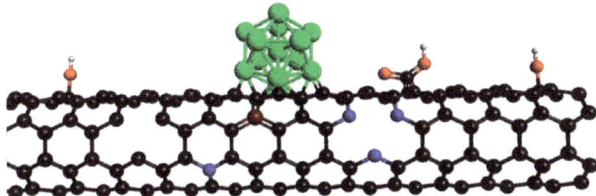

Figure 1. Surface functionalization and defect engineering in CNTs.

The surface functionalization of CNTs is classified as noncovalent or covalent, where noncovalent functionalization is based on supramolecular complexation via wrapping and adsorptive forces (e.g., π-stacking interactions and van der Waals forces). This type of functionalization does not damage the structure of the sidewall of CNTs. Noncovalent functionalization is commonly used by surfactants and polymers due to the interactions of the hydrophobic part of the adsorbed molecules with nanotubes sidewalls through van der Waals, π–π, CH–π, and other interactions, and aqueous solubility is provided by the hydrophilic part of the molecule [59,60]. This can provide an increase in the solubility and the hydrophilicity of CNTs and help reduce the tendency of CNTs to aggregate. Nevertheless, as a result of the surface functionalization of CNTs by surfactants, their interfacial adhesion is weak [59]. In covalent functionalization, the organic molecules, polymers, or metal nanoparticles are covalently bonded on the surface of CNTs, as shown in Figure 2 [61]. Covalent functionalization of the surface of CNTs can be achieved by adding covalently linked oxygen-containing groups, such as hydroxyl (OH), carbonyl (C = O), and carboxyl groups (COOH) [62]. These groups can be added at the ends, defects, and the sidewall. This chemical modification is achieved by chemical treatment with oxidizing agents such as nitric acid (HNO_3), sulfuric acid (H_2SO_4), hydrochloric acid (HCl), and potassium permanganate ($KMNO_4$) [63].

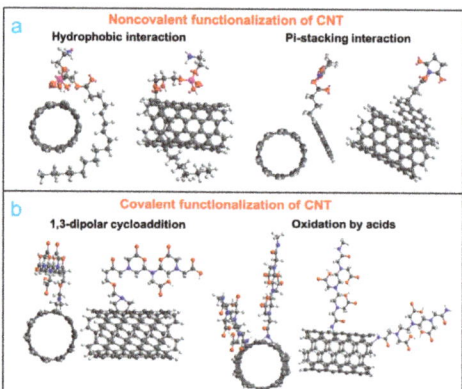

Figure 2. Functionalization modes of CNTs surface. (**a**) noncovalent functionalization and (**b**) covalent functionalization on the CNTs surface [61].

More recently, defect engineering has become an important method to modify the properties of CNTs [64,65]. Different types of defects have been explored to modify the reactivity of CNTs, including vacancies, substitutional defects (i.e., doping), combined vacancies and substitutional defects, and edge defects. Figure 3a shows the types of vacancies most commonly used to modify the properties of CNTs. Another type of defect widely used is the substitutional defect, which is also known as doping. Doping with a single type of atom or combining two types of atoms in doping have been employed, as shown in Figure 3b, and another strategy is the combination of vacancy and doping. A well-known case of this type of structure is pyridine-type nitrogen doping. It has been reported that defect engineering substantially modifies the electronic and structural properties of pristine CNTs, which causes a substantial improvement in the reactivity of CNT [66–69].

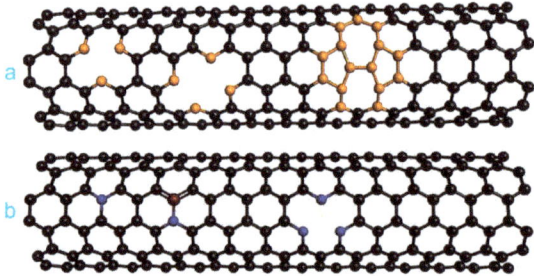

Figure 3. Defect engineering in CNTs. Types of (**a**) vacancies and (**b**) doping.

3. Experimental Studies

CNTs have been widely used for NO_x sensing [70,71]. In this context, the first studies showed the good performance of pristine CNTs for NO_x detection. For instance, Kong et al. demonstrated that the CNTs can be used as chemical gas sensor [25]. In their study, CNTs thin films were deposited onto SiO_2/Si substrates via chemical vapor deposition technique. The measurements were performed under argon or under an air atmosphere at room temperature (RT); the gas sensors showed fast response and high sensitivity when they were exposed to NO_2. In another study, Li et al. fabricated a NO_2 sensor using SWCNTs on gold electrodes [72]. The response of the sensor was up to 0.044 ppm with a recovery time of 10 h. Afterwards, the sensing properties of pristine CNTs on Si_3N_4/Si were reported by Valentini et al. [73]. CNTs were synthetized using plasma-enhanced CVD process. The CNTs/Si_3N_4 sensors were tested at different temperatures and exhibited a higher sensitivity to NO_2 at RT. Piloto et al. sensed NO_2 gas using pristine CNTs films with different thicknesses. They reported a detection limit of 1 ppm and a high sensitivity using a thickness of ≈5 nm, which was tested at RT. This response is attributed to the high density of CNTs [74]. Although these studies have demonstrated the potential of pristine CNTs for NO_x detection, several approaches have been used to improve the pristine CNTs sensing properties (e.g., high sensitivity and low operating temperature, fast response, shorter recovery time, high selectivity, easily scalable for mass production and low cost) toward NO_x gas such as surface functionalization and defect engineering. Through these approaches, the modified CNTs sensing properties toward NO_x gases have been substantially improved. Accordingly, to date, many experimental investigations have been performed using different synthesis methods to improve the NO_x detection by CNTs modified through defect engineering and surface functionalization. Therefore, in this section, in last decade, experimental studies on CNTs modified through surface functionalization and defect engineering to improve the sensitivity and selectivity toward NO_x are analyzed. These studies are mainly focused on chemiresistive gas sensors.

3.1. Functionalized CNTs

A strategy to improve the sensing of NO_2 gases is the use of CNTs functionalized by molecules anchored on their surface. Polyaniline (PANI) is a polymer widely used for the functionalization of CNTs owing to its extraordinary electrical properties, chemical stability, the ease of property modification via inorganic acids, low cost, and the ease of synthesis; when interacting with CNTs, it increases the transportation rate of charge carriers [75–78]. As a result of these properties, PANI is widely used to modify the surface of CNTs. Yun et al. functionalized CNTs using PANI polymerization for NO gas sensing. The samples were tested using TiO_2 as a catalyst in a vacuum chamber at a pressure of 1×10^{-6} mbar, where the sensors were exposed to NO gas at 25 ppm [79]. In another investigation, the PANI and poly(3,4-ethylenedioxythiophene)–polystyrene sulfonic acid (PEDOT:PSS) were used to modify the reactivity of CNTs for NO_2 gas detection; multiwalled CNTs were synthetized on silicon substrates by the CVD method. PANI and PEDOT:PSS were dissolved using different organic solvents to modify their properties and then spin-coated onto CNTs grown on silicon substrates; a high sensitivity (29.8%) to NO_2 at 100 ppm was observed for the PANI–MWCNT composite using dimethyl sulfoxide at RT [80]. Using the same approach, PANI modified with sulfonic acid has been used to functionalize CNTs for the detection of NO_2 [81]. Sensors based on the PANI/MWCNT composite showed very good sensitivity and fast response time of 50 s when were exposed at different concentrations of NO_2 gas and a detection limit of 0.05 ppm. In addition, the PANI/CNT composites changed their electronic properties from a p-type to an n-type semiconductor when the samples underwent heat treatment at 80 °C for 24 h; sensors improved the response time to 5.2 s with a detection limit of 0.0167 ppm [82]. This improvement is due to the high permeability of the PANI/CNT compound, which causes a rapid diffusion of the gas through the polymer passageways, the high mobility of the charge carriers of the composite, and the interaction between amino groups and NO_2 molecules, as shown in Figure 4.

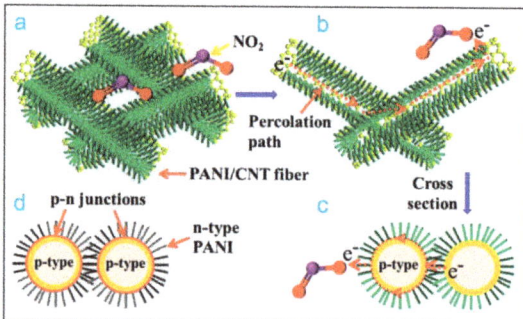

Figure 4. Possible sensing mechanism: (**a**) sketch diagram of conductive network of hierarchical p-PANI/CNT fibers, (**b**) percolation path through conjugate interfaces of PANI and MWCNTs, (**c**) cross-section of PANI/CNT fibers, and (**d**) p-n heterojunction structure of hierarchical n-PANI/CNT fibers [82].

Another research group functionalized CNTs using a carboxylic group to improve the detection of NO_2 gas. The sensing properties were obtained at different temperatures and NO_2 concentrations, which obtained the highest sensitivity of 26.88% at a concentration of 100 ppm tested at RT [83]. Jeon et al. fabricated a NO gas sensor based on CNTs functionalized with an amine group, where the response was 50% at RT [84]. Sensing properties based on sulfuric acid-functionalized CNTs have been reported by Ionete et al. The sensors exhibited good response at RT with high sensitivity when exposed to NO and NO_2 gas at a concentration of 0.04–0.8 ppm [85]. For NO sensing, the sensor showed a response and a recovery time of 255 and 50 s, respectively; whereas, for NO_2, it exhibited a response and a recovery time of 540 and 420 s, respectively. Finally, carbon nanotubes functionalized with manganese porphyrin have been used for the fabrication of NO_2 sensors. The sensors were operated at different temperatures with a high sensitivity at temperature of 100 °C [86].

3.2. Decorated CNTs

Another strategy to improve gas sensing is using metallic nanoparticles deposited or supported on the structure of CNTs [87,88]. This strategy has attracted much interest in sensing applications because the catalytic properties of metallic nanoparticles can modify the electronic properties of CNTs using transition metals supported on their surface. The metallic nanoparticles act as catalysts, promoting more reactive sites on their surface for the adsorption of gas molecules. Furthermore, the metal nanoparticles have demonstrated that can bonds strongly with small gas molecules due to their electronic structure and empty orbitals [89]. For example, gold nanoparticles supported on vertically aligned CNTs (VA-CNTs) have been used as gas sensors to detect NO_2 molecules; CNTs were synthesized using the CVD technique. The tests were carried out at RT using CNTs with lengths of 150, 300, and 500 μm, as shown in Figure 5, to find the best response for 300 μm when the sensors were exposed to NO_2 gas at different concentrations [90]. This high response to NO_2 detection was related with the transport unidirectional of the electrical charges and the high effective surface-area-to-volume ratio for CNTs of 300 μm compared with CNTs of 500 μm and 150 μm. However, longer lengths (e.g., 500 μm) of the VA-CNTs could produce a lofty packaging, and this can make the gas detection difficult. In addition, they evaluated the gas sensors at different humidity and found that gas sensors measured at 50% humidity increased the detection of NO_2 gas. It has reported that relative humidity plays an important role in the electrical conductivity and sensitivity of CNTs [91,92].

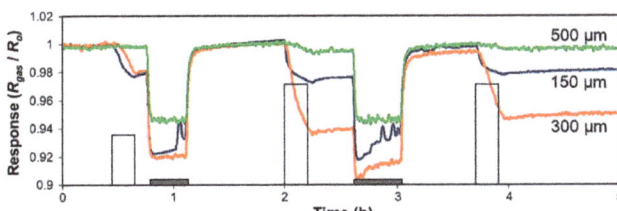

Figure 5. Room temperature detection of NO$_2$ for sensors with different CNT lengths. White pulses indicate the exposure to 0.5 ppm, 1 ppm, and 1 ppm of NO$_2$ (duration: 15 min). Gray bars indicate the periods of heating at 150 °C that help clean the surface of CNT after being exposed to NO$_2$. Heat was not applied after the last exposure cycle, and the baseline was not regained [90].

Ada Fort et al. developed sensors based on CNTs decorated with gold nanoparticles and TiO$_2$, which were operated at temperatures below 250 °C. CNTs decorated with nanoparticles showed more activity at low temperatures than pristine CNTs, thus enabling high sensitivity for CNTs decorated with gold nanoparticles with a value of 10% at 12 ppm NO$_2$ operating at 240 °C [93]. Dilonardo et al. developed sensors based on CNTs with metallic nanoparticles of Au and Pd deposited on its surface. Sensing tests were carried out at different concentrations of NO$_2$ operating at different temperatures (45–200 °C) [38]. Metallic nanoparticles were deposited into CNTs using the electrophoresis technique; these sensors obtained high gas sensitivity, fast response, and low limit of detection, as shown in Figure 6. Using the same strategy, the Pt nanoparticles were supported into CNTs; these sensors were manufactured using the sputtering technique and annealed at a temperature of 500 °C for 1 h under argon atmosphere. Measurements were performed at different concentrations and at various temperatures (25–150 °C), obtaining the best sensing response at a concentration of 2 ppm operating at 100 °C (at least five times higher than pristine CNTs) [94]; in that investigation, it was shown that the performance of sensors based on CNTs were degraded when the sensors were preserved in humid environments. In addition, Mahmood and Naje reported sensors based on Pt nanoparticles deposited into CNTs to detect NO$_2$ molecules, which exhibited high sensitivity tested at different temperatures. Their study showed an increase of 4.1 times compared to that of pristine CNTs with a value of 150% at RT [95]. Furthermore, TiO$_2$ and Au nanoparticles on CNTs were used for NO$_2$ gas sensing. The sensitivity of the sensor was enhanced using pulsed temperature mode, which consisted of variable working temperature using a pulse train [96]. CNTs were decorated with WO$_3$ nanoparticles to form WO$_3$/MWCNT composite by metal organic decomposition method for NO$_2$ gas sensing [97]. The sensors were exposed to NO$_2$ gas at different concentrations and measured at RT; the highest sensitivity obtained was in the range of 0.1–0.2 ppm.

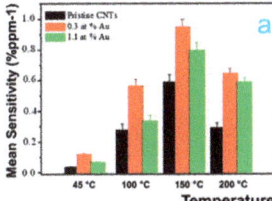 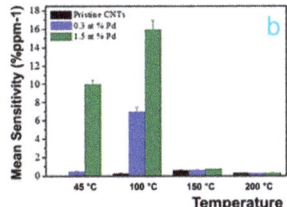

Figure 6. Mean sensitivity of pristine and (**a**) Au- and (**b**) Pd-modified MWCNTs-based sensors toward NO$_2$ gas at different sensor operating temperatures in the range 45–200 °C [38].

Table 1 summarizes the different nanoparticles supported on CNTs that have been used for the sensing of NO$_2$ gases. An analysis of sensing properties presented in this table indicates that the lower detection limit is 0.003 ppm detected by the Pt-SWCNTs system operating at 200 °C [87], while at RT, the detection limit is 0.088 ppm using ZnO-SWCNTs [98]. Table 1 also indicates that the Ag-SWCNTs sensor has the lowest time response of 8 s compared with other systems [95]. All these results demonstrate that decorated CNTs with metal nanoparticles can enhance the response sensor due the metal nanoparticles. Thus, the role of metal nanoparticles on CNTs is to accelerate the surface reaction, increase active sites for the adsorption of gas molecules, and improve the electrical properties, which in turn increase the sensitivity to small gas molecules.

Table 1. Summary of NO_2 gas sensors based on CNTs decorated with nanoparticles.

Sensor Type	Operating Temperature °C	Limit of Detection (ppm)	Response Time	Recovery Time	Reference
Pt-SWCNTs	200	0.003	<600 s	-	[87]
Pt-MWCNTs	25	1.7	-	-	[99]
Pt-SWCNTs	25–150	2	>180 s	849–1411 s	[94]
Pd-SWCNTs	200	0.009	<600 s	-	[87]
Pd-MWCNTs	25	1.7	-	-	[99]
Pd-SWCNTs	45–200	0.2	<300 s	>1300 s	[38]
Au-MWCNTs	RT	0.1	>600 s	-	[100]
Au-MWCNTs	45–200	0.2	<300 s	>1300 s	[38]
Au-MWCNTs	100–250	5	>30 s	7–4 min	[93]
SnO_2-MWCNTs	30–200	0.1	<420 s	>8 min	[40]
SnO_2-SWCNTs	180–380	0.3	<100 s	-	[101]
TiO_2-SWCNTs	100–250	5	>60 s	6–3 min	[93]
ZnTe-SWCNTs	RT	0.5	-	-	[102]
Rh-MWCNTs	RT	0.05	20 min	-	[88]
Cdots-SWCNTs	RT	0.1	381 s	294 s	[103]
ZnO-SWCNTs	RT	0.088	<220 s	-	[98]
ZnO-SWCNTs	25–300	1	300 s	5–8 min	[104]
Ag-SWCNTs	RT	-	8 s	15 s	[95]
WO_3-SWCNTs	250–300	0.05	25 min	-	[105]
WO_3-SWCNTs	RT	0.1	10 min	27 min	[106]

3.3. Doped CNTs

The doping of materials is one of the most used strategies to modify the electrical and electronic properties of CNTs by replacing carbon atoms with heteroatoms. Several elements have been used for the doping of CNTs, which have improved the electrical properties of the CNTs. For example, the detection of NO_2 on pristine double-walled CNTs and doped with N was studied by Muangrat et al. CNTs were synthesized at different temperatures and doped at different concentrations, where the pristine CNTs diameter was slightly larger than nitrogen-doped CNTs. In addition, they showed that a higher concentration of nitrogen (1.6 at %) decreases the crystallinity of the material. For sensor fabrication, the CNTs powders were dispersed in ethanol using ultrasonication and then deposited by drop casting on a hot ceramic substrate with a temperature of 100 °C. Their results showed that the N-doped nanotube synthesized at 900 °C with 1.6 at % of nitrogen exhibited the best response to NO_2 gas with a value of 60% more than the pristine CNTs tested at RT [39] (Figure 7). Thus, the previous study demonstrated that the CNTs doped with heteroatoms of N increased the NO_2 detection, which is related to the high transfer of charge between the CNTs defects and gas molecules.

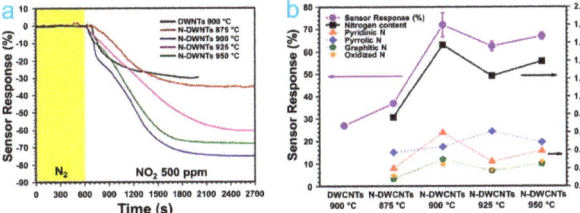

Figure 7. (**a**) sensor responses (%) of all sensors toward 500 ppm of NO$_2$ as a function of time. (**b**) sensor response saturation limit (left axis) and the ratio of different types of nitrogen as a function of synthesized temperature [39].

4. Theoretical Studies

In recent years, the theoretical design of nanomaterials has gained great importance. Among the different levels of theory that have been used to study novel materials, approaches based on quantum mechanics can be discarded, e.g., density functional theory (DFT), which has been widely used as a predictive tool for novel materials because its agreement with the experiment is very good [107,108]. It has been widely used to study novel materials in different fields, such as catalysis, electronics, and sensors [109–112]. In this section, we analyze the theoretical studies developed on CNTs modified through surface functionalization and defect engineering to improve the sensitivity and selectivity to NO$_x$.

4.1. Decorated CNTs

As previously mentioned, a strategy used to modify the reactivity of CNTs is through surface modification by depositing different atoms or nanoparticles on their surface. In this sense, there are several DFT-based theoretical studies on the reactivity of transition metals-decorated CNTs to NO$_x$ gases. For example, the NO adsorption on Pd- and Pt-decorated CNTs was investigated using the PW91 functional [113]. The NO adsorption energy on Pd-decorated CNTs and Pt-decorated CNTs was −1.81 and −2.29 eV, respectively. Very recently, the NO$_2$ interaction on Rh$_3$M alloys (M = Rh, Ag, Ir, Pd, Pt, and Au) deposited on CNTs was computed employing the BLYP functional [114]. The NO$_2$ adsorption energy on RhM-decorated CNTs was between −1.67 and −2.28 eV. These results suggest that the deposition of atoms or nanoparticles on the surfaces of CNTs is a good strategy to increase their surface reactivity. In addition, it can be deduced that transition metals-decorated CNTs can be better candidates for the detection of NO$_x$ molecules.

4.2. Doped CNTs

Doping has been widely used to modify the structural, electronic, and reactive properties of pristine carbon nanomaterials to NO$_x$ gases [115–117]. In this same direction, there are numerous studies on the use of doping as a strategy to improve the properties of CNTs with respect to NO$_x$ gases. At the theoretical level, various strategies have been used to dope the CNTs. One of the most used routes to dope the CNTs is by substituting a C atom for a heteroatom. For NO molecules, the NO interaction on Al-doped SWNT was recently investigated using the Perdew–Burke–Ernzerhof (PBE) generalized gradient approximation (GGA) [118]. The NO adsorption energy on Al-doped SWCNTs (−1.57 eV) was high compared to pristine SWCNT (−0.09 eV). In another study, the NO molecule interaction on Ni-, Pd-, and Pt-doped SWCNTs, employing the WB97XD functional, was analyzed [119]. The calculated adsorption energy values were −2.47, −3.58, and −3.56 eV for Ni-, Pd-, and Pt-doped SWCNTs, respectively. For the NO$_2$ molecule, Table 2 lists the different doping elements used to dope the CNTs by substituting the doping atom for a C atom. All the interaction energies of NO$_2$ on doped CNTs were higher than those on pristine CNTs, because for the pristine CNTs, NO$_2$ adsorption energies of less than 0.25 eV have been reported. These results clearly show that the doped CNTs are better candidates for NO$_2$

detection than pristine CNTs [120–122]. This increase in the NO_2 adsorption energies can be associated to the modification of the electronic and structural properties of doped CNTs with respect to pristine CNTs. In addition, it is clearly demonstrated that when doping occurs in CNTs, the concentration of the doping element substantially determines the properties of the doped CNTs. In this direction, NO and NO_2 adsorption on CNTs doped with different numbers of Al atoms was investigated using the PBE approximation [118]. It was shown that the NO and NO_2 adsorption energies tend to increase as the content of Al in the doped CNTs increased, which can be associated to the enlarged active sites on CNTs provided by Al atoms.

Table 2. NO_2 adsorption on doped CNTs.

Doping Atom	E_{ads} (in eV)	Methodology	Reference
Al	−2.20	B3LYP	[120]
Al	−4.24	BPE	[118]
P	−1.60	B3LYP	[120]
Cr	−2.34	B3LYP	[121]
Mn	−1.82	B3LYP	[121]
Co	−2.36	B3LYP	[122]
Zn	−2.02	B3LYP	[123]
Mo	−3.17	B3LYP	[121]
Tc	−2.06	B3LYP	[121]
Rh	−2.08	B3LYP	[122]
Pd	−2.09	B3LYP	[123]
W	−3.90	B3LYP	[121]
Re	−2.83	B3LYP	[121]
Os	−2.50	B3LYP	[123]
Ir	−2.62	B3LYP	[122]

More complex doping has recently been investigated such as N_4 porphyrin-like CNTs with transition metals. The presence of this structure in CNTs provides an increase in their reactivity compared with pristine CNTs. In this direction, the NO adsorption on CoN_4-CNTs was investigated using the PBE functional. The NO adsorption energy on CoN_4-CNTs (−1.21 eV) was high compared to those reported for pristine CNT [124]. In another more recent study, the NO and NO_2 interaction with MnN_4-CNT was computed employing the PBE approximation [125]. The NO and NO_2 adsorption energies on MnN_4-CNT were −2.41 and −1.74 eV, respectively. As in the previous case, the interaction energies of NO and NO_2 on MnN_4-CNT were higher than those on pristine CNTs, which shows that N_4-CNTs with transition metals are better candidates for the detection of these toxic gases than pristine CNTs.

4.3. Vacancies

Another strategy used to modify the reactivity of pristine CNTs is through vacancies (Figure 1). As previously documented, vacancy defects substantially modify the electronic, mechanical, and chemical properties of CNTs [126,127]. In this direction, there are some theoretical studies on the reactivity of CNTs with vacancies. For example, Vasylenko et al. investigated the NO interaction on metallic SWCNTs (8,0) with vacancy using generalized gradient approximation [128]. The adsorption energy of NO on metallic SWCNTs (8,0) with vacancy was −2.49 eV, which was higher than reported on pristine CNTs. These results show that SWCNTs with vacancy are better candidates for the detection of NO_x gases than pristine CNTs.

5. Combined Theoretical and Experimental Studies

A very interesting route for designing novel materials is combining theory and experiment [129,130]. It has been demonstrated that combining experimental results and DFT calculations is very efficient for designing novel toxic-gas sensors [131–133]. Therefore, in this section, we analyze the coupled theoretical and experimental studies on developed on CNTs modified through surface functionalization and defect engineering to improve the sensitivity and selectivity to NO_x. The sensing properties of CNTs decorated with gold nanoparticles to NO_2 were investigated through coupled theoretical and experimental methods [100]. First, three type of active layers (O_2-MWNTs, Au (5 Å)-decorated MWNTs, and Au (10 Å)-decorated MWNTs) were exposed to various concentrations of NO_2. The Au–O_2-decorated MWNTs sensors detect NO_2 down to 0.1 ppm. In addition, Au-decorated MWCNTs improve the detection of NO_2 compared with that of O_2-functionalized MWCNTs sensors. Finally, the best response to NO_2 is achieved for Au (5 Å)-decorated MWCNTs sensors. To explain the sensing properties obtained experimentally, DFT calculations were carried out on SWCNTs (5,5) decorated with an Au_{13} nanoparticles in the presence of NO_2 gas. The Au_{13} deposited on SWCNTs slightly modified the electronic properties of pristine SWCNTs. In addition, a strong interaction (−3.26 eV) between NO_2 and Au_{13}-decorated SWCNTs was observed, which can be associated with a good sensitivity of Au_{13}-decorated SWCNTs to NO_2 gas. In another study, the adsorption of NO_2 molecules on B- and N-doped CNTs was studied by Adjizian et al. [134]. First, the CNTs doped with B and N were obtained using the CVD technique. The presence of B and N atoms in the structure of CNTs increased the value of the intensity ratio between the D-band and G-band in the Raman spectrum, which increase the density of structural defects modifying the chemical reactivity. The tests were performed using an airtight chamber at concentrations of 0.05, 0.1, 0.2, 0.5, and 1.0 ppm of NO_2 operating at RT and 150 °C (Table 3). The sensors showed a good response to NO_2 for both N- and B-doped CNTs operating at both temperatures compared with pristine CNTs. The sensors showed a response at low concentrations with a detection limit of 0.05 ppm increasing with gas concentration. Furthermore, the experimental results demonstrated that the N-doped CNTs are more stable than B-CNTs, which showed the best sensitivity when exposed to NO_2 gas. Then, they used graphene as a model for the density functional calculations, which demonstrated that the reactivity of pristine graphene is enhanced by doping with B and N.

Table 3. Experimental gas sensing responsiveness, S, for nitrogen- and boron-doped nanotubes at an ambient temperature and 150 °C for different gas concentrations. Republished with permission of Elsevier from [134].

Sensor Type	Operating Temperature	NO_2			
		0.05 ppm	0.2 ppm	0.5 ppm	1.0 ppm
N-CNT	Ambient	−0.75	−2.01	−3.27	−5.5
	150 °C	−0.54	−1.21	−1.87	−2.76
B-CNT	Ambient	0.00	−0.91	−1.39	−1.63
	150 °C	−1.33	−1.98	−3.56	−3.98

6. Conclusions and Future Directions

This review analyzes the progress of modified CNTs as NO_x sensors in the last decade. The different modifications made to the CNTs from the experimental, theoretical, and combined theoretical–experimental perspectives are reviewed. Based on this review, the following conclusions and future directions are proposed.

At the experimental level, CNTs functionalized with conductive polymers, such as PANI and PEDOT, improved the sensor's response to NO_2. In addition, the use of metallic nanoparticles supported on CNTs has achieved great progress in the development of NO_2 gas sensors, which is related to the catalytic spillover effect of the nanoparticles increasing the electron transfer between metal nanoparticles and CNTs. Therefore, sensors based on polymer-modified or metallic nanoparticle-modified CNTs have shown good responses to NO_2 compared with pristine CNTs because these materials have more reactive sites for the adsorption of the gas molecule. Unfortunately, there is little research on the use of CNTs doped with heteroatoms for gas sensing, which may be due to the difficulty of controlling doping concentration experimentally. Although there has been significant advancement in NO_2 sensing with these materials, there are challenges to be overcome, such as improving the sensitivity at low concentrations and RT, selectivity, and industrial manufacturing scalable sensors.

Many theoretical DFT-based studies have been developed on modified CNTs as NO_x sensors. At the DFT level, different modifications have been investigated on CNTs, such as decorated, doping, and vacancy, in which doping is the most explored. These modified CNTs have shown higher reactivity than pristine CNTs; therefore, they are a good strategy to modify the sensitivity of the CNTs to NO_x. However, to date, DFT studies have been mainly focused on the sensitivity of modified CNTs to NO_x. Therefore, it is necessary to investigate the selectivity of the modified CNTs to NO_x gases at the DFT level. In addition, feasible approaches (e.g., applying an electronic field) to facilitate the desorption of NO_x gases on the modified CNTs should be theoretically investigated in detail.

Coupled theoretical–experimental investigations are a good strategy for designing more sensitive and selective NO_x sensors based on modified CNTs. However, in the last decade, these types of investigations have remained scarce. Therefore, investigations combining theory and experiment should be performed to design novel NO_x sensors employing modified CNTs.

Author Contributions: Conceptualization, M.A.V.-M., F.M.-A., A.S.C.-R., H.R.-C., R.R.-D., H.C.-M., D.I.M.; Formal analysis, M.A.V.-M., F.M.-A., A.S.C.-R., H.R.-C., H.C.-M., D.I.M.; investigation, M.A.V.-M., F.M.-A., A.S.C.-R., H.R.-C., H.C.-M., D.I.M.; Data curation, M.A.V.-M., F.M.-A., A.S.C.-R., R.R.-D.; Writing—original draft preparation, M.A.V.-M., F.M.-A., A.S.C.-R., R.R.-D.; Writing—review and editing, H.R.-C., H.C.-M., D.I.M.; Supervision, H.R.-C., H.C.-M., D.I.M.; Funding acquisition, H.C.-M., D.I.M. All authors have read and agreed to the published version of the manuscript.

Funding: Heriberto Cruz-Martínez appreciates the funding sources provided by the Tecnológico Nacional de México (TecNM) through the grant number 10800.21-P. The APC was funded by Tecnologico de Monterrey.

Institutional Review Board Statement: Not applicable.

Informed Consent Statement: Not applicable.

Data Availability Statement: Not applicable.

Conflicts of Interest: The authors declare no conflict of interest.

References

1. Liu, W.D.; Yu, Y.; Dargusch, M.; Liu, Q.; Chen, Z.G. Carbon allotrope hybrids advance thermoelectric development and applications. *Renew. Sustain. Energy Rev.* **2021**, *141*, 110800. [CrossRef]
2. Daniels, S.L. Products of incomplete combustion (O_x, CO_x, HO_x, NO_x, SO_x RO_x, MO_x and PO_x). *J. Hazard. Mater.* **1989**, *22*, 161–173. [CrossRef]
3. Usman, M.; Ma, Z.; Zafar, M.W.; Haseeb, A.; Ashraf, R.U. Are air pollution, economic and non-economic factors associated with per capita health expenditures? Evidence from emerging economies. *Int. J. Environ. Res. Public Health* **2019**, *16*, 1967. [CrossRef]
4. Timilsina, G.R.; Csordás, S.; Mevel, S. When does a carbon tax on fossil fuels stimulate biofuels? *Ecol. Econ.* **2011**, *70*, 2400–2415. [CrossRef]
5. Helm, D. The future of fossil fuels—Is it the end? *Oxf. Rev. Econ. Policy* **2016**, *32*, 191–205. [CrossRef]
6. Skalska, K.; Miller, J.S.; Ledakowicz, S. Trends in NO_x abatement: A review. *Sci. Total Environ.* **2010**, *408*, 3976–3989. [CrossRef]
7. Zhang, Y.Q.; Liu, Y.J.; Liu, Y.L.; Zhao, J.X. Boosting sensitivity of Boron Nitride Nanotube (BNNT) to nitrogen dioxide by Fe encapsulation. *J. Mol. Graph. Model.* **2014**, *51*, 1–6. [CrossRef]
8. Kumar, U.; Yadav, B.C.; Haldar, T.; Dixit, C.K.; Yadawa, P.K. Synthesis of MWCNT/PPY nanocomposite using oxidation polymerization method and its employment in sensing such as CO_2 and humidity. *J. Taiwan Inst. Chem. Eng.* **2020**, *113*, 419–427. [CrossRef]
9. Nundy, S.; Eom, T.-Y.; Kang, J.-G.; Suh, J.; Cho, M.; Park, J.-S.; Lee, H.-J. Flower-shaped ZnO nanomaterials for low-temperature operations in NOX gas sensors. *Ceram. Int.* **2020**, *46*, 5706–5714. [CrossRef]
10. Trung, D.D.; Van Toan, N.; Van Tong, P.; Van Duy, N.; Hoa, N.D.; Van Hieu, N. Synthesis of single-crystal SnO_2 nanowires for NOx gas sensors application. *Ceram. Int.* **2012**, *38*, 6557–6563. [CrossRef]
11. Cantalini, C.; Sun, H.T.; Faccio, M.; Ferri, G.; Pelino, M. Niobium-doped α-Fe_2O_3 semiconductor ceramic sensors for the measurement of nitric oxide gases. *Sens. Actuators B Chem.* **1995**, *25*, 673–677. [CrossRef]

12. Majhi, S.M.; Mirzaei, A.; Kim, H.W.; Kim, S.S.; Kim, T.W. Recent advances in energy-saving chemiresistive gas sensors: A review. *Nano Energy* **2021**, *79*, 105369. [CrossRef]
13. Hirsch, A. The era of carbon allotropes. *Nat. Mater.* **2010**, *9*, 868–871. [CrossRef]
14. Falcao, E.H.; Wudl, F. Carbon allotropes: Beyond graphite and diamond. *J. Chem. Technol. Biotechnol.* **2007**, *82*, 524–531. [CrossRef]
15. Rodríguez-Quintana, R.; Carbajal-Franco, G.; Rojas-Chávez, H. DFT study of the H_2 molecules adsorption on pristine and Ni doped graphite surfaces. *Mater. Lett.* **2021**, *293*, 129660. [CrossRef]
16. Iijima, S. Helical microtubules of graphitic carbon. *Nature* **1991**, *354*, 56–58. [CrossRef]
17. Okpalugo, T.I.T.; Papakonstantinou, P.; Murphy, H.; McLaughlin, J.; Brown, N.M.D. High resolution XPS characterization of chemical functionalised MWCNTs and SWCNTs. *Carbon* **2005**, *43*, 153–161. [CrossRef]
18. Marinho, B.; Ghislandi, M.; Tkalya, E.; Koning, C.E.; de With, G. Electrical conductivity of compacts of graphene, multi-wall carbon nanotubes, carbon black, and graphite powder. *Powder Technol.* **2012**, *221*, 351–358. [CrossRef]
19. Niu, J.J.; Wang, J.N.; Jiang, Y.; Su, L.F.; Ma, J. An approach to carbon nanotubes with high surface area and large pore volume. *Microporous Mesoporous Mater.* **2007**, *100*, 1–5. [CrossRef]
20. Li, Y.; Kang, Z.; Yan, X.; Cao, S.; Li, M.; Guo, Y.; Huan, Y.; Wen, X.; Zhang, Y. A three-dimensional reticulate CNT-aerogel for a high mechanical flexibility fiber supercapacitor. *Nanoscale* **2018**, *10*, 9360–9368. [CrossRef]
21. Liu, Q.; Ren, W.; Li, F.; Cong, A.H.; Cheng, H.-M. Synthesis and high thermal stability of double-walled carbon nanotubes using nickel formate dihydrate as catalyst precursor. *J. Phys. Chem. C* **2007**, *111*, 5006–5013. [CrossRef]
22. Scardamaglia, M.; Struzzi, C.; Aparicio Rebollo, F.J.; de Marco, P.; Mudimela, P.R.; Colomer, J.F.; Amati, M.; Gregoratti, L.; Petaccia, L.; Snyders, R.; et al. Tuning electronic properties of carbon nanotubes by nitrogen grafting: Chemistry and chemical stability. *Carbon* **2015**, *83*, 118–127. [CrossRef]
23. Xiao, K.; Liu, Y.; Hu, P.; Yu, G.; Wang, X.; Zhu, D. High-mobility thin-film transistors based on aligned carbon nanotubes. *Appl. Phys. Lett.* **2003**, *83*, 150. [CrossRef]
24. Sayago, I.; Santos, H.; Horrillo, M.C.; Aleixandre, M.; Fernández, M.J.; Terrado, E.; Tacchini, I.; Aroz, R.; Maser, W.K.; Benito, A.M.; et al. Carbon nanotube networks as gas sensors for NO_2 detection. *Talanta* **2008**, *77*, 758–764. [CrossRef]
25. Kong, J.; Franklin, N.R.; Zhou, C.; Chapline, M.G.; Peng, S.; Cho, K.; Dai, H. Nanotube molecular wires as chemical sensors. *Science* **2000**, *287*, 622–625. [CrossRef]
26. Kumar, D.; Chaturvedi, P.; Saho, P.; Jha, P.; Chouksey, A.; Lal, M.; Rawat, J.S.B.S.; Tandon, R.P.; Chaudhury, P.K. Effect of single wall carbon nanotube networks on gas sensor response and detection limit. *Sens. Actuators B Chem.* **2017**, *240*, 1134–1140. [CrossRef]
27. Nurazzi, N.M.; Harussani, M.M.; Zulaikha, N.D.S.; Norhana, A.H.; Syakir, M.I.; Norli, A. Composites based on conductive polymer with carbon nanotubes in DMMP gas sensors—An overview. *Polimery* **2021**, *66*, 85–97. [CrossRef]
28. Abdulla, S.; Mathew, T.L.; Pullithadathil, B. Highly sensitive, room temperature gas sensor based on polyaniline-multiwalled carbon nanotubes (PANI/MWCNTs) nanocomposite for trace-level ammonia detection. *Sens. Actuators B Chem.* **2015**, *221*, 1523–1534. [CrossRef]
29. Bagheri, H.; Hajian, A.; Rezaei, M.; Shirzadmehr, A. Composite of Cu metal nanoparticles-multiwall carbon nanotubes-reduced graphene oxide as a novel and high performance platform of the electrochemical sensor for simultaneous determination of nitrite and nitrate. *J. Hazard. Mater.* **2017**, *324*, 762–772. [CrossRef]
30. Sharma, A.K.; Mahajan, A.; Bedi, R.K.; Kumar, S.; Debnath, A.K.; Aswal, D.K. Non-covalently anchored multi-walled carbon nanotubes with hexa-decafluorinated zinc phthalocyanine as Ppb level chemiresistive chlorine sensor. *Appl. Surf. Sci.* **2018**, *427*, 202–209. [CrossRef]
31. Kothari, R.; Kundalwal, S.I.; Sahu, S.K. Transversely isotropic thermal properties of carbon nanotubes containing vacancies. *Acta Mech.* **2018**, *229*, 2787–2800. [CrossRef]
32. Wang, C.; Wang, C.Y. Geometry and electronic properties of single vacancies in achiral carbon nanotubes. *Eur. Phys. J. B Condens. Matter Complex. Syst.* **2006**, *54*, 243–247. [CrossRef]
33. Kuzmany, H.; Kukovecz, A.; Simon, F.; Holzweber, M.; Kramberger, C.; Pichler, T. Functionalization of carbon nanotubes. *Synth. Met.* **2004**, *141*, 113–122. [CrossRef]
34. Abdelhalim, A.; Abdellah, A.; Scarpa, G.; Lugli, P. Metallic nanoparticles functionalizing carbon nanotube networks for gas sensing applications. *Nanotechnology* **2014**, *25*, 055208. [CrossRef]
35. Aroutiounian, V.M. Gas sensors based on functionalized carbon nanotubes. *J. Contemp. Phys. Arm. Acad. Sci.* **2015**, *50*, 333–354. [CrossRef]
36. Zhang, W.-S.; Liu, Y.-T.; Yao, T.-T.; Wu, G.-P.; Liu, Q. Oxygen defect engineering toward the length-selective tailoring of carbon nanotubes via a two-step electrochemical strategy. *J. Phys. Chem. C* **2020**, *124*, 27097–27106. [CrossRef]
37. Hoefer, M.A.; Bandaru, P.R. Defect engineering of the electrochemical characteristics of carbon nanotube varieties. *J. Appl. Phys.* **2010**, *108*, 034308. [CrossRef]
38. Dilonardo, E.; Penza, M.; Alvisi, M.; Rossi, R.; Cassano, G.; Di Franco, C.; Palmisano, F.; Torsi, L.; Cioffi, N. Gas sensing properties of MWCNT layers electrochemically decorated with Au and Pd nanoparticles. *Beilstein J. Nanotechnol.* **2017**, *8*, 592–603. [CrossRef]
39. Muangrat, W.; Wongwiriyapan, W.; Yordsri, V.; Chobsilp, T.; Inpaeng, S.; Issro, C.; Domanov, O.; Ayala, P.; Pichler, T.; Shi, L. Unravel the active site in nitrogen-doped double-walled carbon nanotubes for nitrogen dioxide gas sensor. *Phys. Status Solidi Appl. Mater. Sci.* **2018**, *215*, 1–6. [CrossRef]

40. Sharma, A.; Tomar, M.; Gupta, V. Room temperature trace level detection of NO_2 gas using SnO_2 modified carbon nanotubes based sensor. *J. Mater. Chem.* **2012**, *22*, 23608–23616. [CrossRef]
41. Norizan, M.N.; Moklis, M.H.; Demon, S.Z.N.; Halim, N.A.; Samsuri, A.; Mohamad, I.S.; Knight, V.F.; Abdullah, N. Carbon nanotubes: Functionalisation and their application in chemical sensors. *RSC Adv.* **2020**, *10*, 43704–43732. [CrossRef]
42. Rasheed, T.; Nabeel, F.; Adeel, M.; Rizwan, K.; Bilal, M.; Iqbal, H.M.N. Carbon nanotubes-based cues: A pathway to future sensing and detection of hazardous pollutants. *J. Mol. Liq.* **2019**, *292*, 111425. [CrossRef]
43. Dai, H. Carbon nanotubes: Synthesis, integration, and properties. *Acc. Chem. Res.* **2002**, *35*, 1035–1044. [CrossRef]
44. Ibrahim, K.S. Carbon nanotubes-properties and applications: A review. *Carbon Lett.* **2013**, *14*, 131–144. [CrossRef]
45. Beitollahi, H.; Movahedifar, F.; Tajik, S.; Jahani, S. A review on the effects of introducing CNTs in the modification process of electrochemical sensors. *Electroanalysis* **2019**, *31*, 1195–1203. [CrossRef]
46. Bassyouni, M.; Mansi, A.E.; ElGabry, A.; Ibrahim, B.A.; Kassem, O.A.; Alhebeshy, R. Utilization of carbon nanotubes in removal of heavy metals from wastewater: A review of the CNTs' potential and current challenges. *Appl. Phys. A* **2020**, *126*, 1–33. [CrossRef]
47. Khot, L.R.; Sankaran, S.; Maja, J.M.; Ehsani, R.; Schuster, E.W. Applications of nanomaterials in agricultural production and crop protection: A review. *Crop. Prot.* **2012**, *35*, 64–70. [CrossRef]
48. Ma, L.; Dong, X.; Chen, M.; Zhu, L.; Wang, C.; Yang, F.; Dong, Y. Fabrication and water treatment application of carbon nanotubes (CNTs)-based composite membranes: A review. *Membranes* **2017**, *7*, 16. [CrossRef]
49. Xiang, L.; Zhang, H.; Hu, Y.; Peng, L.-M. Carbon nanotube-based flexible electronics. *J. Mater. Chem. C* **2018**, *6*, 7714–7727. [CrossRef]
50. Elias, A.; Uddin, N.; Hossain, A.; Saha, J.K.; Siddiquey, I.A.; Sarker, D.R.; Diba, Z.R.; Uddin, J.; Choudhury, M.H.R.; Firoz, S.H. An experimental and theoretical study of the effect of Ce doping in ZnO/CNT composite thin film with enhanced visible light photo-catalysis. *Int. J. Hydrogen Energy* **2019**, *44*, 20068–20078. [CrossRef]
51. Lehman, J.H.; Terrones, M.; Mansfield, E.; Hurst, K.E.; Meunier, V. Evaluating the characteristics of multiwall carbon nanotubes. *Carbon* **2011**, *49*, 2581–2602. [CrossRef]
52. Belin, T.; Epron, F. Characterization methods of carbon nanotubes: A review. *Mater. Sci. Eng. B* **2005**, *119*, 105–118. [CrossRef]
53. Prasek, J.; Drbohlavova, J.; Chomoucka, J.; Hubalek, J.; Jasek, O.; Adam, V.; Kizek, R. Methods for carbon nanotubes synthesis. *J. Mater. Chem.* **2011**, *21*, 15872–15884. [CrossRef]
54. Andrews, R.; Jacques, D.; Qian, D.; Rantell, T. Multiwall carbon nanotubes: Synthesis and application. *Acc. Chem. Res.* **2002**, *35*, 1008–1017. [CrossRef]
55. Tasis, D.; Tagmatarchis, N.; Bianco, A.; Prato, M. Chemistry of carbon nanotubes. *Chem. Rev.* **2006**, *106*, 1105–1136. [CrossRef]
56. Dilonardo, E.; Penza, M.; Alvisi, M.; Di Franco, C.; Rossi, R.; Palmisano, F.; Torsi, L.; Cioffi, N. Electrophoretic deposition of Au NPs on MWCNT-based gas sensor for tailored gas detection with enhanced sensing properties. *Sens. Actuators B Chem.* **2016**, *223*, 417–428. [CrossRef]
57. Sinha, N.; Ma, J.; Yeow, J.T. Carbon nanotube-based sensors. *J. Nanosci. Nanotechnol.* **2006**, *6*, 573–590. [CrossRef]
58. Meyyappan, M. Carbon nanotube-based chemical sensors. *Small* **2016**, *12*, 2118–2129. [CrossRef]
59. Hu, C.Y.; Xu, Y.J.; Duo, S.W.; Zhang, R.F.; Li, M.S. Non-covalent functionalization of carbon nanotubes with surfactants and polymers. *J. Chin. Chem. Soc.* **2009**, *56*, 234–239. [CrossRef]
60. Kocharova, N.; Ääritalo, T.; Leiro, J.; Kankare, J.; Lukkari, J. Aqueous dispersion, surface thiolation, and direct self-assembly of carbon nanotubes on gold. *Langmuir* **2007**, *23*, 3363–3371. [CrossRef]
61. Kotagiri, N.; Kim, J.W. Stealth nanotubes: Strategies of shielding carbon nanotubes to evade opsonization and improve biodistribution. *Int. J. Nanomed.* **2014**, *9*, 85–105. [CrossRef]
62. Tang, R.; Shi, Y.; Hou, Z.; Wei, L. Carbon nanotube-based chemiresistive sensors. *Sensors* **2017**, *17*, 882. [CrossRef]
63. Jun, L.Y.; Mubarak, N.M.; Yee, M.J.; Yon, L.S.; Bing, C.H.; Khalid, M.; Abdullah, E.C. An overview of functionalised carbon nanomaterial for organic pollutant removal. *J. Ind. Eng. Chem.* **2018**, *67*, 175–186. [CrossRef]
64. Pomoell, J.A.V.; Krasheninnikov, A.V.; Nordlund, K.; Keinonen, J. Ion ranges and irradiation-induced defects in multiwalled carbon nanotubes. *J. Appl. Phys.* **2004**, *96*, 2864–2871. [CrossRef]
65. Lehtinen, O.; Nikitin, T.; Krasheninnikov, A.V.; Sun, L.; Banhart, F.; Khriachtchev, L.; Keinonen, J. Characterization of ion-irradiation-induced defects in multi-walled carbon nanotubes. *New J. Phys.* **2011**, *13*, 073004. [CrossRef]
66. Da Silva, L.B.; Fagan, S.B.; Mota, R. Ab initio study of deformed carbon nanotube sensors for carbon monoxide molecules. *Nano Lett.* **2004**, *4*, 65–67. [CrossRef]
67. Peng, S.; Cho, K. Ab initio study of doped carbon nanotube sensors. *Nano Lett.* **2003**, *3*, 513–517. [CrossRef]
68. Liu, Y.; Zhang, H.; Zhang, Z.; Jia, X.; An, L. CO adsorption on Fe-doped vacancy-defected CNTs–A DFT study. *Chem. Phys. Lett.* **2019**, *730*, 316–320. [CrossRef]
69. Tabtimsai, C.; Rakrai, W.; Phalinyot, S.; Wanno, B. Interaction investigation of single and multiple carbon monoxide molecules with Fe-, Ru-, and Os-doped single-walled carbon nanotubes by DFT study: Applications to gas adsorption and detection nanomaterials. *J. Mol. Model.* **2020**, *26*, 1–13. [CrossRef]
70. Yeow, J.T.W.; Wang, Y. A review of carbon nanotubes-based gas sensors. *J. Sens.* **2009**, *2009*, 493904. [CrossRef]
71. Han, T.; Nag, A.; Mukhopadhyay, S.C.; Xu, Y. Carbon nanotubes and its gas-sensing applications: A review. *Sens. Actuators A Phys.* **2019**, *291*, 107–143. [CrossRef]

72. Li, J.; Lu, Y.; Ye, Q.; Cinke, M.; Han, J.; Meyyappan, M. Carbon nanotube sensors for gas and organic vapor detection. *Nano Lett.* **2003**, *3*, 929–933. [CrossRef]
73. Valentini, L.; Cantalini, C.; Armentano, I.; Kenny, J.M.; Lozzi, L.; Santucci, S. Highly sensitive and selective sensors based on carbon nanotubes thin films for molecular detection. *Diam. Relat. Mater.* **2004**, *13*, 1301–1305. [CrossRef]
74. Piloto, C.; Mirri, F.; Bengio, E.A.; Notarianni, M.; Gupta, B.; Shafiei, M.; Pasquali, M.; Motta, N. Room temperature gas sensing properties of ultrathin carbon nanotube films by surfactant-free dip coating. *Sens. Actuators B* **2016**, *227*, 128–134. [CrossRef]
75. Ray, A.; Sadhukhan, P.; Naskar, K.; Lal, G.; Bhar, R.; Sinha, C.; Das, S. Polyaniline-multiwalled carbon nanotube (PANI-MWCNT): Room temperature resistive carbon monoxide (CO) sensor. *Synth. Met.* **2018**, *245*, 182–189. [CrossRef]
76. Zhang, Z.; Wei, Z.; Wan, M. Nanostructures of polyaniline doped with inorganic acids. *Macromolecules* **2002**, *35*, 5937–5942. [CrossRef]
77. Fei, J.; Cui, Y.; Yan, X.; Yang, Y.; Wang, K.; Li, J. Controlled fabrication of polyaniline spherical and cubic shells with hierarchical nanostructures. *ACS Nano* **2009**, *3*, 3714–3718. [CrossRef]
78. Cho, S.; Kwon, O.S.; You, S.A.; Jang, J. Shape-controlled polyaniline chemiresistors for high-performance DMMP sensors: Effect of morphologies and charge-transport properties. *J. Mater. Chem. A* **2013**, *1*, 5679–5688. [CrossRef]
79. Yun, J.; Jeon, S.; Kim, H.I. Improvement of NO gas sensing properties of polyaniline/MWCNT composite by photocatalytic effect of TiO_2. *J. Nanomater.* **2013**, *2013*, 184345. [CrossRef]
80. Mangu, R.; Rajaputra, S.; Singh, V.P. MWCNT-polymer composites as highly sensitive and selective room temperature gas sensors. *Nanotechnology* **2011**, *22*, 215502. [CrossRef]
81. Zhang, Y.; Kim, J.J.; Chen, D.; Tuller, H.L.; Rutledge, G.C. Electrospun polyaniline fibers as highly sensitive room temperature chemiresistive sensors for ammonia and nitrogen dioxide gases. *Adv. Funct. Mater.* **2014**, *24*, 4005–4014. [CrossRef]
82. Zhang, W.; Cao, S.; Wu, Z.; Zhang, M.; Cao, Y.; Guo, J.; Zhong, F.; Duan, H.; Jia, D. High-performance gas sensor of polyaniline/carbon nanotube composites promoted by interface engineering. *Sensors* **2019**, *20*, 149. [CrossRef] [PubMed]
83. Pisal, S.H.; Harale, N.S.; Bhat, T.S.; Deshmukh, H.P.; Patil, P.S. Functionalized multi-walled carbon nanotubes for nitrogen sensor. *IOSR J. Appl. Chem.* **2014**, *7*, 49–52. [CrossRef]
84. Jeon, J.Y.; Kang, B.C.; Byun, Y.T.; Ha, T.J. High-performance gas sensors based on single-wall carbon nanotube random networks for the detection of nitric oxide down to the ppb-level. *Nanoscale* **2019**, *11*, 1587–1594. [CrossRef] [PubMed]
85. Ionete, E.I.; Spiridon, S.I.; Monea, B.F.; Stratulat, E. A room temperature gas sensor based on sulfonated SWCNTs for the detection of NO and NO_2. *Sensors* **2019**, *19*, 1116. [CrossRef]
86. Popescu, M.; Simandan, I.D.; Sava, F.; Velea, A.; Fagadar-Cosma, E. Sensor of nitrogen dioxide based on single wall carbon nanotubes and manganese-porphyrin. *Dig. J. Nanomater. Biostructures* **2011**, *6*, 1253–1256.
87. Penza, M.; Rossi, R.; Alvisi, M.; Cassano, G.; Signore, M.A.; Serra, E.; Giorgi, R. Pt- and Pd-nanoclusters functionalized carbon nanotubes networked films for sub-ppm gas sensors. *Sens. Actuators B Chem.* **2008**, *135*, 289–297. [CrossRef]
88. Leghrib, R.; Dufour, T.; Demoisson, F.; Claessens, N.; Reniers, F.; Llobet, E. Gas sensing properties of multiwall carbon nanotubes decorated with rhodium nanoparticles. *Sens. Actuators B Chem.* **2011**, *160*, 974–980. [CrossRef]
89. Kim, S.J.; Park, Y.J.; Ra, E.J.; Kim, K.K.; An, K.H.; Lee, Y.H.; Choi, J.Y.; Park, C.H.; Doo, S.K.; Park, M.H.; et al. Defect-induced loading of Pt nanoparticles on carbon nanotubes. *Appl. Phys. Lett.* **2007**, *90*, 94–97. [CrossRef]
90. Mudimela, P.R.; Scardamaglia, M.; González-León, O.; Reckinger, N.; Snyders, R.; Llobet, E.; Bittencourt, C.; Colomer, J.F. Gas sensing with gold-decorated vertically aligned carbon nanotubes. *Beilstein J. Nanotechnol.* **2014**, *5*, 910–918. [CrossRef]
91. Na, P.S.; Kim, H.; So, H.M.; Kong, K.J.; Chang, H.; Ryu, B.H.; Choi, Y.; Lee, J.O.; Kim, B.K.; Kim, J.J.; et al. Investigation of the humidity effect on the electrical properties of single-walled carbon nanotube transistors. *Appl. Phys. Lett.* **2005**, *87*, 10–13. [CrossRef]
92. Loghin, F.C.; Falco, A.; Albrecht, A.; Salmerón, J.F.; Becherer, M.; Lugli, P.; Rivandeneyra, A. A handwriting method for low-cost gas sensors. *ACS Appl. Mater. Interfaces* **2018**, *10*, 34683–34689. [CrossRef]
93. Fort, A.; Panzardi, E.; Al-Hamry, A.; Vignoli, V.; Mugnaini, M.; Addabbo, T.; Kanoun, O. Highly sensitive detection of NO_2 by au and TiO_2 nanoparticles decorated SWCNTs sensors. *Sensors* **2020**, *20*, 12. [CrossRef] [PubMed]
94. Choi, S.W.; Kim, J.; Byun, Y.T. Highly sensitive and selective NO_2 detection by Pt nanoparticles-decorated single-walled carbon nanotubes and the underlying sensing mechanism. *Sens. Actuators B Chem.* **2017**, *238*, 1032–1042. [CrossRef]
95. Mahmood, W.K.; Naje, A.N. Fabrication of room temperature NO_2 gas sensor based on silver nanoparticles-decorated carbon nanotubes. *J. Nano- Electron. Phys.* **2018**, *10*, 1–6. [CrossRef]
96. Panzardi, E.; Lo Grasso, A.; Vignoli, V.; Mugnaini, M.; Lupetti, P.; Fort, A. NO_2 Sensing with SWCNT decorated by nanoparticles in temperature pulsed mode: Modeling and characterization. *Sensors* **2020**, *20*, 4729. [CrossRef]
97. Su, P.G.; Pan, T.T. Fabrication of a room-temperature NO_2 gas sensor based on WO_3 films and WO_3/MWCNT nanocomposite films by combining polyol process with metal organic decomposition method. *Mater. Chem. Phys.* **2011**, *125*, 351–357. [CrossRef]
98. Park, S.; Byoun, Y.; Kang, H.; Song, Y.-J.; Choi, S.-W. ZnO nanocluster-functionalized single-walled carbon nanotubes synthesized by microwave irradiation for highly sensitive NO_2 detection at room temperature. *ACS Omega* **2019**, *4*, 10677–10686. [CrossRef]
99. Baccar, H.; Thamri, A.; Clément, P.; Llobet, E.; Abdelghani, A. Pt- and Pd-decorated MWCNTs for vapour and gas detection at room temperature. *Beilstein J. Nanotechnol.* **2015**, *6*, 919–927. [CrossRef] [PubMed]
100. Zanolli, Z.; Leghrib, R.; Felten, A.; Pireaux, J.J.; Llobet, E.; Charlier, J.C. Gas sensing with au-decorated carbon nanotubes. *ACS Nano* **2011**, *5*, 4592–4599. [CrossRef]

101. Choi, K.Y.; Park, J.S.; Park, K.B.; Kim, H.J.; Park, H.D.; Kim, S.D. Low power micro-gas sensors using mixed SnO_2 nanoparticles and MWCNTs to detect NO_2, NH_3, and xylene gases for ubiquitous sensor network applications. *Sens. Actuators B Chem.* **2010**, *150*, 65–72. [CrossRef]
102. Kim, D.; Park, K.M.; Shanmugam, R.; Yoo, B. Electrochemically decorated ZnTe nanodots on single-walled carbon nanotubes for room-temperature NO_2 sensor application. *J. Nanosci. Nanotechnol.* **2014**, *14*, 8248–8252. [CrossRef] [PubMed]
103. Lim, N.; Lee, J.S.; Byun, Y.T. Negatively-doped single-walled carbon nanotubes decorated with carbon dots for highly selective NO2 detection. *Nanomaterials* **2020**, *10*, 2509. [CrossRef]
104. Albiss, B.A.; Sakhaneh, W.A.; Jumah, I.; Obaidat, I.M. NO_2 gas sensing properties of ZnO/single-wall carbon nanotube composites. *IEEE Sens. J.* **2010**, *10*, 1807–1812. [CrossRef]
105. Evans, G.P.; Buckley, D.J.; Skipper, N.T.; Parkin, I.P. Single-walled carbon nanotube composite inks for printed gas sensors: Enhanced detection of NO_2, NH_3, EtOH and acetone. *RSC Adv.* **2014**, *4*, 51395–51403. [CrossRef]
106. Yaqoob, U.; Phan, D.T.; Uddin, A.S.M.I.; Chung, G.S. Highly flexible room temperature NO_2 sensor based on MWCNTs-WO_3 nanoparticles hybrid on a PET substrate. *Sens. Actuators B Chem.* **2015**, *221*, 760–768. [CrossRef]
107. Jones, R.O. Density functional theory: Its origins, rise to prominence, and future. *Rev. Mod. Phys.* **2015**, *87*, 897. [CrossRef]
108. Mardirossian, N.; Head-Gordon, M. Thirty years of density functional theory in computational chemistry: An overview and extensive assessment of 200 density functionals. *Mol. Phys.* **2017**, *115*, 2315–2372. [CrossRef]
109. Cruz-Martínez, H.; Tellez-Cruz, M.M.; Solorza-Feria, O.; Calaminici, P.; Medina, D.I. Catalytic activity trends from pure Pd nanoclusters to M@PdPt (M = Co, Ni, and Cu) core-shell nanoclusters for the oxygen reduction reaction: A first-principles analysis. *Int. J. Hydrogen Energy* **2020**, *45*, 13738–13745. [CrossRef]
110. Zeng, Y.; Lin, S.; Gu, D.; Li, X. Two-dimensional nanomaterials for gas sensing applications: The role of theoretical calculations. *Nanomaterials* **2018**, *8*, 851. [CrossRef] [PubMed]
111. Piras, A.; Ehlert, C.; Gryn'Ova, G. Sensing and sensitivity: Computational chemistry of graphene-based sensors. *Wiley Interdiscip. Rev. Comput. Mol. Sci.* **2021**, *11*, e1526. [CrossRef]
112. Bakar, A.; Afaq, A.; Ahmed, M.; Bashir, A.; Asif, M. Optoelectronic properties of CuCoMnZ (Z = Si, Sn, Sb): A DFT study. *J. Electron. Mater.* **2021**, *50*, 4006–4015. [CrossRef]
113. Li, K.; Wang, W.; Cao, D. Metal (Pd, Pt)-decorated carbon nanotubes for CO and NO sensing. *Sens. Actuators B Chem.* **2011**, *159*, 171–177. [CrossRef]
114. Dutta, A.; Pradhan, A.K.; Qi, F.; Mondal, P. Computation-led design of pollutant gas sensors with bare and carbon nanotube supported rhodium alloys. *Monatsh. Chem.* **2020**, *151*, 159–171. [CrossRef]
115. Cruz-Martínez, H.; Rojas-Chávez, H.; Montejo-Alvaro, F.; Peña-Castañeda, Y.A.; Matadamas-Ortiz, P.T.; Medina, D.I. Recent developments in graphene-based toxic gas sensors: A theoretical overview. *Sensors* **2021**, *21*, 1992. [CrossRef] [PubMed]
116. Esrafili, M.D.; Janebi, H. B-, N-doped and BN codoped C60 heterofullerenes for environmental monitoring of NO and NO_2: A DFT study. *Mol. Phys.* **2020**, *118*, e1631495. [CrossRef]
117. Ramirez-de-Arellano, J.M.; Canales, M.; Magaña, L.F. Carbon nanostructures doped with transition metals for pollutant gas adsorption systems. *Molecules* **2021**, *26*, 5346. [CrossRef] [PubMed]
118. Jia, X.; An, L.; Chen, T. Adsorption of nitrogen oxides on Al-doped carbon nanotubes: The first principles study. *Adsorption* **2020**, *26*, 587–595. [CrossRef]
119. Demir, S.; Fellah, M.F. Carbon nanotubes doped with Ni, Pd and Pt: A density functional theory study of adsorption and sensing NO. *Surf. Sci.* **2020**, *701*, 121689. [CrossRef]
120. Azizi, K.; Karimpanah, M. Computational study of Al-or P-doped single-walled carbon nanotubes as NH_3 and NO_2 sensors. *Appl. Surf. Sci.* **2013**, *285*, 102–109. [CrossRef]
121. Tabtimsai, C.; Wanno, B.; Utairueng, A.; Promchamorn, P.; Kumsuwan, U. First principles investigation of NH_3 and NO_2 adsorption on transition metal-doped single-walled carbon nanotubes. *J. Electron. Mater.* **2019**, *48*, 7226–7238. [CrossRef]
122. Tabtimsai, C.; Wanno, B.; Ruangpornvisuti, V. Theoretical investigation of CO_2 and NO_2 adsorption onto Co-, Rh-and Ir-doped (5, 5) single-walled carbon nanotubes. *Mater. Chem. Phys.* **2013**, *138*, 709–715. [CrossRef]
123. Tabtimsai, C.; Keawwangchai, S.; Wanno, B.; Ruangpornvisuti, V. Gas adsorption on the Zn–, Pd–and Os–doped armchair (5, 5) single–walled carbon nanotubes. *J. Mol. Model.* **2012**, *18*, 351–358. [CrossRef]
124. Zhang, X.; Wang, Y.; Wang, Z.; Ma, S. Small gas adsorption on Co-N_4 porphyrin-like CNT for sensor exploitation: A first-principles study. *Carbon Lett.* **2020**, *30*, 177–187. [CrossRef]
125. Luo, M.; Liang, Z.; Peera, S.G.; Chen, M.; Liu, C.; Yang, H.; Liu, J.; Kumar, U.P.; Liang, T. Theoretical study on the adsorption and predictive catalysis of MnN4 embedded in carbon substrate for gas molecules. *Appl. Surf. Sci.* **2020**, *525*, 146480. [CrossRef]
126. Hankins, A.; Willard, T.C.; Liu, A.Y.; Paranjape, M. Role of defects in the sensing mechanism of CNTFET gas sensors. *J. Appl. Phys.* **2020**, *128*, 084501. [CrossRef]
127. Zanolli, Z.; Charlier, J.C. Defective carbon nanotubes for single-molecule sensing. *Phys. Rev. B* **2009**, *80*, 155447. [CrossRef]
128. Vasylenko, A.I.; Tokarchuk, M.V.; Jurga, S. Effect of a vacancy in single-walled carbon nanotubes on He and NO adsorption. *J. Phys. Chem. C* **2015**, *119*, 5113–5116. [CrossRef]
129. Cruz-Martínez, H.; Rojas-Chávez, H.; Matadamas-Ortiz, P.T.; Ortiz-Herrera, J.C.; López-Chávez, E.; Solorza-Feria, O.; Medina, D.I. Current progress of Pt-based ORR electrocatalysts for PEMFCs: An integrated view combining theory and experiment. *Mater. Today Phys.* **2021**, *19*, 100406. [CrossRef]

130. Cruz-Martínez, H.; Guerra-Cabrera, W.; Flores-Rojas, E.; Ruiz-Villalobos, D.; Rojas-Chávez, H.; Peña-Castañeda, Y.A.; Medina, D.I. Pt-free metal nanocatalysts for the oxygen reduction reaction combining experiment and theory: An overview. *Molecules* **2021**, *26*, 6689. [CrossRef]
131. Pineda-Reyes, A.M.; Herrera-Rivera, M.R.; Rojas-Chávez, H.; Cruz-Martínez, H.; Medina, D.I. Recent advances in ZnO-based carbon monoxide sensors: Role of doping. *Sensors* **2021**, *21*, 4425. [CrossRef]
132. Chikate, P.R.; Sharma, A.; Rondiya, S.R.; Cross, R.W.; Dzade, N.Y.; Shirage, P.M.; Devan, R.S. Hierarchically interconnected ZnO nanowires for low-temperature-operated reducing gas sensors: Experimental and DFT studies. *New J. Chem.* **2021**, *3*, 1404–1414. [CrossRef]
133. Santucci, S.; Picozzi, S.; Di Gregorio, F.; Lozzi, L.; Cantalini, C.; Valentini, L.; Kenny, J.M.; Delley, B. NO_2 and CO gas adsorption on carbon nanotubes: Experiment and theory. *J. Chem. Phys.* **2003**, *119*, 10904–10910. [CrossRef]
134. Adjizian, J.-J.; Leghrib, R.; Koos, A.A.; Suarez-Martinez, I.; Crossley, A.; Wagner, P.; Grobert, N.; Llobet, E.; Ewels, C.P. Boron-and nitrogen-doped multi-wall carbon nanotubes for gas detection. *Carbon* **2014**, *66*, 662–673. [CrossRef]

MDPI
St. Alban-Anlage 66
4052 Basel
Switzerland
Tel. +41 61 683 77 34
Fax +41 61 302 89 18
www.mdpi.com

International Journal of Molecular Sciences Editorial Office
E-mail: ijms@mdpi.com
www.mdpi.com/journal/ijms

www.ingramcontent.com/pod-product-compliance
Lightning Source LLC
LaVergne TN
LVHW070406100526
838202LV00014B/1402